T0217224

Lecture Notes in Computer Science   5070

Commenced Publication in 1973
Founding and Former Series Editors:
Gerhard Goos, Juris Hartmanis, and Jan van Leeuwen

Editorial Board

Małgorzata Marciniak
Agnieszka Mykowiecka (Eds.)

Aspects
of Natural Language
Processing

Essays Dedicated to Leonard Bolc
on the Occasion of His 75th Birthday

 Springer

Volume Editors

Małgorzata Marciniak
Agnieszka Mykowiecka
Institute of Computer Science
Polish Academy of Sciences
ul. J.K. Ordona 21, 01-237 Warsaw, Poland
E-mail: {Malgorzata.Marciniak, Agnieszka.Mykowiecka}@ipipan.waw.pl

The cover illustration, showing the Rosetta Stone, was retrieved from
`http://de.wikipedia.org/wiki/Datei:Rosetta_Stone_BW.jpeg`

Library of Congress Control Number: 2009935055

CR Subject Classification (1998): I.2, H.3-5, J.5, F.4

LNCS Sublibrary: SL 3 – Information Systems and Application, incl. Internet/Web and HCI

ISSN   0302-9743

ISBN   978-3-642-04734-3 Springer Berlin Heidelberg New York

springer.com

Typesetting: Camera-ready by author, data conversion by Markus Richter, Heidelberg
Printed on acid-free paper   SPIN: 12766055   06/3180   5 4 3 2 1 0

Leonard Bolc

Preface

This book is dedicated to Professor Leonard Bolc on the occasion of his 75th birthday, and contains essays written by his friends, former students and colleagues to celebrate his scientific career. For many years Leonard Bolc has played an important role in the Polish computer science community. He was especially known for his clear vision in the development of artificial intelligence, inspiring research, organizational and editorial achievements.

The papers included in this volume, present research in the areas which Leonard Bolc and his colleagues investigated during his long scientific career, i.e., logic, automatic reasoning, natural language processing, and computer applications of natural language or human-like reasoning.

Part I of the book is devoted to logic—the domain which was one of the most explored by Leonard Bolc himself. The first paper in this section describes a new approach to making correct judgments keeping in mind real-world constraints. The second paper addresses the problem of qualitative interpretation of fuzzy-like paraconsistent reasoning with natural language words.

Part II contains papers focusing on different aspects of computational linguistics. This part begins with papers devoted to morphology. Two of them describe morphological aspects of the Polish and Uzbek languages, and the third paper discusses morphological properties of Polish multi-word proper names. The subsequent paper presents a new definition of Polish nominal phrases defined within the metamorphosis grammar formalism, which was first described in one of the numerous publications edited by Professor Bolc. The following four papers are devoted to automatic acquisition of linguistic knowledge: extraction of semantic relations, valence dictionaries, and semantic restrictions on verb arguments.

Part III comprises papers describing different applications in which natural language processing or automatic reasoning plays an important role. The first paper is devoted to speech technology and describes the issues of voice portal implementation. Speech analysis was a subject of interest to Professor Bolc in the mid-1970s, which is commemorated in the following paper. The next two papers are devoted to different aspects of information extraction. The first one presents real-time event extraction, whereas the next one describes the creation of a domain model for information extraction applications in the medical domain. The next topics addressed are methods of searching for exact and approximate patterns in texts, and decision making in the context of a support system that is capable of dynamically adjusting its prediction model. The last paper presents an environment that allows a group of professional analysts to work together on complex problems and exchange their results in active and tacit ways.

By preparing this volume we would like to express our gratitude to our mentor Professor Leonard Bolc, for the many years of his generous and friendly leadership. While preparing this volume we had the unique opportunity to experience the type of work to which Leonard Bolc devoted so much time and effort, and in which he continues to be very active. We would like to thank all the authors for their positive response to the idea of creating the volume, and for their help in reviewing the papers. We would also like to thank Maciej Korzeniowski for help in proofreading the volume.

We would like to express our thanks for the scientific encouragement Professor Bolc gave us, on behalf of all members of the Linguistic Engineering Group in the Institute of Computer Science, which he organized and led for many years, his former students who were inspired by him to devote their life to scientific research, and also on behalf of those who decided to work in industry. Finally, our colleagues, Piotr Rychlik, Paweł Stefański, Anna Ostaszewska, Agata Wrzos-Kamińska, Jacek Wrzos-Kamiński, Marek Wójcik, Agata Sawicka-Wójcik, and Łukasz Dębowski, who worked in the Group in the space of nearly 25 years, join the authors of the papers included in this volume in wishing Leonard Bolc all the best!

August 2009   Małgorzata Marciniak
                                                 Agnieszka Mykowiecka

A Scientific Portrait of Professor Leonard Bolc

Leonard Bolc started his scientific career studying at the Philological Department of the Adam Mickiewicz University in Poznań. From 1961 to 1987 Leonard Bolc worked at the Warsaw University (UW) in the Institute of Computer Science from the position of assistant to associate professor. He received his PhD in applied linguistics from the Adam Mickiewicz University in 1964. In the same year he began his postdoctoral studies at the Institute of Mathematics of the University of Münster. He spent four years there, expanding his knowledge of mathematics and computer science under the supervision of Professor H. Hermes, whose mentor was one of the founders of modern logic, G. Frege. During his stay at Münster, Leonard Bolc attended the lectures of Noam Chomsky, who sparked his interest in computational linguistics, especially in formal grammars and natural language parsers. In 1969, he received his habilitation at the Adam Mickiewicz University in using a context-free grammar formalism for natural language text analysis. The next year was spent in Dresden working on digital processing of speech signals. When he returned to Warsaw, he was nominated the Head of Department of Information Analysis and Synthesis in the Institute of Computer Science UW where he worked on artificial intelligence, focusing on various aspects of natural language processing [2]. From 1987 to 2007 he worked at the Institute of Computer Science, Polish Academy of Sciences, where he organized the Man-Machine Communication Group, that later changed its name to the Linguistics Engineering Group.

Professor Bolc focused his research on logic and computational linguistics, with the aim of creating a computer system with a natural language user interface. To achieve this goal, it is necessary to solve various problems from many fields—speech technologies, language engineering, knowledge representation and reasoning. Professor Leonard Bolc himself explored some of them and inspired his students and colleagues to work on many others. He worked in the fields of formal grammar [8], semantic interpretation of texts [1], searching in large data sets [7], information retrieval [3], [5], non-classical logic [10, 11], uncertain and incomplete knowledge processing [6] and automatic deduction [4].

The wide range of Leonard Bolc's research was the result of his conviction that only by taking into account all aspects of natural language communication— speech understanding, syntactic, semantic and pragmatic analysis of utterances, and automatic deduction based on human reasoning—can we implement a system that processes data in natural language.

Professor Leonard Bolc was very active in the dissemination of many important scientific results. He has been on the editorial board of many journals published by Academic Press, Springer-Verlag, Kluwer Academic Publishing, and Blackwell Scientific Publications. Many important monographs concerning new achievements in various fields of computer science have been edited by him. He has been a co-founder of the series "Symbolic Computation" published by

Springer-Verlag. He has edited more than 40 volumes in world-famous publishing houses, and he continues to be a very active editor. Leonard Bolc is also a very well-known editor in Poland. More than 120 volumes have been published in various book series edited by him. On the basis of many of these publications, Polish scientists received their habilitation or were nominated professors. Over the last few years, he has been the editor-in-chief for many academic handbooks of the Polish-Japanese Institute of Information Technology Press.

During his long scientific career, first at the Warsaw University, then at the Institute of Computer Science PAS, he was a scientific mentor to many students and young scientists. He encouraged many young people to devote their life to research in various aspects of artificial intelligence. All papers in this volume are co-authored by scientists who worked with Professor Bolc at various stages of their scientific careers; eight authors worked in the Group he created.

Selected Publications

1. Bolc, L., Strzalkowski, T.: Transformation Of Natural Language Into Logical Formulas. COLING 1982, 29–36 (1982)
2. Bolc, L., Cichy, M., Różańska, L.: Przetwarzanie języka naturalnego. WNT Warszawa (1982)
3. Bolc, L., Kochut, K., Lesniewski, A., Strzalkowski, T.: Natural Language Information Retrieval System Dialog. EACL 1983, 196–203 (1983)
4. Bolc, L., Rychlik, P.: The Use of Modal Default Reasoning in a Medical Diagnostic System with Natural Language Interface, Proc. of the 6th ECAI, Pisa (1984)
5. Bolc, L., Kowalski, A., Kozlowska, M., Strzalkowski, T.: A Natural Language Information Retrieval System with Extensions Towards Fuzzy Reasoning. International Journal of Man-Machine Studies 23(4), 335–367 (1985)
6. Bolc, L., Borodziewicz, W., Wójcik, M.: Podstawy przetwarzania informacji niepewnej i niepełnej. PWN, Warszawa (1991)
7. Bolc, L., Cytowski, J.: Search Methods for Artificial Intelligence. Academic Press, London (1992)
8. Bolc, L., Mykowiecka, A.: Podstawy przetwarzania języka naturalnego. Wybrane metody formalnego zapisu składni, Akademicka Oficyna Wydawnicza (1992)
9. Bolc, L., Zaremba, J.: Wprowadzenie do uczenia się maszyn. PWN, Warszawa (1993)
10. Bolc, L., Borowik, P.: Many-Valued Logics. Vol. 1: Theoretical Foundations. Springer, Heidelberg (2000)
11. Bolc, L., Borowik, P.: Many-Valued Logics 2: Automated Reasoning and Practical Applications. Springer, Heidelberg (2003)

Springer Acknowledgements

It was a great pleasure for Springer to receive the proposal to publish a book to be dedicated to Professor Leonard Bolc on the occasion of his 75th birthday – after a quick round of evaluation of the proposal in discussion with the series editors of the Lecture Notes in Artificial Intelligence series, there was unanimous agreement to publish this book as part of the LNCS/LNAI Festschrift subline.

The extraordinary scientific achievements and merits of Professor Bolc during his scientific career, now spanning almost half a century, are dignified in an impressive way by all those contributing to this Festschrift. However, beyond being an outstanding scientist and organizer of the computer science community, in Poland as well as internationally, Professor Bolc was also a very prolific author and was successfully involved in editing several books and entire book series and reviewing for various journals. Starting in the 1970s, a substantial part of Professor Bolc's editorial activities resulted in publications by Springer-Verlag and we are very proud to enjoy his cooperation up to the present day.

Starting out in 1978 when acting as volume editor of LNCS 63, *Natural Language Communication with Computers*, Professor Bolc developed close contacts and excellent cooperation with several generations of staff of the Heidelberg-based computer science editorial team in Springer-Verlag. From Springer's point of view, the book series *Artificial Intelligence and Symbolic Computation* initiated by Professor Bolc was a very successful venue for the publication of monographs and edited collections of chapters on timely topics in artificial intelligence; also his own twin-monographs on Many-Valued Logics, coauthored by Piotr Borowik, are landmark publications for the logics community. Overall, in his editorial activities Professor Bolc patiently and considerately encouraged many scientists around the world to write down their consolidated results for monograph publications. Beyond publications in AI&SC and LNCS/LNAI, Professor Bolc brought to Springer quite a few remarkable standalone publications, including the bestseller *Genetic Algorithms + Data Structures = Evolution Programs* by Zbigniew Michalewicz.

As head of the Springer Heidelberg computer science editorial team, it is my privilege to be in close contact and cooperation with Professor Bolc up to the present – at occasional personal meetings and during periodic phone conversations he generously shares his scientific knowledge and community connections and many of the leads he has offered in these exchanges have been turned into successful projects.

On behalf of Springer, in particular on behalf of the present and former members of the computer science editorial team, I would like to express my sincere gratitude to Professor Bolc for the excellent cooperation he has offered over decades and which we hope he will offer for many more years. We wish Professor Bolc all the best for the years to come!

August 2009   Alfred Hofmann

Table of Contents

I Logic

Wisdom Technology: A Rough-Granular Approach 3
Andrzej Jankowski and Andrzej Skowron

Paraconsistent Reasoning with Words 43
Alicja S. Szalas and Andrzej Szałas

II Language

On the Root-Based Lexicon for Polish 61
Joanna Rabiega-Wiśniewska

Representation of Uzbek Morphology in Prolog 83
Gayrat Matlatipov and Zygmunt Vetulani

Inflection of Polish Multi-Word Proper Names with Morfeusz
and Multiflex 111
Agata Savary, Joanna Rabiega-Wiśniewska, and Marcin Woliński

A New Formal Definition of Polish Nominal Phrases 143
Marek Świdziński and Marcin Woliński

Morphosyntactic Constraints in the Acquisition of Linguistic Knowledge
for Polish 163
Maciej Piasecki and Adam Radziszewski

Towards the Automatic Acquisition of a Valence Dictionary for Polish ... 191
Adam Przepiórkowski

Semantic Annotation of Verb Arguments in Shallow Parsed
Polish Sentences by Means of the EM Selection Algorithm 211
Elżbieta Hajnicz

Adjectives: Constructions vs. Valence 241
Anna Kupść

III Applications

User-Centered Design for a Voice Portal 273
*Krzysztof Marasek, Łukasz Brocki, Danijel Koržinek,
Krzysztof Szklanny, and Ryszard Gubrynowicz*

Speech Understanding System SUSY—A New Version of the Speech
Synthesis Program 295
    Jerzy Cytowski

Exploring Curvature-Based Topic Development Analysis for Detecting
Event Reporting Boundaries...................................... 311
    Jakub Piskorski

Domain Model for Medical Information Extraction—The LightMedOnt
Ontology.. 333
    Agnieszka Mykowiecka and Malgorzata Marciniak

A Survey of Text Processing Tools for the Automatic Analysis of
Molecular Sequences 359
    Andrzej Polański, Rafał Pokrzywa, and Marek Kimmel

Intelligent Decision Support: A Fuzzy Stock Ranking System 379
    Adam Ghandar, Zbigniew Michalewicz, and Ralf Zurbruegg

COLLANE: An Experiment in Computer-Mediated Tacit Collaboration.. 411
    *Tomek Strzalkowski, Sarah Taylor, Samira Shaikh, Ben-Ami Lipetz,
    Hilda Hardy, Nick Webb, Tony Cresswell, Ting Liu, Min Wu,
    Yu Zhan, and Song Chen*

Author Index 451

Part I

Logic

Wisdom Technology:
A Rough-Granular Approach

Andrzej Jankowski[1] and Andrzej Skowron[2]

[1] Institute of Decision Processes Support
and
AdgaM Solutions Sp. z o.o.
Wąwozowa 9 lok. 64, 02-796 Warsaw, Poland
andrzejj@adgam.com.pl
[2] Institute of Mathematics,
Warsaw University
Banacha 2, 02-097 Warsaw, Poland
skowron@mimuw.edu.pl

Abstract. We discuss foundations for modern intelligent systems in the framework of Wisdom Technology (Wistech). The approach is based on the rough-granular approach.

Key words: wisdom technology, judgment, interaction, rough sets, granular computing, rough-granular computing

> *66. Consider for example the proceedings that we call "games"*
> *... we see a complicated network of similarities overlapping and criss-crossing: sometimes overall similarities, sometimes similarities of detail.*

> *67. I can think of no better expression to characterize these similarities than "family resemblances"; for the various resemblances between members of family: build, features, colour of eyes, gait, temperament, etc., etc. overlap and criss-cross in the same way. - And I shall say: games form a family.*

> *569. Language is an instrument. Its concepts are instruments. Now perhaps one thinks that it can make no great difference which concepts we employ. As, after all, it is possible to do physics in feet and inches as well as in metres and centimetres; the difference is merely one of convenience. But even this is not true, if for instance, calculations in some system of measurement demand more time and trouble than it is possible for us to give them.*

> *570. Concepts lead us to make investigations; are the expression of our interest, and direct interest.*
> – Ludwig Wittgenstein [97]

M. Marciniak and A. Mykowiecka (Eds.): Bolc Festschrift, LNCS 5070, pp. 3–41, 2009.

1 Introduction: Definition of the Subject and Its Importance

There are many indications that we are currently witnessing the onset of an era of radical technological changes. These radical changes depend on the further advancement of technology to acquire, represent, store, process, discover, communicate and learn wisdom. In this paper, we call this technology *wisdom technology* (or Wistech, for short). The term *wisdom* commonly means *rightly judging* [29]. This common notion can be refined. By *wisdom*, we understand an adaptive ability to make judgments correctly to a satisfactory degree (in particular, correct decisions) having in mind real-life constraints. The intuitive nature of wisdom understood in this way can be expressed by the so called *wisdom equation* [26], metaphorically as shown in (1).

$$wisdom = knowledge + adaptive\ judgment + interactions. \qquad (1)$$

Wisdom can be treated as a special type of knowledge processing. In order to explain the specificity of this type of knowledge processing, let us assume that a control system of a given agent Ag consists of a society of agent control components interacting with the other agent Ag components and with the agent Ag environments. Moreover, there are special agent components, called as the agent coordination control components which are responsible for the coordination of control components. Any agent coordination control component mainly searches for answers for the following question: *What to do next?* or, more precisely: *Which of the agent Ag control components should be activated now?* Of course, any agent control component has to process some kind of knowledge representation. In the context of agent perception, the agent Ag itself (by using, e.g., interactions, memory, and coordination among control components) is processing a very special type of knowledge reflecting the agent perception of the hierarchy of needs (objectives, plans, etc.) and the current agent or the environment constraints. This kind of knowledge processing mainly deals with complex vague concepts (such as risk or safety) from the point of view of the *selfish* agent needs. Usually, this kind of knowledge processing is not necessarily logical reasoning in terms of proving statements (i.e., labeling statements by truth values such as TRUE or FALSE). This knowledge processing is rather analogous to the judgment process in a court aiming at recognition of evidence which could be used as an argument *for* or *against*. Arguments *for* or *against* are used in order to make the final decision which one of the solutions is the best for the agent in the current situation (i.e., arguments are labeling statements by judgment values expressing the action priorities). The evaluation of currents needs by agent Ag is realized from the point of view of hierarchy of agent Ag *life* values/needs). Wisdom type of knowledge processing by the agent Ag is characterized by the ability to improve quality of the judgment process based on the agent Ag experiences. In order to emphasize the importance of this ability, we use the concept of *adaptive judgment* in the wisdom equation instead of just *judgment*. An agent who is able to perform adaptive judgment in the above sense, we simply call as a *judge*.

The adaptivity aspects are also crucial from the point of view of interactions [25,51,54,79]. The need for adaptation follows, e.g., from the fact that complex vague concepts on the basis of which the judgment is performed by the agent Ag are approximated by classification algorithms (classifiers) which should drift in time following changes in data and represented knowledge.

An important aspect of Wistech is that the complexity and uncertainty of real-life constraints mean that in practice we must reconcile ourselves to the fact that our judgments are based on non-crisp concepts (i.e., concepts with border-line cases) and also do not take into account all the knowledge accumulated and available to us. This is why our judgments are usually imperfect. But as a consolation, we also learn to improve the quality of our judgments via observation and analysis of our experience during interaction with the environment. Satisfactory decision-making levels can be achieved as a result of improved judgments.

Thus wisdom is directly responsible for the focusing of an agents attention (see Aristotle tetrahedron in Figure 10) on problems and techniques of their solution which are important in terms of the agent judgment mechanism. This mechanism is based on the Maslow hierarchy of needs (see Figure 8) and agent perception of ongoing interactions with other agents and environments. In particular, the agent's wisdom can be treated, as the control at the highest level of hierarchy of the agent's actions and reactions and is based on concept processing in the metaphoric Aristotle tetrahedron (Figure 10). One can use the following conceptual simplification of agent wisdom. Agent wisdom is an efficient and an on-line agent judgment mechanism making it possible for an agent to answer the following questions: (i) How to currently construct the most important priority list of problems to be solved? (ii) How to solve the top priority problems under real life constraints? (iii) What to do next?

One of the main barriers hindering an acceleration in the development of Wistech applications lies in developing satisfactory computational models implementing the functioning of *adaptive judgment*. This difficulty primarily consists in overcoming the complexity of integrating the local assimilation and processing of changing non-crisp and incompletely specified concepts necessary to make correct judgments. In other words, we are only able to model tested phenomena using local (subjective) models and interactions between them. In practical applications, usually, we are not able to give perfect global models of analyzed phenomena. However, we can only approximate global models by integrating the various incomplete perspectives of problem perception.

Wisdom techniques include approximate reasoning by agents or teams of agents about vague concepts concerning real-life dynamically changing, usually distributed, systems in which these agents are operating. Such systems consist of other autonomous agents operating in highly unpredictable environments and interacting with each other.

Wistech is based on techniques of reasoning about knowledge, information and data which helps apply the current knowledge in problem solving in real-life highly unpredictable environments and autonomous multiagent systems. This includes such methods as identification of the current situation on the basis of interactions or dialogs, extraction of relevant fragments of knowledge from knowledge networks, judgment for prediction for relevant actions or plans in the current situation, or judgment of the current plan reconfiguration.

In [26,27,28] Wisdom Technology (Wistech) is discussed as one of the main paradigms for development of new applications in intelligent systems.

Let us summarize the concepts surrounding Wistech.

Wisdom technology (Wistech) is a collection of techniques aimed at the further advancement of technology to acquire, represent, store, process, discover, communicate, and learn *wisdom* in the design and implementation of intelligent systems. These techniques include approximate reasoning by agents or teams of agents about vague concepts concerning real-life dynamically changing, usually distributed, systems in which these agents are operating. Such systems consist of other autonomous agents operating in highly unpredictable environments and interacting with each other. Wistech can be treated as the successor of database technology, information technology, and knowledge management technologies. Wistech is the combination of the technologies represented in (1) and offers an intuitive starting point for a variety of approaches to designing and implementing computational models for Wistech in intelligent systems.

Knowledge technology in Wistech is based on techniques for reasoning about knowledge, information and data which helps apply the current knowledge in problem solving. This includes, e.g., extracting of relevant fragments of knowledge from knowledge networks for making decisions or reasoning by analogy.

Judgment technology in Wistech covers the representation of agent perception and adaptive judgment strategies based on results of the perception of real life scenes in environments and their representations in the agent's mind. The role of judgment is crucial, e.g., in adaptive planning relative to the Maslov Hierarchy of agent needs or goals. Judgment also includes techniques used for perception, analysis of perceived facts, learning, and adaptive improving of approximations of vague complex concepts (from different levels of concept hierarchies in real-life problem solving) applied to modeling interactions in dynamically changing environments (in which cooperating, communicating, and competing agents exist) by using uncertain and insufficient knowledge or resources.

Interaction technology includes techniques for performing and monitoring actions by agents and environments. Techniques for planning and controlling actions are result of the combination of judgment and interaction technologies.

There are many ways to build Wistech computational models. In this paper, the focus is on rough-granular computing (RGC).

Gottfried Wilhelm Leibniz should be considered a precursor of modern Granular Computing (GC) understood as a calculus of human thoughts:

> If controversies were to arise, there would be no more need of disputation between two philosophers than between two accountants. For it would suffice to take their pencils in their hands, and say to each other: 'Let us calculate'.
>
> *Gottfried Wilhelm Leibniz [36]*

... Languages are the best mirror of the human mind, and that a precise analysis of the signification of words would tell us more than anything else about the operations of the understanding.

Gottfried Wilhelm Leibniz [37]

Through centuries since Leibniz formulated the above sentences, mathematicians have been developing tools to deal with this issue. Unfortunately, the tools developed in *crisp* mathematics, in particular, in classical mathematical logic do not yet allow for the understanding natural language used by humans to express thoughts and reasoning about theses thoughts, an understanding which will allow us to construct truly intelligent systems.

One of the reasons is that humans, capable of efficiently solving many real-life problems, are able to express their thoughts by means of vague, uncertain, imprecise concepts and reason with such concepts. Lotfi Zadeh proposed to base the calculus of thoughts using fuzzy logic to move from computing with numbers to computing with words, from manipulations of measurements to manipulations of perceptions, and further to granular computing (GC). This idea has been developed by Lotfi Zadeh himself in a number of papers (see, e.g., [99,100,101,104,105]) and by other researchers, also using rough set methods (see, e.g., [56,62]). The label Granular Computing was suggested by T.Y. Lin in 1998.

The basic objects on which computations are performed in GC are granules. Lotfi Zadeh proposed the term *information granule* as *a clump of objects of some sort, drawn together on the basis of indistinguishability, similarity or functionality. In this definition, being general enough to comprise a large number of special cases, the stress is laid on the reasons for clustering objects into clumps, and three such motives are suggested: indistinguishability, similarity, and functionality.*

Solving complex problems, e.g., by multi-agent systems requires new approximate reasoning methods based on new computing paradigms. One such recently emerging computing paradigm is Rough Granular Computing (RGC) (see, e.g., [40,67,66,74,78,26,27,28,75]).

RGC is an approach to the constructive definition of computations over objects, called granules, aiming at searching for solutions of problems which are specified using vague concepts. Computations in RGC are performed on granules representing often vague, partially specified, and compound concepts delivered by agents engaged in tasks such as knowledge representation, communication with other agents, and reasoning. Granules are obtained through the process of granulation. Granulation can be viewed as a human way of achieving data compression and it plays a key role in implementing the divide-and-conquer strategy in human problem-solving. The approach combines rough set methods with other soft computing methods, and methods based on granular computing (GC). RGC is used for developing one of the possible Wistech foundations based on approximate reasoning using vague concepts.

The research on the foundations on RGC is based on the rough set approach. The rough set concept, due to Pawlak [58,59,61] is based on classical two valued logic. The rough set approach has been developed to deal with uncertainty and vagueness. The approach makes it possible to reason precisely about the approximations of vague concepts. These approximations are temporary, subjective, and change adaptively with changes in environments [7,70,76].

2   Vague Concepts and Approximate Reasoning About Vague Concepts

One of the issues discussed in connection with the notion of a concept is vagueness. Vagueness is usually associated with the existence of borderline objects which cannot be uniquely classified relative to a concept or its complement. Borderline cases of concepts create their boundaries (boundary regions). *Vague concepts* lack sharp (crisp) boundaries and they are susceptible to *sorites paradoxes* [32]. Almost all concepts we use in natural language are vague (e.g., *tall, young, safe*). Such concepts are often used in task specifications or domain knowledge description for intelligent systems. Vague concepts can be complex, e.g., *dangerous situation on the road, tax fraud, financial risk.*

Approximate reasoning about vague concepts concerns the approximation of vague concepts and approximate vague concept reasoning techniques based on their approximations. Techniques for approximate reasoning about complex vague concepts are basic tools in the design of intelligent systems. Approximations of vague concepts are constructed through the fusion of local models based on different subjective views. Such local models can lead to contradictory conclusions because they are built using incomplete knowledge about concepts from different sources and different generalization strategies in inducing local models. Hence, techniques for resolving such contradictory conclusions are important (e.g., voting strategies).

Mathematics requires that all mathematical notions (including sets) must be exact, otherwise precise reasoning would be impossible. However, philosophers (see, e.g., [32]) and recently computer scientists (see references in [61]) as well as other researchers have become interested in *vague* (imprecise) concepts.

In classical set theory a set is uniquely determined by its elements. In other words, this means that every element must be uniquely classified as belonging to the set or not. That is to say the notion of a set is a *crisp* (precise) one. For example, the set of odd numbers is crisp because every number is either odd or even. In contrast to odd numbers, the notion of a beautiful painting is vague, because we are unable to classify uniquely all paintings into two classes: beautiful and not beautiful. Some paintings cannot be classified whether they are beautiful or not and thus they remain in the doubtful (grey) area. Thus, *beauty* is not a precise but a vague concept. Almost all concepts in natural language are vague. Therefore, common sense reasoning based on natural language must be based on vague concepts and not on classical logic.

The idea of vagueness can be traced back to the ancient Greek philosopher Eubulides of Megara (ca. 400BC) who first formulated the so called *sorites* (heap) and *falakros* (bald man) paradoxes (see, e.g., [32]). The bald man paradox goes as follows: suppose a man has 100,000 hairs on his head. Removing one hair from his head surely cannot make him bald. Repeating this step we arrive at the conclusion that a man with no hair is not bald. Similar reasoning can be applied to a heap of stones.

Vagueness is usually associated with the boundary region approach (i.e., existence of objects which cannot be uniquely classified relative to a set or its complement) which was first formulated in 1893 by the father of modern logic, German logician, Gottlob Frege (1848-1925) (see [19]). According to Frege the concept must have a sharp boundary. To a concept without a sharp boundary there would correspond an area that would not have any sharp boundary–line all around. It means that mathematics must use crisp (precise), not vague concepts, otherwise it would be impossible to reason precisely.

We would like to note that modern understanding of the notion of a vague (imprecise) concept has a quite firmly established meaning context including the following issues [32]:

1. The presence of borderline cases.
2. Boundary regions of vague concepts are not crisp.
3. Vague concepts are susceptible to sorites paradoxes.

Moreover, it is usually assumed that the understanding (approximation) of vague concepts (their semantics is determined by the satisfiability relation) depends on the agent's knowledge, which is often changing. Hence, the approximation of vague concepts by an agent should also be considered changing with time (this is known as the concept drift). In the 20th century, it became obvious that new specialized logic tools needed to be developed to investigate and implement practical problems involving vague concepts. One of such tool is that of rough sets. It was shown that rough set tools are making it possible to deal with vague concepts (see, e.g., [61,7,70]).

Vague complex concepts are very often related to one another through a hierarchy induced by abstraction levels of these concepts. Such hierarchies occur, for instance, when some concepts are components of other concepts. In such a context, there is a great interest in investigating the relation *being a part-of*. It is a different approach from the ontology of modern mathematics (also known as *Cantor ontology*), which is based on the relation *being an element-of*. Such alternative ontology for mathematics was proposed by Stanisław Leśniewski [38,39] in 1929 and became the inspiration for an important research within *rough mereology* (see, e.g., [67,56]). Within this approach, the notion *rough inclusion relation* plays a central role. It describes to what degree some concepts are parts of other concepts. A rough mereological approach is based on the relation to be a part to a degree.

Certainly, the mereological approach is not the only attempt at establishing links between vague concepts and rough sets. Among others are links based on treating vague concepts by means of logical values. In this case, one can build an algebra of such vague concepts as an algebra of logical values. Usually, this kind of algebra is a pseudo-Boolean algebra and the relationships between vague concepts can be expressed as relationships of logical values of an intermediate logic (see, e.g. [27]).

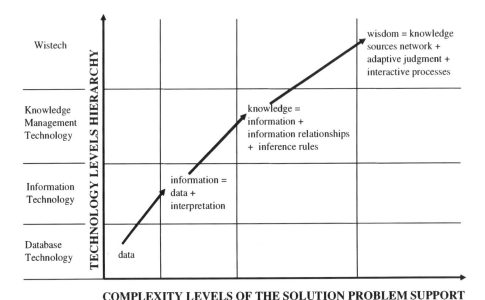

COMPLEXITY LEVELS OF THE SOLUTION PROBLEM SUPPORT

Fig. 1. Wisdom equation context

3   Wisdom Technology (Wistech)

On the one hand, the idea expressed by (1) (the wisdom equation paradigm) is a step in the direction of a new philosophy for the use of computing machines in our daily life, referred to as ubiquitous computing. This paradigm is strongly connected with various applications of autonomic computing [43]. On the other hand, it should be emphasized that the idea of integrating many basic AI concepts (e.g., interaction, knowledge, network, adaptation, assessment, simulation of behavior in an uncertain environment, planning and problem solving) is as old as the history of AI itself. Many examples of such an approach adopted by researchers in the middle of the 20th century can be found in [18]. This research was intensively continued in the second half of the 20th century and at the beginning of this century (see, e.g., [106,107,108,109]).

The experiences of the authors of these articles, gained from numerous real-life projects, are summarized by the authors in the metaphoric wisdom equation (1). This equation can also be illustrated using the following diagram presented in Figure 1. In this figure the term *data* is understood as a stream of symbols without any interpretation of their meaning. From the perspective of the metaphor expressed in the wisdom equation (1), Wistech can be perceived as the integration of three technologies (corresponding to three components in the wisdom equation (1)). At the current stage two of them, i.e., *knowledge* (presented in a bit more detailed form in Figure 1 as *knowledge sources network*) and *interactions* (presented in a bit more detailed form in Figure 1 as *interactive processes*) seem to be conceptually more clear, than the third one, namely:

1. *knowledge sources network*—by knowledge we traditionally understand every organized set of information along with the inference rules; in this context one can easily imagine the following examples illustrating the concept of a knowledge sources network:
    - representation of states of reality perceived by our senses (or observed by the *receptors* of another observer) are integrated as a whole in our minds in a network of sources of knowledge and then stored in some part of our additional memory,
    - a network of knowledge levels represented by agents in some multi-agent system and the level of knowledge about the environment registered by means of receptors;
2. *interactive processes*—interaction understood as a sequence of stimuli and reactions over time; examples are:
    - the dialogue of two people,
    - a sequence of actions and reactions between an unmanned aircraft and the environment in which the flight takes place, or
    - a sequence of movements during some multi-player game.

Let us note that, while some aspects of interactions are quite clear, the methods for modeling of interactive computations necessary for developing Wistech need a lot of investigations (see, e.g., [25,51,54,79]).

The concept of adaptive judgment distinguishing wisdom from the general concept of problem solving conceptually seems to be far more difficult. Intuitively this concept can be expressed as follows:

- *adaptive judgment*—understood here as mechanisms in a metalanguage (meta-reasoning) which on the basis of selection of available sources of knowledge, and on the basis of understanding of history of interactive processes and their current status, enable us to perform the following activities under real-life constraints:
    - identification and judgment of importance (for future judgment) of phenomena, available for observation, in the surrounding environment;
    - planning current priorities for actions to be taken (in particular, on the basis of an understanding of the history of interactive processes and their current status) toward making optimal judgments;
    - selection of fragments of ordered knowledge (hierarchies of information and judgment strategies) satisfactory for making a decision at the planned time (a decision here is understood as a commencing interaction with the environment or as selecting the future course to make judgments);
    - prediction of important consequences of the planned interaction of processes;
    - learning and, in particular, reaching conclusions from experience leading to adaptive improvement in the adaptive judgment process.

The difficulty in developing satisfactory computational models implementing the functioning of *adaptive judgment* primarily consists in overcoming the complexity of integrating the local assimilation and processing of changing non-crisp and incomplete concepts necessary to make correct judgments. In other words, we are only able to model tested phenomena using local (subjective) models and interactions between them. In practical applications, usually, we are not able to give global models of analyzed phenomena (see, e.g., [89,41,43,16,14]). However, we can approximate global models by integrating the various incomplete perspectives of problem perception. One of the potential computation models for *adaptive judgment* might be the *rough-granular approach*.

In natural language, the concept of wisdom is used in various semantic contexts. In particular, it is frequently semantically associated with such concepts as inference, reasoning, deduction, problem solving, common sense reasoning, reasoning by analogy, and others. As a consequence this semantic proximity may lead to misunderstandings. For example, one could begin to wonder what the difference is between the widely known and applied concept in AI of *inference engine* and the concept of *wisdom engine* defined here? In order to avoid this type of misunderstanding it is worth explaining the basic difference between the understanding of wisdom and such concepts as inference, reasoning, deduction and others.

Above all, let us start with an explanation of how we understand the difference between problem solving and wisdom.

An attempt at explaining the concept of wisdom can be made using the concept of *problem solving* in the following manner: wisdom is the ability to identify important problems, search for sufficiently correct solutions to them, having in mind real life, available knowledge sources, personal experience, constraints, etc. Having in mind this understanding of wisdom, at once we get the first important difference. Namely, in the problem solving process we do not have the following important wisdom factor: *Identify important problems and problem solution constraints.*

Certainly, this is not the only difference. Therefore, one can illustrate the general difference between the concept of problem solving and wisdom as the difference between the concept of flying in an artificially controlled environment (e.g., using a flight simulator and problem solving procedures) and the concept of flying a Boeing 767 aeroplane in a real-life dangerous environment (wisdom in a particular domain).

One can therefore think that wisdom is very similar to *the ability of problem solving in a particular domain of application*, which in the context of the world of computing machines is frequently understood as an *inference engine*. The commonly accepted definition of the concept of an inference engine can be found for example in Wikipedia (http://en.wikipedia.org/wiki/Inference_engine). It reads as follows:

> An inference engine is a computer program that tries to derive answers from a knowledge base. It is the *brain* that expert systems use to reason about the information in the knowledge base, for the ultimate purpose of formulating new conclusions.

An inference engine has three main elements. They are:

1. An interpreter. The interpreter executes the chosen agenda items by applying the corresponding base rules.
2. A scheduler. The scheduler maintains control over the agenda by estimating the effects of applying inference rules in light of item priorities or other criteria on the agenda.
3. A consistency enforcer. The consistency enforcer attempts to maintain a consistent representation of the emerging solution.

In other words, the concept of an inference engine relates to generating strategies for the inference planning from potentially varied sources of knowledge which are in interaction together. So this concept is conceptually related to the following two elements of the wisdom equation:

1. knowledge sources network,
2. interactive processes.

However, it should be remembered that wisdom in our understanding is not only some general concept of *inference*. The basic characteristic of wisdom, distinguishing this concept from the general understanding of inference, is *adaptive ability to make correct judgments having in mind real-life constraints*. The significant characteristic differentiating wisdom from the general understanding of such concepts as problem solving or inference engines, is adaptive judgment.

In analogy to what we did in the case of *problem solving* we can now attempt to explain the concept of wisdom using the concept of *inference engine* in the following manner: Wisdom is an inference engine interacting with the real-life environment, which is able to identify important problems and to find sufficiently correct solutions for them having in mind real-life constraints, available knowledge sources and personal experience. In this case, one can also illustrate the difference between the concept of an inference engine and the concept of wisdom using the metaphor of flying a plane.

One could ask the question of which is the more general concept: wisdom or problem solving? Wisdom is a concept carrying a certain additional structure of adaptive judgment which in a continuously improving manner assists us in identifying the most important problem to resolve in a given set of constraints, and what an acceptable compromise between the quality of the solution and the possibility of achieving a better solution is. Therefore, the question of which is the more general concept, closely resembles the question from mathematics: What is the more general concept in mathematics: the concept of a field (problem solving), or the concept of the vector space over a field (wisdom understood as problem solving + adaptive judgment)? The vector space is a richer mathematical structure due to the possible operations on vectors. Analogously to wisdom it is the richer process (it includes adaptive judgment—a kind of meta-judgment to include judgments relating to problem solving). On the other hand, research into single-dimensional space can be treated as the research of fields. In this sense, the concept of vector space over a field is more general than the concept of a field.

Today nobody doubts that technologies based on computing machines are among the most important technology groups of the 20th century, and, to a considerable degree, have been instrumental in the progress of other technologies. Analyzing the stages in the development of computing machines, one can quite clearly distinguish the following three stages in their development in the 20th century:

1. Database Technology (gathering and processing of transaction data).
2. Information Technology (understood as adding to the database technology the ability to automate analysis, processing and visualization of information).
3. Knowledge Management Technology (understood as systems supporting organization of large data sets and the automatic support for knowledge processing and discovery (see, e.g., [33])).

The three stages of development in computing machine technology show us the trends for the further development in applications of these technologies. These trends can be easily imagined using the further advancement of complexity of information processing (Shannon Dimension) and the advancement of complexity of dialogue intelligence (Turing Dimension), viz.

- *Shannon Dimension* level of information processing complexity (representation, search, use);
- *Turing Dimension* the complexity of queries that a machine is capable of understanding and answering correctly. One of the objectives of AI is for computing machines to reach the point in the Turing Dimension that is well-known as the Turing Test (see [93]).

In this framework, the development trends in the application of computing machines technology can be illustrated in Figure 2.

Immediately from the beginning of the new millennium one can see more and more clearly the following new application of computing machine technology, viz., wisdom technology (Wistech) which put simply can be presented in a table (see Figure 3, being an extension of the table presented in Figure 2).

In other words, the trends in the development of the technology of computing machines can be presented using the so-called DIKW hierarchy (i.e., Data, Information, Knowledge, Wisdom). Intuitively speaking, each level of the DIKW hierarchy adds certain attributes over and above the previous one. The hierarchy is presented graphically in Figure 4.

DIKW hierarchy can be traced back to the well-known poem by T.S. Eliot, *The Rock*, written in 1932. He wrote:

Where is the life we have lost in living?
Where is the wisdom we have lost in knowledge?
Where is the knowledge we have lost in information?

Technology	Additional attributes	Shannon Dimensions	Turing Dimensions
Database Technology	data is the most basic level	How to represent information?	SQL
Information Technology	information = data + interpretation	Where to find information?	Who? What? When? Where? How much?
Knowledge Management Technology	knowledge = information + information relationships + inference rules	How to use information?	How? Why? What if?

Fig. 2. Computing machines technology

Technology	Addional attributes	Shannon Dimensions	Turing Dimensions
Wisdom Technology (Wistech)	Wisdom equation, i.e. wisdom = knowledge sources network + adaptive judgment + interactive processes	Learn when to use information Learn how to get important information	How to make correct judgments (in. particular correct decisions) keeping in mind real life constrains?

Fig. 3. Computing machine technology (continued)

It is a truism to state that the effects of any activity depend to a decisive degree on the wisdom of the decisions taken, at the start, during the implementation, improvement and completion of the activity. The main objective of Wistech is to automate support for the processes leading to wise actions. These activities cover all areas of man's activities, from the economy, through medicine, education, research, development, etc.

In this context, one can clearly see how important a role Wistech development may have in the future. The following comment from G.W. Leibniz on the idea to automate the processing of concepts representing thoughts should not surprise us either:

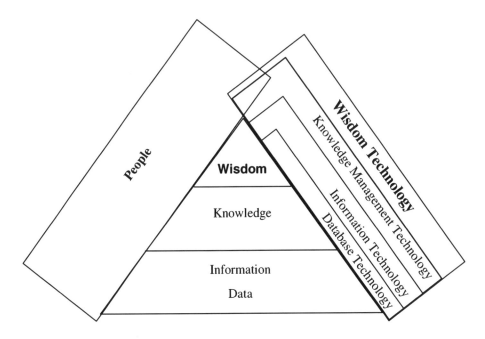

Fig. 4. DIKW hierarchy

No one else, I believe, has noticed this, because if they had ... they would have dropped everything in order to deal with it; because there is nothing greater that man could do.

In order to create general Wistech computational models let us start with an analysis of the concept of adaptive judgment.

For better familiarization with adaptive judgment we shall use the visualization of processes based on the IDEFO standard. This means of visualisation is described in the diagram presented in Figure 5.

An intrinsic part of the concept of judgment is its relation to the entity implementing the judgment. Intuitively this can be a person, animal, machine, abstract agent, society of agents, etc. In general, we shall call the entity making a judgment the *judge*. We shall also assume that the knowledge sources network is divided into external sources (i.e., sources of knowledge that are also available to other judges) and internal sources (which are only available to the specific judge in question).

The first level of the model is presented in Figure 6. Of course, successive levels of the model are more complex. Its details may depend on the assumed paradigms for the implementation of adaptive judgment. However, these details should include such elements as:

1. Learning of the *External Communication Language* understood as a language based on concepts used to communicate and process knowledge with a network of external sources of knowledge;

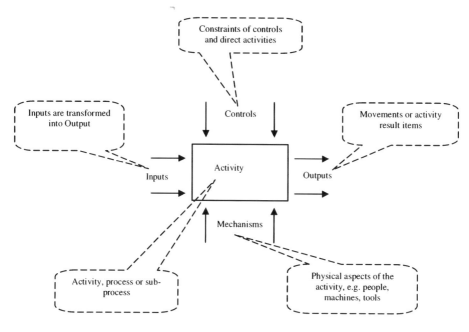

Fig. 5. Activity

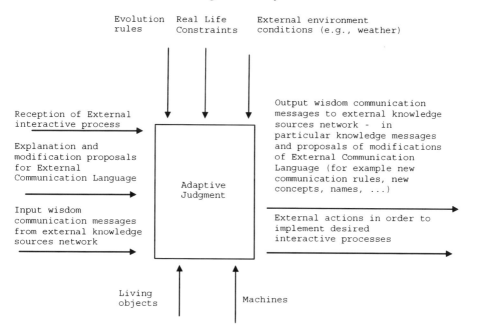

Fig. 6. The first level of the model

2. Learning of the *Internal Communication Language* understood as a hierarchy of meta-languages based on concepts used to process and improve the External Communication Language and a language based on concepts used to communicate and process knowledge with a network of internal sources of knowledge;

3. Receiving in memory signals from signal receptors and interactive processes and expressing their significance in the *External Communication Language* and the *Internal Communication Language*;

4. Planning the current priorities for internal actions (mainly related to the processing of wisdom) on the basis of an assessment in relation to the hierarchy of values controlling the adaptive judgment process;

5. Selection of fragments of ordered knowledge (hierarchies of information and judgment strategies) sufficient to take a decision at the planned time (a decision is understood here as commencing an interaction with the environment or selecting a future course to resolve the problem);

6. Outputting wisdom communication messages to external knowledge sources network, in particular, knowledge messages and proposals of modifications of the *External Communication Language* (e.g., new communication rules, new concepts, names);

7. External actions in order to implement the desired interactive processes.

All elements occurring in the above list are very complex and important, but the following two problems are particularly important for adaptive judgment computational models:

1. *Concept learning and integration*—this is the problem of designing computational models for the implementation of learning concepts important to the representation, processing and communication of wisdom and, (in particular, this relates to the learning of concepts) improving the quality of approximation of the integration of incomplete local perceptions of a problem (arising during local assimilation and processing of fuzzy and incomplete concepts).

2. *Judge hierarchy of habits controls*—this is the problem of building of computational models for the implementation of a functioning hierarchy of habit controls by a judge controlling the judgment process in an adaptive way.

Now, we outline the idea of a framework for a solution to the problem of implementing a judge's hierarchy of habit controls. In this paper, we treat the concept of a *habit* as an elementary and repeatable part of a behavioral pattern. In this context, the meaning of elementary should be considered, by comparison to the required reasoning (knowledge usage) complexity necessary to the behavioral pattern implementation. In other words, by a habit we mean any regularly repeated behavioral pattern that requires little or no reasoning effort (knowledge usage). In general, any behavioral pattern could be treated as a sequence of habits and other activities which use knowledge intensively. Among such activities especially important are activities leading to new habits. We assume that such habit processing is controlled by so-called *habit controls* which

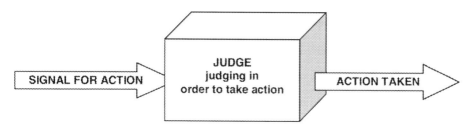

Fig. 7. Judge diagram

support the following aspects of an adaptive judgment process for a situation considered by a *judge*:

1. *Continuous habit prioritization* to be used in a particular situation after identification of habits. This is a prioritization from the point of view of the following three criteria:
   - The predicted consequences of the phenomena observed in a considered situation;
   - Knowledge available to a judge;
   - The actual plans of a judge's action.
2. *Knowledge prioritization* is used if we do not identify any habit to be used given a considered situation, then we have to perform a prioritization of the pieces of available knowledge which could be used for identification of the best habit or for the construction of a new habit to be used in the considered situation.
3. *Habit control assessment* for continuous improvement of the adaptive judgment process and for the construction of new habits and habit controls.

One of the key components of Wistech is the judge (or agent) hierarchy of habit control[3], essential for optimal decision making and closely correlated with the knowledge held and interactions with the environment. Judge hierarchy also means the desire of the judge to satisfy their needs in interactions with their environment. Put very simply, the judge receives and sends out signals according to the diagram presented in Figure 7.

The interior of the box is the place for the judge to process signals and to take an action.

For judgment the following two interacting adaptive processes are of great importance:

1. *adaptation of the environment*, in which the judge *lives* with their *needs and objectives* so as to best fit the *needs and objectives* of the environment,
2. *adaptation of the internal processes taking place in a judge* in such a way as to best realize their *needs and objectives* based on the resources available in the environment.

---

[3] Habits are understood as elementary and repeatable parts of behavioral patterns.

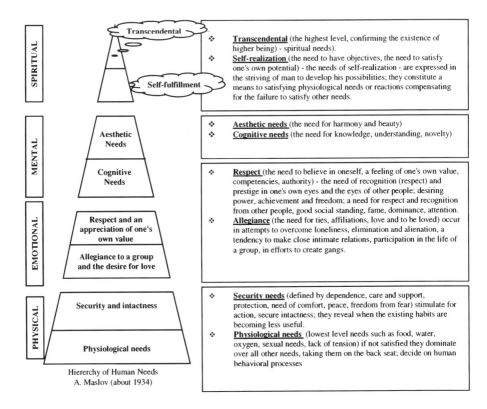

Fig. 8. The Maslov Hierarchy of human needs (about 1934) as an example of judge hierarchy of habit controls

The judge's environment adaptation is the basis for computational models of judge learning. A key part of this, is the evolution of a judge hierarchy of habit controls. The judge hierarchy of habits controls constitutes a catalyst for evolutionary processes in the environment, and also constitutes an approach to expressing various paradigms of computational models to be used in the machine implementation of this concept. For example, these paradigms can be based on the metaphorically understood principle of Newtonian dynamics (e.g., action = reaction), thermodynamics (e.g., increase in entropy of information), quantum mechanics (the principle of it being impossible to determine *location and speed simultaneously*), psychology (e.g., based on metaphorical understanding of Maslow's hierarchy of needs). Particularly worthy of attention in relation to Wistech is the metaphoric approach to Maslow's hierarchy of needs in reference to the abstractly understood community of agents. This hierarchy is presented in Figure 8. It could be used for the direct construction of computational models of judge's hierarchy of habit controls.

4 Wisdom Technology & Rough-Granular Computing

Rough-Granular Computing (RGC) is used for developing one of the possible Wistech foundations based on approximate reasoning about vague concepts.

One of the main branches of Granular Computing (GC) is Computing with Words and Perceptions (CWP). *GC derives from the fact that it opens the door to computation and reasoning with information which is perception—rather than measurement—based. Perceptions play a key role in human cognition, and underlie the remarkable human capability to perform a wide variety of physical and mental tasks without any measurements and any computations. Everyday examples of such tasks are driving a car in city traffic, playing tennis and summarizing a story* [103].

We consider the optimization tasks in which we are searching for optimal solutions satisfying some constraints. These constraints are often vague, imprecise, and the specifications of concepts and their dependencies which constitute the constraints, are often incomplete. Decision tables [59] are examples of such constraints. Another example of constraints can be found, e.g., [5,4,50,74] where a specification is given by domain knowledge and data sets. Domain knowledge is represented by an ontology of vague concepts and the dependencies between them. In a more general case, the constraints can be specified in a simplified fragment of a natural language [103].

Granules are constructed using information calculi (see, e.g., [66,74,78,76,75]). Granules are objects constructed in computations aiming at solving the mentioned above optimization tasks. Among examples of granules one may distinguish different kinds of patterns, decision rules or sets of decision rules, clusters, classifiers, coalitions of agents [74,76,75]. In RGC the general optimization criteria based on the minimal length principle [61] are used. In searching for (sub)optimal solutions it is necessary to construct many compound granules using some specific operations such as generalization, specification or fusion. Granules are labeled with parameters. By tuning these parameters we optimize the granules relative to their description size and the quality of data description, i.e., two basic components on which the optimization measures are defined.

From this general description of tasks in RGC, it follows, that together with the specification of elementary granules and operation on them it is necessary to define measures of granule quality (e.g., measures of their inclusion such as the rough inclusion measures [67,74], covering or closeness) and tools for measuring the sizes of granules. Optimization strategies of already constructed (parameterized) granules are also very important.

Developing methods for the approximation of compound concepts expressing the result of perception is one of the main challenges of Perception Based Computing (PBC). The perceived concepts are expressed in natural language. We discuss the rough-granular approach to approximation of such concepts from sensory data and domain knowledge. This additional knowledge, represented by an ontology of concepts, is used to make it feasible to search for features (condition attributes) relevant to the approximation of concepts on different levels of the concept hierarchy defined by a given ontology. We report several experiments of the proposed methodology for the approximation of compound concepts from sensory data and domain knowledge. The approach is illustrated by examples relative to interactions of agents, ontology approximation, adaptive hierarchical learning of compound concepts and skills, behavioral pattern identification, planning, conflict analysis and negotiations, and perception-based reasoning. The presented results seem to justify the following claim of Lotfi A. Zadeh:

> In coming years, granular computing is likely to play an increasingly important role in scientific theories—especially in human-centric theories in which human judgment, perception and emotions are of pivotal importance.

The question of how concept ontologies can be discovered from sensory data remains as one of the greatest challenges for many interdisciplinary projects on the learning of concepts.

The concept approximation problem is the basic problem investigated in machine learning, pattern recognition and data mining [20]. It is necessary to induce approximations of concepts (models of concepts) consistent (or almost consistent) with some constraints. In the most typical case, the constraints are defined by a training sample. For more complex concepts, we consider constraints defined by a domain ontology consisting of vague concepts and dependencies between them. Information about the classified objects and concepts is incomplete. In the most general case, the adaptive approximation of concepts is performed under interaction with the dynamically changing environment. In all these cases, searching for sub-optimal models relative to the minimal length principle (MLP) is performed. Notice that in adaptive concept approximation, one of the components of the model should be the adaptation strategy. Components involved in the construction of concept approximations which are tuned in searching for sub-optimal models relative to MLP are called information granules. In rough granular computing (RGC), information granule calculi are used in the construction of classifier components and classifiers themselves (see, e.g., [76]) satisfying given constraints. An important mechanism in RGC is related to generalization schemes making it possible to construct complex patterns from simpler patters. Generalization schemes are tuned using, e.g., some evolutionary strategies.

Rough set theory due to Zdzisław Pawlak [59,61], is a mathematical approach to imperfect knowledge. The problem of imperfect knowledge has been tackled for a long time by philosophers, logicians and mathematicians. Recently has also become a crucial issue for computer scientists, particularly in the area of artificial intelligence. There are many approaches to the problem of how to understand and manipulate imperfect knowledge. The most successful one is, no doubt, the fuzzy set theory proposed by Lotfi A. Zadeh [98]. Rough set theory presents another attempt to solve this problem. It is based on an assumption that objects and concepts are perceived by partial information about them. Due to this some objects can be indiscernible. From this fact, it follows, that some sets can not be exactly described by available information about objects; they are rough and not crisp. Any rough set is characterized by its (lower and upper) approximations. The difference between the upper and lower approximation of a given set is called its boundary. Rough set theory expresses vagueness by employing a boundary region of a set. If the boundary region of a set is empty it means that the set is crisp, otherwise the set is rough (inexact). A nonempty boundary region of a set indicates that our knowledge about the set is insufficient to define the set precisely. One can recognize that rough set theory is, in a sense, a formalization of the idea presented by Gotlob Frege [19].

One of the consequences of perceiving objects using only available information about them is that for some objects one cannot decide if they belong to a given set or not. However, one can estimate the degree to which objects belong to sets. This is another crucial observation in building the foundations for approximate reasoning. In dealing with imperfect knowledge one can only characterize satisfiability of relations between objects to a degree, and not precisely. Among relations between objects, the rough inclusion relation, which describes to what degree objects are parts of other objects, plays a special role. A rough mereological approach (see, e.g., [67,56]) is an extension of the Leśniewski mereology [38] and is based on the relation *to be a part to a degree*.

A very successful technique for rough set methods was Boolean reasoning (see, e.g., [61]). The idea of Boolean reasoning is based on the construction for a given problem P and a corresponding Boolean function f_P with the following property: the solutions for the problem P can be decoded from prime implicants of the Boolean function f_P . It is worth mentioning that to solve real-life problems it is necessary to deal with Boolean functions having a large number of variables.

A successful methodology based on the discernibility of objects and Boolean reasoning has been developed in rough set theory for computing of many key constructs like reducts and their approximations, decision rules, association rules, discretization of real value attributes, symbolic value grouping, searching for new features defined by oblique hyperplanes or higher order surfaces, pattern extraction from data as well as conflict resolution or negotiation [69,49,61]. Most of the problems involving the computation of these entities are NP-complete or NP-hard. However, we have been successful in developing efficient heuristics yielding sub-optimal solutions for these problems. The results of experiments on many data sets are very promising. They show a very good quality of solutions generated by the heuristics in comparison with other methods reported in literature (e.g., with respect to the classification quality of unseen objects). Moreover, they are very time-efficient. It is important to note that the methodology makes it possible to construct heuristics having a very important approximation property. Namely, expressions generated by heuristics (i.e., implicants) close to prime implicants define approximate solutions for the problem.

The rough set approach offers tools for approximate reasoning in multiagent systems (MAS). A typical example is one agent's approximation of concepts of another agent. The approximation of a concept is based on a decision table representing information about objects perceived by both agents.

The strategies for data model inducing developed so far are often unsatisfactory for the approximation of compound concepts that occur in the perception process. Researchers from different areas have recognized the necessity to work on new methods of concept approximation (see, e.g., [13,95]). The main reason for this is that these compound concepts are, in a sense, too far from measurements which makes the searching for relevant features in a very large space unfeasible. There are several research directions aiming to overcome this difficulty. One of them is based on the interdisciplinary research where knowledge pertaining to perception in psychology and neuroscience is used to help to deal with compound concepts (see, e.g., [46]). There is a great effort in neuroscience towards understanding the hierarchical structures of neural networks in living organisms [46]. Also mathematicians are recognizing the problems of learning as the main problem of the current century [65]. These problems are closely related to complex system modeling as well. In such systems again the problem of concept approximation and its role in reasoning about perceptions is one of the challenges nowadays. One should take into account that modeling complex phenomena entails the use of local models (captured by local agents, if one would like to use the multi-agent terminology [89]) that should be fused afterwards. This process involves negotiations between agents [89] to resolve contradictions and conflicts in local modeling. This kind of modeling is becoming more and more important in dealing with complex real-life phenomena which we are unable to model using traditional analytical approaches. The latter approaches lead to exact models. However, the necessary assumptions used to develop them result in solutions that are too far from reality to be accepted. New methods or even a new science should therefore be developed for such modeling [23].

Let us consider some illustrative examples of granule modeling. In this example we would like to emphasize the interactions between modeling of syntax and semantics. One can find here an analogy to the Aristotle triangle and the interactions between the left-brain and the right brain [55,85].

Any object $x \in U$, in a given information system $IS_1 = (U, A)$, is perceived by means of its signature $Inf_A(x) = \{(a, a(x)) : a \in A\}$ [61]. On the first level, we consider objects with signatures represented by the information system $IS_1 = (U, A)$. Objects with the same signature are indiscernible. On the next level of modeling we consider as objects some relational structures over the signatures of objects from the first level. For example, for any signature u one can consider as a relational structure a neighborhood defined by a similarity relation between signatures of objects from the first level. Attributes of objects on the second level describe properties of relational structures. Hence, indiscernibility classes defined by such attributes are sets of relational structures; in our example sets of neighborhoods. One can continue this process of hierarchical modeling on the third level by considering signatures of objects from the second level as objects. In our example, the third level of modeling represents modeling of clusters of neighborhoods defined by the similarity relation. Observe that it is possible to link objects from a higher level with objects from a lower level. In our example, any object from the second level is a neighborhood of similar object descriptions. Any element from this neighborhood defines an elementary granule (indiscernibility class) of objects on the first level. Hence, any such neighborhood defines, on the first level, a family of elementary granules corresponding to signatures from the neighborhood. One can consider as a granule quality measure a function assigning to the neighborhood a degree to which the union of its elementary granules is included into a given concept.

The idea of interactions between syntax and semantics can be expressed on a more abstract level using adjoint functors [27]. Now, we would like to outline the basic idea of this approach.

Alfred Tarski, in his research on satisfiability relation and the concept of truth, investigated features of a class $Mod(A)$ of models satisfying some set A of expressions of the language as well as the set of features $Th(M)$ of a class of models M [92]. Clearly, by definition, these functions Mod and Th are *adjoint*, i.e., they satisfy the condition:

$$M \subseteq Mod(A) \text{ if and only if } Th(M) \vdash A, \tag{2}$$

where

- M denotes models, worlds, memorized sequences of receptors receiving stimuli from the environment; sometimes subclasses of M are called *scenes*,
- A denotes a set of expressions of the language; the expressions are used to represent or denote concepts,
- $M_1 \subseteq M_2$ asserts inclusion of the model M_1 in M_2; for instance, a *scene reasoning* process may involve, for some reasons, certain models (e.g., when planning a trip by car, we consider all possible access paths at the beginning; then, depending on some other conditions, constraints, and other criteria, we gradually rule out irrelevant models and, as a consequence, we consider only a class of models M_1 with actual prospective access route models containing only the necessary information to make the trip),
- $Mod(A)$ denotes the class of all models satisfying expressions A,
- $Th(M)$ denotes the class of all language expressions that hold true in all models belonging to the class M,
- $A_1 \vdash A_2$ states that the all expressions belonging to A_2 can be derived in accordance with considered deduction rules from the set of formulas A_1.

Condition (2) is usually expressed in the formalism of Galois connections or, more generally, adjoint functors as follows:

$$\frac{M \subseteq Mod(A)}{Th(M) \vdash A}, \tag{3}$$

and we say that Th is the left adjoint to functor Mod, whereas Mod is the right adjoint to functor Th. Using the language of category theory [45,44], we can also say that Th and Mod are adjoint to each other, and simply write

$$Th \dashv Mod. \tag{4}$$

Intuitively, $Th(M)$ can be regarded as a verbal description in the language of features of the models belonging to the class M. On the other hand, $Mod(A)$ can be intuitively considered as a projection of an agent's understanding of the expressions A about the class of models satisfying these expressions. In this way, we obtain the following metaphor of the adjoint functor above mentioned:

$$\begin{array}{c} \textit{Linguistic description in a language} \\ \textit{of a class of models } M \end{array} \dashv \begin{array}{c} \textit{Imaginated models of a given set of} \\ \textit{language expressions } A \end{array}$$

Similarly, a metaphoric evaluation of the aforementioned conjugation allows us to obtain, for a class of *perceived worlds* M and *expressions* A describing M, the following metaphors in natural language:

$$\textit{Description of } M \textit{ in a language} \dashv \textit{Imaginated interpretation of } A,$$

$$\textit{Symbolic reasoning about } M \dashv \begin{array}{l} \textit{Scene reasoning described by} \\ \textit{expressions from } A, \end{array}$$

$$\textit{Judgments concerning } M \dashv \begin{array}{l} \textit{Action plans derived from expressions} \\ A, \end{array}$$

$$\textit{Analysis of features of models in } M \dashv \begin{array}{l} \textit{Synthesis of classes of models} \\ \textit{satisfying given features from } A, \end{array}$$

$$\textit{Judgments concerning } M \dashv \begin{array}{l} \textit{Emotions concerning expressions} \\ \textit{from } A, \end{array}$$

$$\begin{array}{c} \textit{Logical functions of the left} \\ \textit{hemisphere (e.g., word computing) as} \\ \textit{perception effects by sensory organs} \\ \textit{of models in } M \end{array} \dashv \begin{array}{l} \textit{Imaginative functions of the right} \\ \textit{hemisphere (scene calculus) as effects} \\ \textit{of understanding of propositions from} \\ A \textit{ by the brain.} \end{array}$$

It can be seen from the metaphors above that adjoint functors can be regarded in a natural way as a generalization of the concept of semantics understood as a binary relation between model and language. It particularly concerns the Cartesian-closed categories, intensively investigated by Joachim Lambek and others [35]. In these categories, morphisms are considered as deductive reasoning operators. Hence, adjoint categories can be considered as pairs of categories corresponding to language and models. This kind of approach to semantics by means of adjoint functors could be a starting point to research about formula-less and model-less semantics through an analogy to point-less topology where *points* are represented by open sets including points (see, [30]). In other words, this kind of semantics deals with approximations of formulas and approximations of models instead of dealing directly with formulas and models only.

The above metaphors are only to draw attention to the adjoint relations between the symbolic reasoning category and imaginative-scene reasoning category. It is worth stressing that it has become common in cognitive studies and multi-agent interaction [86] to pay more and more attention to the role of the duality, illustrated above, in processes involving cognition and intelligence

[85]. The intuitions illustrated here are basic for the implementation of granular computing [2,62,26,27,28,74,75,76,78,80], where an *intelligent* agent is equipped with two hemispheres (*of a brain*)—the left one is used for describing things in a symbolic language and for symbolic reasoning, while the right hemisphere deals with imagining by the agent the acceptable models satisfying certain features and with reasoning on possibilities of traversing from one model to the other (scene reasoning). Such an agent communicates with the world through sensors attached to both the brain's hemispheres. Intuitively, the essence of granular computing is to *construct* the best possible ontological tools that are helpful in the discovery of information granules which may help us wisely solve practical problems. In the context of rough sets and approximation spaces, granulation is rooted in the discovery of elementary sets that identify neighborhoods of objects of interest and give substance to our search for instances of classes of objects (i.e., concepts) as well as which serve to establish a ground for perception [103]. A classical example illustrating these intuitions by our civilization for nearly 2000 years since the ancient Greeks, is the history of solving the problem of geometric constructions. This problem has become relatively easy due to a granulation method initiated by Évariste Galois. Using the Galois theory, certain problems in field theory may be reduced to group theory, which is in some sense simpler and better understood. In order to demonstrate the impossibility of such geometric constructions as squaring the circle, angle trisection or doubling the cube, we can express the problem in an algebraic language (instead of a geometric one) by means of the following observation: A given number can be constructed using a ruler and a compass if and only if its rank over the field of rationals is a natural power of 2. At the same time, using the Galois theory, one can express an algebraic problem by means of group theory, which can be solved relatively more easily [22].

In some sense, the Galois idea has been generalized by Garrett Birkhoff. He noticed in 1940 [11] that any binary relation yields two inverse dual isomorphisms called *polarities*. He introduced this name because they also generalize the dual isomorphism between *polars* in analytic and projective geometry. This concept has been further generalized to any partial order, and was called *Galois connections*. Next, the concept of Galois connections has been generalized to category theory by Daniel M. Kan [31], who introduced the concept of *adjoint functors*. In computer science, Galois connections are associated with several other names such as classification [3], contexts [21], or the Chu spaces [3].

One of the key research directions within the Rasiowa–Pawlak school [27] is the investigation of algebraic features of various kinds of logics. Especially important is the method of proving completeness by means of the so-called *canonical models* (which are built of terms, language's expressions and a filter representing a given theory), described in the following section. The method shows how, having at ones disposal a description, to imagine models for it (to use the aforementioned metaphor, it is a cycle of back and forth shifts from the left hemisphere to the right one, and reverse). On the other hand, having models that can be transformed by suitable functors into models expressible in another language, one can build an approximation language for new models (a cycle of back and forth shifts from the right hemisphere to the left one, and reverse).

It is worth remembering that in modeling of concept approximations using rough sets, it is frequently necessary to discover relevant semantic granular structures (in right-hemisphere), syntactic granular structures (in left-hemisphere), and plans of interactions with the environment and other agents. Next, based on the results of interactions it is necessary to upgrade the world perception, estimate the distance to the planned goals, and reconstruct plans. Intuitively speaking, this approach to adaptive granular computing is very similar to the quality improvement cycle known as PDCA cycle (Plan Do Check Act) or the Deming cycle. This is a fundamental challenge for methods concerning the approximate (inductive) defining of concepts from acquired information (e.g., sensor measurements represented by a given information system). This approach could also be applied to data mining, text mining, machine learning, and pattern recognition.

Relationships between concepts such as thinking, imagination, judgment, perception and psyche were already investigated by the ancient Greeks. In particular, Aristotle—the father of formal logic and its basic concepts like deduction and induction—has dedicated many papers to clarify this issue. In particular, in [83] one can find a sentence:

> Thinking is different from perceiving and is held to be in part imagination, in part judgment: we must therefore first mark off the sphere of imagination and then speak of judgment.

If we treat any interpreter of soul (psyche) as the most complex concept we can illustrate this statement as it is show in Figure 9.

Aristotle emphasized the importance of links between symbols of spoken words (which are the basis for judgment and syllogistics), mental experiences represented in our imagination, and things of which our experiences are the images.

In [17], one can read:

> Spoken words are the symbols (symbola) of mental experience (pathemata) and written words are the symbols of spoken words. Just as all men have not the same writing, so all men have not the same speech sounds, but the mental experiences, which these directly symbolize, are the same for all, as also are those things (pragmata) of which our experiences are the images (homoiomata).

In some sense, this is an extension of the idea presented in Figure 9 which can now be represented in the *Aristotle Tetrahedron* (AT) (see Figure 10). AT can be treated as a metaphoric starting point for constructions of computational models for wisdom technology based on granular computing. However, AT is a very complex concept, even just its sides and walls represent very complex relationships. For example, one of the greatest challenges is to learn how to

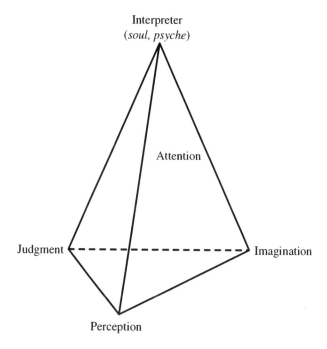

Fig. 9. Relationships between imagination, judgment, perception and psyche

connect words, phrases, and sentences to the perception of objects and events in the world [48]. Notice also that modern linguistics is mainly based on the exploration of relationships represented by projections of the tetrahedron. This is the main relation studied in the well known book [55]. The corresponding illustration is presented in Figure 11.

Notice that the Tarski satisfiability relation $(M, L, models)$, where M is any class of abstract models, L any set of language expressions, and $models \subseteq M \times L$ is a satisfiability relation, could be treated as an exploration of links represented by sides of AT. More generally, having in mind that any adjoint functors $Th \dashv Mod$ can be treated as a generalization of the Tarski satisfiability (for formula-less and model-less reasoning), one can conclude that the research on adjoint functors is related to understanding of the sides of AT.

One can consider the following situation of three adjoint functor pairs:

$$
\begin{array}{ccccccc}
& Th' & & Th & & Th'' & \\
U' & \rightleftarrows & U & \rightleftarrows & L & \rightleftarrows & L' \\
& Mod' & & Mod & & Mod'' &
\end{array}
$$

where $Th' \dashv Mod'$ is defined as a *left granulation* of $Th \dashv Mod$ and $Th'' \dashv Mod''$ is defined as a *right granulation* of $Th \dashv Mod$.

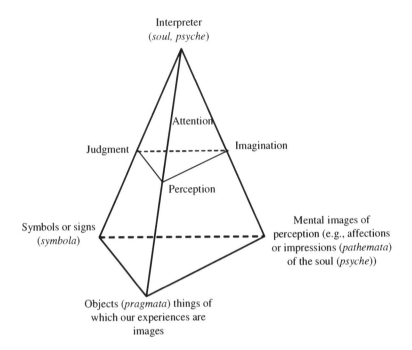

Fig. 10. Aristotle tetrahedron (AT)

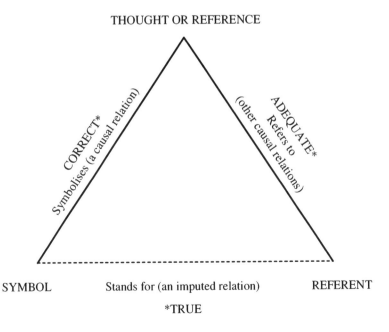

Fig. 11. Semiotic triangle

It could be shown that the fundamental concepts for rough set theory could be generalized using a concept of *adjoint approximation* represented by a tuple

$$(U, L, Mod, Lower, Upper), \tag{5}$$

where $Mod : L \longrightarrow U$, $Lower : U \longrightarrow L$; $Upper : U \longrightarrow L$, $Upper \dashv Mod$, $Mod \dashv Lower$.

5 Examples of Applications

One of the possible approaches in developing methods for compound concept approximations can be based on layered (hierarchical) learning [84]. Inducing concept approximation should be developed hierarchically, starting from concepts that can be directly approximated using sensor measurements, and aiming toward compound target concepts related to perception. This general idea can be realized using additional domain knowledge represented in natural language. For example, one can use some rules of behavior on the streets, expressed in natural language, to assess from recordings (made, e.g., by camera and other sensors) of actual traffic situations, if a particular situation is safe or not (see, e.g., [50,6,4,15]). Hierarchical learning has also been used for identification of risk patterns in medical data and extended to therapy planning (see, e.g. [8,9,10]). Another application of hierarchical learning for sunspot classification is reported in [52]. To deal with such problems one should develop methods for concept approximations together with methods aiming at the approximation of reasoning schemes (over such concepts) expressed in natural language. The foundations of such an approach, creating a core of perception logic, are based on rough set theory [59,61,15] and its extension rough mereology [67,56]. Boolean (approximate) reasoning methods can be scaled to the case of compound concept approximation.

The prediction of behavioral patterns of a compound object evaluated over time is usually based on some historical knowledge representation used to store information about changes in relevant features or parameters. This information is usually represented as a data set and has to be collected during long-term observation of a complex dynamic system. For example, in case of road traffic, we associate the object-vehicle parameters with the readouts of different measuring devices or technical equipment placed inside the vehicle or in the outside environment (e.g., alongside the road, in a helicopter observing the situation on the road, in a traffic patrol vehicle). Many monitoring devices serve as informative sensors such as GPS, laser scanners, thermometers, range finders, digital cameras, radar, image and sound converters. Hence, many vehicle features serve as models of physical sensors. Here are some example sensors: location, speed, current acceleration or deceleration, visibility, humidity (slipperiness) of the road. By analogy to this example, many features of compound objects are often dubbed sensors. Some rough set tools have been developed (see, e.g., [4]) for perception modeling that make it possible to recognize behavioral patterns of objects and their parts changing over time. More complex behavior of compound objects or groups of compound objects can be presented in the form of *behavioral graphs*. Any behavioral graph can be interpreted as a *behavioral pattern* and can be used as a complex classifier for the recognition of complex behaviors. The complete approach to the perception of behavioral patterns, based on behavioral graphs and the dynamic elimination of behavioral patterns, is presented in [4,5]. The tools for dynamic elimination of behavioral patterns are used for excluding many objects from analysis by the *system attention* procedures searching for identification of some behavioral patterns. The developed rough set tools for perception modeling are used to model networks of classifiers. Such networks make it possible to recognize behavioral patterns of objects changing over time. They are constructed using an ontology of concepts provided by experts that engage in approximate reasoning on concepts embedded in such an ontology. Experiments on data from a vehicular traffic simulator [4,5,6,50] show that the developed methods are useful in the identification of behavioral patterns.

The following example concerns human computer-interfaces that allow for a dialog with experts to transfer to the system their knowledge about structurally compound objects. For pattern recognition systems [20], e.g., for Optical Character Recognition (OCR) systems it would be helpful to transfer to the system a certain knowledge about the expert view on border line cases. The central issue in such pattern recognition systems is the construction of classifiers within vast and poorly understood search spaces, which is a very difficult task. Nonetheless, this process can be greatly enhanced with knowledge about the investigated objects provided by a human expert. A framework was developed for the transfer of such knowledge from the expert and for incorporating it into the learning process of a recognition system using methods based on rough mereology (see, e.g., [53]). Is was also demonstrated how this knowledge acquisition can be conducted in an interactive manner, with a large dataset of handwritten digits as an example.

The next two examples are related to approximation of compound concepts in reinforcement learning and planning.

In reinforcement learning [88,76,63,64], the main task is to learn the approximation of the function $Q(s, a)$, where s, a denotes a global state of the system and an action performed by an agent $ag \in Ag$ and the real value of $Q(s, a)$ describes the reward for executing the action a in the state s. In approximation of the function $Q(s, a)$ probabilistic models are used. However, for compound real-life problems it may be hard to build such models for such a compound concept as $Q(s, a)$ [95]. We propose another approach to the approximation of $Q(s, a)$ based on ontology approximation. The approach is based on the assumption that in a dialog with experts additional knowledge can be acquired, making it possible to create a ranking of values $Q(s, a)$ for different actions a in a given state s. In the explanation given by an expert about possible values of $Q(s, a)$, concepts from a special ontology are used. Then, using this ontology one can follow hierarchical learning methods to learn the approximations of concepts from the ontology. Such concepts can have a temporal character too. This means that the ranking of actions may depend not only on the actual action and the state but also on actions performed in the past and changes caused by these actions.

In [8,9,10] a computer tool based on rough sets for supporting the automated planning of medical treatment (see, e.g., [24]) is discussed. In this approach, a given patient is treated as an investigated complex dynamical system, whilst diseases of this patient (RDS, PDA, sepsis, Ureaplasma and respiratory failure) are treated as compound objects changing and interacting over time. As a measure of planning success (or failure) in experiments, we use a special hierarchical classifier that can predict the similarity between two plans as a number between 0.0 and 1.0. This classifier has been constructed on the basis of the special ontology specified by human experts and data sets. It is important to mention that beside the ontology, experts provided the example data (values of attributes) for the purpose of concept approximation from the ontology. The methods of construction such classifiers are based on approximate reasoning schemes (AR schemes, for short) and were described, e.g., in [6,50,74,80,81,4]. We applied this method for the approximation of similarity between plans generated in automated planning and plans proposed by human experts during realistic clinical treatment.

One can observe an analogous problem in learning discernibility between compound objects. For example, let us consider the state of a complex system in two close moments of time: $x(t + \Delta t)$ and $x(t)$. The *difference* between these two states cannot be described using numbers but *words* [102], i.e., it can be expressed in terms of concepts from a special ontology of (vague) concepts (e.g., acquired from a domain expert). These concepts should be again approximated using a relevant language and then in terms of the induced approximations the *difference* between $x(t + \Delta t)$ and $x(t)$ can be characterized and next used in decision making.

6   Future Directions

One of the RGC challenges is to develop approximate reasoning techniques for reasoning about the dynamics of distributed systems of judges, i.e., agents performing rightly judging. These techniques should be based on systems of evolving local perception logics (i.e., logics of agents or teams of agents) rather than on a global logic [71,72]. The approximate reasoning on judge's global behavior system is unfeasible without methods for approximation of compound vague concepts and approximate reasoning about them. One can observe here an analogy to phenomena related to the emergent patters in complex adaptive systems [14]. Let us observe that judges can be organized into a hierarchical structure, i.e., one judge can represent a coalition of judges in interaction with other agents existing in the environment [1,34,41]. Such judges representing coalitions play an important role in hierarchical reasoning about the behavior of judge populations. Strategies for coalition formation and cooperation [1,41,43] are of critical importance in designing systems of judges with dynamics adhering to the given specification to a satisfactory degree. Developing strategies for the discovery of information granules representing relevant coalitions and cooperation protocols is a challenge for RGC.

The discussed above problems can be treated as problems of searching for information granules satisfying vague requirements (constraints, specification). The strategies for construction of information granules should be adaptive. It means that the adaptive strategies should make it possible to construct information granules satisfying constraints under dynamically changing environments. This requires the reconstruction or tuning of already constructed information granules which are used as components of data models or the whole models, e.g., classifiers. In the adaptive process, the construction of information granules generalizing some previously constructed ones plays a special role. The mechanism for relevant generalization is crucial. One can imagine many different strategies for this task, e.g., based on adaptive feedback control for tuning the generalization. Cooperation with specialists from different areas such as neuroscience (see, e.g., [46] for visual objects recognition), psychology (see, e.g., [65] for discovery of mechanisms for hierarchical perception), biology (see, e.g., [12] for cooperation based on swarm intelligence) or social science (see, e.g., [41] for modeling of agents behavior) can help to discover such adaptive strategies for extracting sub-optimal (relative to the minimal length principle) data models satisfying vague constraints. This research may also help us to develop strategies for the discovery of ontologies relevant to compound concept approximation.

Another challenge for RGC concerns developing methods for the discovery of structures of complex processes from data (see, e.g., [60,87]), in particular, from temporal data (see, e.g., [94,4]). One of the important problems in discovering structures of complex processes from data is to correctly identify the type of the relevant process, e.g., continuous type of processes modelled by differential equations with some parameters, discrete models represented by behavioral graphs (see, e.g., [4,6]) or models of concurrent processes, e.g. Petri nets (see, e.g., [87]). When a type is selected, usually some optimization of parameters is performed to obtain a relevant structure model. For example, structures of behavioral graphs (see, e.g., [6,4,9,10]) are obtained by composition of temporal patterns discovered from data. So far, in our experiments this process was facilitated by domain experts. However, we plan to develop heuristics searching for such behavioral graphs directly from data without such expert support. The behavioral graphs are granules related to a given level of hierarchy of granules. Properties of such granules representing behavioral patterns are used to define indiscernibility classes or similarity classes of behavioral graphs (processes). From these granules and relations between them (induced from relational structures on value sets of attributes defined by behavioral patterns [75,10]), new behavioral graphs are modelled. In this way, new granules on a higher level of granule hierarchy are obtained. They may represent, e.g., behavioral graphs of more compound groups of objects [10]. Patterns (properties) defined by these more compound behavioral graphs (on the basis of their components and interaction between them) are important in recognition and prediction of properties of complex processes. Learning interactions which lead to the relevant behavioral patterns for compound processes is a challenge, in particular.

Let us consider one more challenge for RGC. We assume that there is given a distributed system of locally interacting agents (parts). These interactions lead to the global pattern represented by means of *emergent patterns* (see, e.g., [14,47]). One can model such a system as a special kind of game in which each agent can only interact with agents from its local neighborhood. The strategy of each agent in the game is defined by its rules of interaction with neighbors. It is well known that emergent patters are very hard to predict. There are many complex real-life systems in which such patterns are observed, e.g., in ecological systems, immune systems, economies, global climate, ant colonies [68]. The challenge we consider is related to developing strategies for learning local interactions among agents leading to a given emergent pattern. Evolutionary strategies are the candidates here (see, e.g., [47,68]). We would like to emphasize the role of RGC in learning such strategies. One possible approach can be based on the assumption that the learning process should be organized in such a way that gradually learned interactions lead to granules of agents represented by coalitions of agents, rather than by particular agents. The behavior of each coalition is determined by some specific interactions of its agents which lead to the interaction of the coalition as a whole with other agents or coalitions. Learning a hierarchy of coalitions is hard but can be made feasible by using domain knowledge, analogously to hierarchical learning (see, e.g., [4,9,10]). Domain knowledge can help to discover languages for expressing behavioral patterns (properties) of coalitions on each level of coalition hierarchy and heuristics for identifying such behavioral patterns from sensor measurements. Without such *hints* based on domain knowledge learning seems to be unfeasible for real-life problems. Using this approach based on hierarchy of coalitions, evolutionary strategies should allow us to learn interactions on each level of hierarchy between existing on this level coalitions and to discover on the next level relevant new coalitions and their properties. Let us observe that the coalition discovered though the relevant granulation of lower level coalitions will usually satisfy the specification for this higher coalition level only to a degree. The discovered coalitions, i.e., new granules can be treated as approximate solutions on a given level of hierarchy. It is worthwhile to mention another possibility related to learning relevant hierarchical coalitions for gradually changing tasks for the whole system, from simple to more compound tasks.

We summarize the above discussion as follows. One may try to describe emergent patterns in terms of interactions between some high level granules—coalitions. We assume that the construction of relevant compound coalitions can be supported by a domain ontology. Then, coalitions are discovered gradually using the hierarchical structure defined by the domain ontology. On each level of hierarchy, a relevant language should be discovered. In this language, it should be possible to express properties of coalitions much easier (e.g., using interactions between teams of agents) than using the language accessible on the preceding level where more detailed descriptions are only accessible (e.g., interactions between single agents). Hence, the same coalition can be described using different languages from two successive levels. However, the language from a higher level is more relevant to further modeling of more compound coalitions. Discovery of a relevant language for each level of hierarchy, from languages on the proceeding levels may be feasible. However, discovery of a relevant language for modeling of compound coalitions on the highest hierarchy level directly from the very basic level (where, e.g., only interactions between single agents are directly expressible) may be unfeasible from the computational complexity point of view.

Finally, let us observe that the following four principles of adaptive information processing in decentralized systems presented in [47] are central for RGC in searching for relevant granules by using evolutionary techniques:

1. Global information is encoded as statistics and dynamics of patterns over the system's components.
2. Randomness and probabilities are essential.
3. The system carries out a fine-grained, parallel search of possibilities.
4. The system exhibits a continual interplay of bottom-up and top-down processes.

Acknowledgments. We thank the reviewers for the valuable comments and suggestions which helped us to improve the presentation.

The research has been partially supported by the grants N N516 368334 and N N516 077837 from Ministry of Science and Higher Education of the Republic of Poland.

References

1. Axaelrod, R.M.: The Complexity of Cooperation. Princeton University Press, Princeton (1997)
2. Bargiela, A., Pedrycz, W.: Granular Computing: An Introduction. Kluwer Academic Publishers, Boston (2003)
3. Barwise, J., Seligman, J.: Information Flow: The Logic of Distributed Systems. Cambridge University Press, Cambridge (1997)
4. Bazan, J., Peters, J.F., Skowron, A.: Behavioral pattern identification through rough set modelling. In: Ślęzak, D., Yao, J., Peters, J.F., Ziarko, W.P., Hu, X. (eds.) RSFDGrC 2005. LNCS (LNAI), vol. 3642, pp. 688–697. Springer, Heidelberg (2005)
5. Bazan, J., Skowron, A.: On-line elimination of non-relevant parts of complex objects in behavioral pattern identification. In: Pal, S.K., Bandyopadhyay, S., Biswas, S. (eds.) PReMI 2005. LNCS, vol. 3776, pp. 720–725. Springer, Heidelberg (2005)
6. Bazan, J., Skowron, A.: Classifiers based on approximate reasoning schemes. In: [16], pp. 191–202. Springer, Heidelberg (2005)
7. Bazan, J.G., Skowron, A., Świniarski, R.W.: Rough sets and vague concept approximation: From sample approximation to adaptive learning. In: Peters, J.F., Skowron, A. (eds.) Transactions on Rough Sets V. LNCS, vol. 4100, pp. 39–62. Springer, Heidelberg (2006)
8. Bazan, J., Kruczek, P., Bazan-Socha, S., Skowron, A., Pietrzyk, J.J.: Risk pattern identification in the treatment of infants with respiratory failure through rough set modeling. In: Proceedings of IPMU'2006, Paris, France, July 2–7, 2006, pp. 2650–2657. Éditions E.D.K, Paris (2006)
9. Bazan, J., Kruczek, P., Bazan-Socha, S., Skowron, A., Pietrzyk, J.J.: Automatic planning of treatment of infants with respiratory failure through rough set modeling. In: Greco, S., Hata, Y., Hirano, S., Inuiguchi, M., Miyamoto, S., Nguyen, H.S., Słowiński, R. (eds.) RSCTC 2006. LNCS (LNAI), vol. 4259, pp. 418–427. Springer, Heidelberg (2006)

10. Bazan, J.: Rough sets and granular computing in behavioral pattern identification and planning. In: [62], pp. 777–822 (2007)
11. Birkhoff, G.: Lattice Theory, 3rd edn. AMS Colloquium Publications, vol. 25. American Mathematical Society, Providence (1967)
12. Bonabeau, E., Dorigo, M., Theraulaz, G.: Swarm Intelligence. In: From Natural to Artificial Systems, Oxford University Press, Oxford (1999)
13. Breiman, L.: Statistical modeling: The two Cultures. Statistical Science 16(3), 199–231 (2001)
14. Desai, A.: Adaptive complex enterprises. Communications ACM 48(5), 32–35 (2005)
15. Doherty, P., Łukaszewicz, W., Skowron, A., Szałas, A.: Knowledge Representation Techniques: A Rough Set Approach. Studies in Fuzziness and Soft Computing, vol. 202. Springer, Heidelberg (2006)
16. Dunin-Kęplicz, B., Jankowski, A., Skowron, A., Szczuka, M.: Monitoring, Security, and Rescue Tasks in Multiagent Systems (MSRAS2004). Series in Soft Computing. Springer, Heidelberg (2005)
17. Edghill, E.M.: On Interpretation by Aristotle (Trans.). eBooks@Adelaide (2007)
18. Feigenbaum, E., Feldman, J. (eds.): Computers and Thought. McGraw Hill, New York (1963)
19. Frege, G.: Grundgesetzen der Arithmetik vol. 2. Verlag von Hermann Pohle, Jena (1903)
20. Friedman, J.H., Hastie, T., Tibshirani, R.: The Elements of Statistical Learning: Data Mining, Inference, and Prediction. Springer, Heidelberg (2001)
21. Ganter, B., Stumme, G., Wille, R. (eds.): Formal Concept Analysis. LNCS (LNAI), vol. 3626. Springer, Heidelberg (2005)
22. Garling, D.J.H.: A Course in Galois Theory. Cambridge University Press, New York (1987)
23. Gell-Mann, M.: The Quark and the Jaguar—Adventures in the Simple and the Complex. Brown and Co., London (1994)
24. Ghallab, M., Nau, D., Traverso, P.: Automated Planning: Theory and Practice. Morgan Kaufmann, San Francisco (2004)
25. Goldin, D., Smolka, S., Wegner, P.: Interactive Computation: The New Paradigm. Springer, Heidelberg (2006)
26. Jankowski, A., Skowron, A.: A wistech paradigm for intelligent systems. In: Peters, J.F., Skowron, A., Düntsch, I., Grzymała-Busse, J.W., Orłowska, E., Polkowski, L. (eds.) Transactions on Rough Sets VI. LNCS, vol. 4374, pp. 94–132. Springer, Heidelberg (2007)
27. Jankowski, A., Skowron, A.: Logic for artificial intelligence: The Rasiowa-Pawlak school perspective. In: Ehrenfeucht, A., Marek, V., Srebrny, M. (eds.) Andrzej Mostowski and Foundational Studies, pp. 106–143. IOS Press, Amsterdam (2007)
28. Jankowski, A., Skowron, A.: Wisdom Granular Computing. In: [62], pp. 329–345 (2007)
29. Johnson, S.: Dictionary of the English Language in Which the Words are Deduced from Their Originals, and Illustrated in their Different Significations by Examples from the Best Writers, 2 Volumes. F.C. and J. Rivington, London (1816)
30. Johnstone, P.: Stone Spaces. Cambridge University Press, Cambridge (1986)
31. Kan, D.M.: Adjoint functors. Trans. Am. Math. Soc., Soc. 87, 294–329 (1958)
32. Keefe, R.: Theories of Vagueness. Cambridge Studies in Philosophy. Cambridge (2000)
33. Kloesgen, W., Żytkow, J.: Handbook of Knowledge Discovery and Data Mining. Oxford University Press, New York (2002)

34. Kraus, S.: Strategic Negotiations in Multiagent Environments. The MIT Press, Massachusetts (2001)
35. Lambek, J., Scott, P.J.: Introduction to Higher-Order Categorical Logic. Cambridge Studies in Advanced Mathematics, vol. 7. Cambridge University Press, Cambridge (1986)
36. Leibniz, G.W.: Dissertio de Arte Combinatoria. Leipzig (1666)
37. Leibniz, G.W.: New Essays on Human Understanding (1705), Translated and edited by Peter Remnant and Jonathan Bennett. Cambridge University Press, Cambridge (1982)
38. Leśniewski, S.: Grundzüge eines neuen Systems der Grundlagen der Mathematik. Fundamenta Mathematicae 14, 1–81 (1929)
39. Leśniewski, S.: On the foundations of mathematics. Topoi 2, 7–52 (1982)
40. Lin, T.Y.: Neighborhood systems and approximation in database and knowledge base systems. In: Emrich, M.L., Phifer, M.S., Hadzikadic, M., Ras, Z.W. (eds.) Proceedings of the Fourth International Symposium on Methodologies of Intelligent Systems (Poster Session), October 12–15, 1989, pp. 75–86. Oak Ridge National Laboratory, Charlotte (1989)
41. Liu, J.: Autonomous Agents and Multi-Agent Systems: Explorations in Learning, Self-Organization and Adaptive Computation. World Scientific Publishing, Singapore (2001)
42. Liu, J., Daneshmend, L.K.: Spatial Reasoning and Planning: Geometry, Mechanism, and Motion. Springer, Berlin (2003)
43. Liu, J., Jin, X., Tsui, K.C.: Autonomy Oriented Computing: From Problem Solving to Complex Systems Modeling. Kluwer Academic Publisher/Springer, Heidelberg (2005)
44. MacLane, S.: Categories for the Working Mathematicians. Graduate Texts in Mathematics. Springer, Berlin (1997)
45. MacLane, S., Moerdijk, I.: Sheaves in Geometry and Logic: A First Introduction to Topos Theory. Universitext. Springer, Berlin (1994)
46. Miikkulainen, R., Bednar, J.A., Choe, Y., Sirosh, J.: Computational Maps in the Visual Cortex. Springer, Heidelberg (2005)
47. Mitchell, M.: Complex systems: Network thinking. Artificial Intelligence 170(18), 1194–1212 (2006)
48. Mooney, R.: Learning to connect language and perception. In: Proceedings of the 23rd AAAI Conference on Artificial Intelligence (AAAI), Senior Member Paper, Chicago, IL, July 13–17, 2008, pp. 1598–1601 (2008)
49. Nguyen, H.S.: Approximate Boolean Reasoning: Foundations and Applications in Data Mining. In: Peters, J.F., Skowron, A. (eds.) Transactions on Rough Sets V. LNCS, vol. 4100, pp. 334–506. Springer, Heidelberg (2006)
50. Nguyen, S.H., Bazan, J.G., Skowron, A., Nguyen, H.S.: Layered learning for concept synthesis. In: Peters, J.F., Skowron, A., Grzymała-Busse, J.W., Kostek, B.z., Świniarski, R.W., Szczuka, M.S. (eds.) Transactions on Rough Sets I. LNCS, vol. 3100, pp. 187–208. Springer, Heidelberg (2004)
51. Nguyen, H.S., Skowron, A., Stepaniuk, J.: Discovery of changes along trajectories generated by process models induced from data and domain knowledge. In: Lindemann, G., Burkhard, H.-D., Czaja, L., Penczek, W., Salwicki, A., Schlingloff, H., Skowron, A., Suraj, Z. (eds.) Proceedings of the Workshop on Concurrency, Specification and Programming (CS&P 2008), September 29–October 1, 2008. Informatik-Berichte, Humboldt Universitaet zu Berlin, vol. 3, pp. 350–362. Gross Vaeter, Germany (2008)

52. Nguyen, S.H., Nguyen, T.T., Nguyen, H.S.: Rough set approach to sunspot classification problem. In: [91], pp. 263–272 (2005)
53. Nguyen, T.T.: Eliciting domain knowledge in handwritten digit recognition. In: [57], pp. 762–767 (2005)
54. Nguyen, T.T., Skowron, A.: Rough-granular computing in human-centric information processing. In: Bargiela, A., Pedrycz, W. (eds.) Human-Centric Information Processing Through Granular Modelling, Springer, Heidelberg (to appear 2009)
55. Ogden, C.K., Richards, I.A.: The Meaning of Meaning. A Study of the Influence of Language upon Thought and of the Science of Symbolism. Kegan Paul, Trench, Trubner and Co., Ltd, London (1923) see also this book: With an Introduction by Postgate, J. P., and Supplementary Essays by Malinowski, B., Crookshank, F.G., Harcourt Brace Jovanovich, Inc., New York (1989)
56. Pal, S.K., Polkowski, L., Skowron, A. (eds.): Rough-Neural Computing: Techniques for Computing with Words. Cognitive Technologies. Springer, Heidelberg (2004)
57. Pal, S.K., Bandyopadhyay, S., Biswas, S. (eds.): PReMI 2005. LNCS, vol. 3776. Springer, Heidelberg (2005)
58. Pawlak, Z.: Rough sets. International Journal of Computer and Information Sciences 11, 341–356 (1982)
59. Pawlak, Z.: Rough Sets: Theoretical Aspects of Reasoning about Data. System Theory, Knowledge Engineering and Problem Solving, vol. 9. Kluwer Academic Publishers, Dordrecht (1991)
60. Pawlak, Z.: Concurrent versus sequential the rough sets perspective. Bulletin of the EATCS 48, 178–190 (1992)
61. Pawlak, Z., Skowron, A.: Rudiments of rough sets. Information Sciences 177(1), 3–27 (2007); Rough sets: Some extensions. Information Sciences 177(1): 28–40; Rough sets and boolean reasoning. Information Sciences 177(1): 41–73 (2007)
62. Pedrycz, W., Skowron, A., Kreinovich, V. (eds.): Handbook of Granular Computing. John Wiley & Sons, New York (2008)
63. Peters, J.F.: Approximation spaces for hierarchical intelligent behavioural system models. In: Kęplicz, B.D., Jankowski, A., Skowron, A., Szczuka, M. (eds.) Monitoring, Security and Rescue Techniques in Multiagent Systems. Advances in Soft Computing, pp. 13–30. Physica-Verlag, Heidelberg (2004)
64. Peters, J.F.: Rough ethology: Towards a biologically-inspired study of collective behavior in intelligent systems with approximation spaces. In: Peters, J.F., Skowron, A. (eds.) Transactions on Rough Sets III. LNCS, vol. 3400, pp. 153–174. Springer, Heidelberg (2005)
65. Poggio, T., Smale, S.: The mathematics of learning: Dealing with data. Notices of the AMS 50(5), 537–544 (2003)
66. Polkowski, L.: Rough Sets: Mathematical Foundations. Advances in Soft Computing. Physica-Verlag, Heidelberg (2002)
67. Polkowski, L., Skowron, A.: Rough mereology: A new paradigm for approximate reasoning. International Journal of Approximate Reasoning 15(4), 333–365 (1996)
68. Segel, L.A., Cohen, I.R. (eds.): Design Principles for the Immune System and Other Distributed Autonomous Systems. Oxford University Press, New York (2001)
69. Skowron, A.: Rough sets in KDD (plenary talk). In: Shi, Z., Faltings, B., Musen, M. (eds.) 16-th World Computer Congress (IFIP'2000): Proceedings of Conference on Intelligent Information Processing (IIP'2000), pp. 1–14. Publishing House of Electronic Industry, Beijing (2000)

70. Skowron, A.: Rough sets and vague concepts. Fundamenta Informaticae 64(1-4), 417–431 (2005)
71. Skowron, A.: Perception logic in intelligent systems (plenary talk). In: Blair, S., et al. (eds.) Proceedings of the 8th Joint Conference on Information Sciences (JCIS 2005), Salt Lake City, Utah, USA, July 21-26, 2005, X-CD Technologies: A Conference & Management Company, Toronto (2005)
72. Skowron, A.: Rough sets in perception-based computing. In: Pal, S.K., Bandyopadhyay, S., Biswas, S. (eds.) PReMI 2005. LNCS, vol. 3776, pp. 21–29. Springer, Heidelberg (2005)
73. Skowron, A., Stepaniuk, J.: Tolerance approximation spaces. Fundamenta Informaticae 27(2–3), 245–253 (1996)
74. Skowron, A., Stepaniuk, J.: Information granules and rough-neural computing. In: [56], pp. 43–84 (2004)
75. Skowron, A., Stepaniuk, J.: Rough sets and granular computing: Toward rough-ranular computing. In: [62], pp. 425–448 (2008)
76. Skowron, A., Stepaniuk, J., Peters, J.F., Swiniarski, R.: Calculi of approximation spaces. Fundamenta Informaticae 72(1–3), 363–378 (2006)
77. Skowron, A., Świniarski, R.W.: Rough sets and higher order vagueness. In: [90], pp. 33–42 (2005)
78. Skowron, A., Świniarski, R.W., Synak, P.: Approximation spaces and information granulation. In: Peters, J.F., Skowron, A. (eds.) Transactions on Rough Sets III. LNCS, vol. 3400, pp. 175–189. Springer, Heidelberg (2005)
79. Skowron, A., Szczuka, M.: Toward Interactive Computations: A Rough-Granular Approach. In: Koronacki, J., Wierzchon, S.T., Ras, Z.W., Kacprzyk, J. (eds.) Commemorative Volume to Honor Ryszard Michalski, Springer, Heidelberg (in preparation 2009)
80. Skowron, A., Synak, P.: Complex patterns. Fundamenta Informaticae 60(1–4), 351–366 (2004)
81. Skowron, A., Synak, P.: Reasoning in information maps. Fundamenta Informaticae 59(2–3), 241–259 (2004)
82. Skowron, A., Peters, J.F.: Rough-granular computing. In: [62], pp. 285–327 (2008)
83. Smith, J.A, (Trans.): On the soul by Aristotle. eBooks@Adelaide (2007), http://etext.library.adelaide.edu.au/a/aristotle/
84. Stone, P.: Layered Learning in Multi-Agent Systems: A Winning Approach to Robotic Soccer. The MIT Press, Cambridge (2000)
85. Sun, R.: Duality of the Mind: A Bottom-up Approach Toward Cognition. Lawrence Erlbaum, Mahwah (2001)
86. Sun, R. (ed.): Cognition and Multi-Agent Interaction. From Cognitive Modeling to Social Simulation. Cambridge University Press, New York (2006)
87. Suraj, Z.: Rough set methods for the synthesis and analysis of concurrent processes. In: Polkowski, L., Tsumoto, S., Lin, T.Y. (eds.) Rough Set Methods and Applications. Studies in Fuzziness and Soft Computing, vol. 56, pp. 379–488. Physica-Verlag, Heidelberg (2000)
88. Sutton, R.S., Barto, A.G.: Reinforcement Learning: An Introduction. The MIT Press, Cambridge (1998)
89. Sycara, K.: Multiagent systems. AI Magazine 19(2), 79–93 (1998)
90. Ślęzak, D., Wang, G., Szczuka, M.S., Düntsch, I., Yao, Y. (eds.): RSFDGrC 2005. LNCS (LNAI), vol. 3641. Springer, Heidelberg (2005)
91. Ślęzak, D., Yao, J., Peters, J.F., Ziarko, W.P., Hu, X. (eds.): RSFDGrC 2005. LNCS (LNAI), vol. 3642. Springer, Heidelberg (2005)

92. Tarski, A.: The Collected Papers of Alfred Tarski, 4 vols. In: Givant, S.R., McKenzie, R.N. (eds.), Birkhäuser, Basel (1986)
93. Turing, A.: Computing machinery and intelligence. Mind LIX(236), 433–460 (1950)
94. Unnikrishnan, K.P., Ramakrishnan, N., Sastry, P.S., Uthurusamy, R.: 4th KDD Workshop on Temporal Data Mining: Network Reconstruction from Dynamic Data. The Twelfth ACM SIGKDD International Conference on Knowledge Discovery and Data (KDD 2006), Philadelphia, USA, August 20–23, 2006 (2006), http://people.cs.vt.edu/~ramakris/kddtdm06/cfp.html
95. Vapnik, V.: Statistical Learning Theory. John Wiley & Sons, New York (1998)
96. Ulam, S.M.: Analogies Between Analogies: The Mathematical Reports of S. M. Ulam and His Los Alamos Collaborators. University of California Press, Berkeley (1990)
97. Wittgenstein, L.: Philosophical Investigations. The German text, with revised English translation (Translated by Anscombe, G.E.M.). Blackwell, Oxford (2001)
98. Zadeh, L.A.: Fuzzy sets. Information and Control 8, 333–353 (1965)
99. Zadeh, L.A.: Outline of a new approach to the analysis of complex systems and decision processes. IEEE Trans. on Systems, Man and Cybernetics SMC 3, 28–44 (1973)
100. Zadeh, L.A.: Fuzzy sets and information granularity. In: Gupta, M., Ragade, R., Yager, R. (eds.) Advances in Fuzzy Set Theory and Applications, Amsterdam: North-Holland Publishing Co, pp. 3–18. North-Holland, Amsterdam (1979)
101. Zadeh, L.A.: Outline of a computational approach to meaning and knowledge representation based on the concept of a generalized assignment statement. In: Thoma, M., Wyner, A. (eds.) Proceedings of the International Seminar on Artificial Intelligence and Man-Machine Systems, pp. 198–211. Springer, Heidelberg (1986)
102. Zadeh, L.A.: From computing with numbers to computing with words—from manipulation of measurements to manipulation of perceptions. IEEE Transactions on Circuits and Systems 45, 105–119 (1999)
103. Zadeh, L.A.: A new direction in AI—toward a computational theory of perceptions. AI Magazine 22(1), 73–84 (2001)
104. Zadeh, L.A.: Foreword. In: Pal, S.K., Polkowski, L., Skowron, A. (eds.) Rough-Neural Computing: Techniques for Computing with Words. Cognitive Technologies Series, pp. IX–XI. Springer, Berlin (2004)
105. Zadeh, L.A.: Generalized theory of uncertainty (GTU)-principal concepts and ideas. Computational Statistics and Data Analysis 51, 15–46 (2006)
106. Zhong, N., Liu, J., Yao, Y.Y.: Envisioning intelligent information technologies (iIT) from the stand-point of Web intelligence (WI). Communications of the ACM 50(3), 89–94 (2007)
107. Zhong, N., Liu, J. (eds.): Intelligent Technologies for Information Analysis. Springer, Berlin (2004)
108. Zhong, N., Liu, J., Yao, Y.Y. (eds.): Web Intelligence. Springer, Berlin (2003)
109. Zhong, N., Liu, J., Yao, Y.Y.: In search of the Wisdom Web. IEEE Computer 35(11), 27–31 (2002)

Paraconsistent Reasoning with Words

Alicja S. Szalas[1] and Andrzej Szałas[2,3]

[1] School of Biological Sciences, Royal Holloway, University of London,
Cooper's Hill Lane Kingswood 1 Hall - C151, Egham, Surrey TW200LG, UK,
`A.S.Szalas@rhul.ac.uk`
[2] Institute of Informatics, Warsaw University
02-097 Warsaw, Poland, `andsz@mimuw.edu.pl`
[3] Department of Computer and Information Science, Linköping University
581 83 Linköping, Sweden

Abstract. Fuzzy logics are one of the most frequent approaches to model uncertainty and vagueness. In the case of fuzzy modeling, degrees of belief and disbelief sum up to 1, which causes problems in modeling the lack of knowledge and inconsistency. Therefore, so called paraconsistent intuitionistic fuzzy sets have been introduced, where the degrees of belief and disbelief are not required to sum up to 1. The situation when this sum is smaller than 1 reflects the lack of knowledge and its value greater than 1 models inconsistency.

In many applications there is a strong need to guide and interpret fuzzy-like reasoning using qualitative approaches. To achieve this goal in the presence of uncertainty, lack of knowledge and inconsistency, we provide a framework for qualitative interpretation of the results of fuzzy-like reasoning by labeling numbers with words, like *true, false, inconsistent, unknown*, reflecting truth values of a suitable, usually finitely valued logical formalism.

Key words: fuzzy logics, four-valued logics, paraconsistent reasoning, reasoning with words

1 Introduction

Computing with words, as described by Zadeh in [33], is

> a methodology in which the objects of computation are words and propositions drawn from a natural language, e.g., small, large, far, heavy, not very likely [...] etc. Computing with words is inspired by the remarkable human capability to perform a wide variety of physical and mental tasks without any measurements and any computations.

In the current paper, instead of "computing with words" we rather concentrate on a more narrow subject of reasoning with words, which is present in everyday activities, in particular those related to expert decision making.

M. Marciniak and A. Mykowiecka (Eds.): Bolc Festschrift, LNCS 5070, pp. 43–58, 2009.

Consider, for example, a medical diagnosis problem. Here a doctor or another health care professional examines symptoms in order to determine the patient's disease. In such cases rules like the following one are used (see [26]):

> IF: 1) the stain of the organism is grampos, and
>   2) the morphology of the organism is coccus, and
>   3) the growth confirmation of the organism is clumps   (1)
> THEN: there is suggestive evidence (0.7) that the identity
>   of the organism is staphylococcus.

Another motivation is related to the Bayesian diagnosis (see [27]), which is based on conditional probabilities of the form $P(A \mid B) = p$. Such probabilities give rise to rules of the form

if B is known to be true then conclude that A is true with probability p. (2)

In order to construct an expert system, one has to encode such rules in a symbolic form which can be processed by computers in interaction with human users (see [6, 20, 24, 26, 8, 19, 29]). Information provided by users is often uncertain and imprecise. It can also be inconsistent, especially if the context includes the fusing of knowledge from different sources.

Quantitative approaches to reasoning in expert systems are usually concentrated around models involving probability, credibility and plausibility, possibility and necessity, degrees of belief and disbelief (mass distributions) (see [20,24]). In such contexts fuzzy modeling of uncertainty and vagueness [34,36,13,14,11,10] is also one of the most frequent approaches. In this setting, degrees of belief and disbelief sum up to 1, as in the case of majority of infinitely many-valued logics (see [5]). Such strong constraints cause problems with modeling the lack of knowledge. Therefore, so called, intuitionistic fuzzy sets have been introduced [2]. In intuitionistic fuzzy sets, the constraints on degrees of belief and disbelief are relaxed, as their sum is required to be no greater than 1. Intuitively, the case when our belief as to A together with the belief as to *not A* is strictly smaller than 1 corresponds to lack of (a part of) knowledge. In [28], paraconsistent intuitionistic fuzzy sets are applied to model uncertainty, lack of knowledge as well as inconsistency by further relaxing the constraint as to belief and disbelief. It is assumed there that the sum of belief and disbelief may be greater than 1. Such a situation models inconsistent knowledge and reflects "overdefined propositions" (the term "overdefined" has been introduced in [7] in the context of fusing Kripke-like worlds but we use it here due to an analogy).

In the current paper we consider the model proposed in [28]. However, we observe that it is often the case that fuzzy values are interpreted qualitatively. On the other hand, even if qualitative (usually finitely-valued) logics developed for similar interpretations are known, the fuzzy reasoning does not employ them and often provides incompatible results. Therefore there is a strong need to guide and interpret the fuzzy-like reasoning using suitable qualitative approaches. In order to guide reasoning in the presence of uncertainty, lack of knowledge and inconsistency, we combine paraconsistent intuitionistic fuzzy sets with four-valued logics, in particular those of [21, 31], developed for handling incomplete and inconsistent knowledge. The logics provided in [21, 31] are based on knowledge ordering considered in [4, 3, 16, 17] and truth ordering considered in [1, 31, 22]. Such a fusion of infinitely many-valued and four-valued logics allows us to combine qualitative and quantitative reasoning in a uniform framework, where the lack of knowledge and inconsistency are handled in a natural and intuitive setting. We show that the use of qualitative logics to guide fuzzy-like reasoning simplifies the problem of results' interpretation.

The paper is structured as follows. First, in Section 2, we discuss and motivate the proposed methodology. In Section 3, we introduce the apparatus we found suitable to deal with the considered phenomena. Then we proceed with Section 4, where we discuss a logic combining the paraconsistent intuitionistic fuzzy approach with that of [31, 22] as well as illustrate the introduced concepts using examples based on medical diagnoses. Finally, Section 5 concludes the paper.

2   The Methodology

Linguistic variables whose values are words (linguistic terms), introduced by Zadeh in [35], are the basis of applications of fuzzy reasoning (see [35, 36, 33]). For example, instead of a numerical variable *temperature* with value of 80°C, *temperature* is treated as a linguistic variable that may assume linguistic values, e.g., of *low, moderate, hot*. In fuzzy reasoning such values are characterized by real numbers from the interval $[0, 1]$, representing the degree of truth. For example, temperature 80°C might be *low* to the degree 0.01, *moderate* to the degree of 0.8 and *hot* to the degree of 0.7.

Fuzzy reasoning is often based on "if-then" rules (see [32, 14, 15]) which take fuzzified scalar values as inputs and produce a fuzzy output. To make the reasoning meaningful, such fuzzy outputs need to be converted into scalar outputs. In practical applications the mentioned "if-then" rules are constructed as qualitative rules reflecting expert reasoning. The following example illustrates the idea.

Example 1. Consider the rule

$$(hot\ temperature \wedge high\ pressure) \rightarrow danger. \tag{3}$$

Suppose that the scalar value of temperature is 80°C and of pressure is 4atm. Whether such temperature is hot and pressure is high depends on a particular application. Let, in our case, the temperature be *hot* to the degree $t = 0.7$ and pressure be *high* to the degree $p = 0.6$. Then the conjunction in the antecedent of implication (3) is evaluated to the $\min\{0.7, 0.6\} = 0.6$. Thus the value of *danger* could also be evaluated to $d = 0.6$.

Let us now analyze this reasoning. Note that fuzzy values attached to linguistic variables like *hot temperature* have their application dependent interpretation. For example, we might have the following interpretation of our linguistic variables:

$$hot\ temperature = \begin{cases} false & \text{when } t \in [0, 0.5] \\ unknown & \text{when } t \in (0.5, 0.8] \\ true & \text{when } t \in (0.8, 1.0] \end{cases}$$

$$high\ pressure = \begin{cases} false & \text{when } p \in [0, 0.4] \\ unknown & \text{when } p \in (0.4, 0.5] \\ true & \text{when } p \in (0.5, 1.0] \end{cases}$$

$$danger = \begin{cases} false & \text{when } d \in [0, 0.7] \\ unknown & \text{when } d \in (0.7, 0.8] \\ true & \text{when } d \in (0.8, 1.0]. \end{cases}$$

When such an interpretation is considered, the implication (3) reduces to

$$\underbrace{(hot\ temperature}_{0.7\ (unknown)} \wedge \underbrace{high\ pressure)}_{0.6\ (true)} \rightarrow \underbrace{danger}_{0.6\ (false)} \ . \tag{4}$$

Observe that in (4) from the value *unknown* of hot temperature and the value *true* of *high pressure* we deduce that the value of *danger* is *false*. In real word applications of such reasoning could be very risky.

One might argue here that the interpretations of considered linguistic variables should be revised. On the other hand, when one deals with many rules, such a revision might result in other unintuitive outputs. For example, in the three-valued logic of Kleene, the resulting value of *danger* should be one of those which are interpreted as *unknown*.   ◁

Consider another example.

Example 2. Consider a choice of meals in a restaurant. Assume that one is interested in a meal which is both tasty and inexpensive which is expressed by the conjunction:

$$tasty \wedge inexpensive \tag{5}$$

Evaluating (5) as the fuzzy conjunction, we obtain that its value is 0.4 for both meals described in Table 1. On the other hand, when both meals have comparable prices, one would rather chose the first one, since it is more tasty. Fuzzy conjunction loses this point.

Assume that the value 0.8 is interpreted for *tasty* as *true* and 0.4 as *false*, while for *inexpensive* 0.4 is interpreted as *unknown*. Observe that the interpretation

Table 1. Table considered in Example 2.

meal	1	2
tasty	0.8	0.4
inexpensive	0.4	0.4

of 0.4 is different for *tasty* and *inexpensive*. Such a situation is not surprising when attributes range over different domains.

Evaluating the conjunction (5) using, for example, the three-valued logic of Kleene, we obtain that its value is *unknown* for the first meal and *false* for the second meal, i.e., it is possible that the first meal is both tasty and inexpensive, while the second one is definitely not. ◁

Concluding the above discussion, the methodology we propose depends on:

1. interpreting real values as truth values in a many-valued logic reflecting the considered truth degrees of linguistic variables and providing qualitative representations for fuzzy-like truth degrees[4]
2. guide and interpret the fuzzy-like reasoning using the principles of the chosen finite-valued logic.

The idea of interpreting real values qualitatively is known and already present (see [26]), where *certainty factor* taking real values from the interval $[-1, 1]$ is, in some cases, interpreted qualitatively as follows:

- $[-1, -0.2)$—known not to hold
- $[-0.2, 0.2]$—unknown
- $(0.2, 1]$—known to hold.

The need for such interpretations has also been observed in [18, 9]. However, the idea of guiding the fuzzy-like reasoning by a finitely-valued logic is, to our knowledge, original.

3 Extending Fuzzy-Like Reasoning for Handling Inconsistency and Incomplete Knowledge

In various infinitely-valued logics [5, 25, 30], including fuzzy reasoning [34, 36, 13, 14, 11, 10] it is assumed that degrees of truth and falsity of any formula A sum up to 1,

$$(\text{truth degree of } A) + (\text{truth degree of } \neg A) = 1. \tag{6}$$

This principle is relaxed in intuitionistic fuzzy sets [2, 8], where it is assumed that

$$0 \leq (\text{truth degree of } A) + (\text{truth degree of } \neg A) \leq 1. \tag{7}$$

Such a generalization allows us to model incomplete knowledge, in this context also called the *hesitation part* (see [8]).

The principle (7) is still further relaxed in [28] by allowing that

$$0 \leq (\text{truth degree of } A) \leq 1 \quad \text{and} \quad 0 \leq (\text{truth degree of } \neg A) \leq 1. \tag{8}$$

Intuitively, the case when (truth degree of A)+(truth degree of $\neg A$) < 1 reflects the lack of knowledge and the case when (truth degree of A) + (truth degree of $\neg A$) > 1 reflects inconsistency.

---

[4] For the purpose of the current paper we consider a four-valued logic, allowing us to interpret propositions as *true, false, unknown* or *inconsistent*.

3.1   Logics $\mathcal{C}(\mathcal{L})$

In order to guide quantitative (e.g., fuzzy-like) reasoning using more qualitative calculus $\mathcal{L}$ we introduce a logic $\mathcal{C}(\mathcal{L})$ whose semantics depends on $\mathcal{L}$. The following sections provide details of the proposed approach.

Syntax of $\mathcal{C}(\mathcal{L})$. Syntax of $\mathcal{C}(\mathcal{L})$ is independent of $\mathcal{L}$. Formulas of $\mathcal{C}(\mathcal{L})$ are those of the classical propositional calculus. More precisely, we first assume that a set $\mathcal{P}$ of *propositional variables* is given. Formulas of $\mathcal{C}(\mathcal{L})$, denoted by $\mathcal{F}$, are then defined inductively by assuming that $\mathcal{P} \subseteq \mathcal{F}$ and that

$$\neg A, A \vee B, A \wedge B, A \to B \text{ are formulas when } A, B \in \mathcal{F}.$$

Semantics of $\mathcal{C}(\mathcal{L})$. In what follows we assume that $\mathcal{L}$ is a four-valued logic with truth values f (standing for *false*), u (standing for *unknown*), i (standing for *inconsistent*) and t (standing for *true*). We assume that $\mathcal{L}$ provides a four-valued semantics of connectives $\neg, \vee, \wedge, \to$. Examples of such logics are given in $[1, 3, 4, 21, 22]$.

In the proposed approach the semantics of each formula A is given by a pair of reals $\langle \beta, \delta \rangle$, where $\beta, \delta \in [0, 1]$. In such cases we use notation $A = \langle \beta, \delta \rangle$ with the intuitive meaning that

  - β provides the *degree of belief* that A holds
  - δ provides the *degree of disbelief* that A holds.

Example 3. Let *red* be a propositional variable. Then:

  - $red = \langle 0.3, 0.7 \rangle$ means that *red* holds to the degree 0.3 and does not hold to the degree 0.7; this corresponds to the situation typical in various infinitely-valued logics, including fuzzy logics, where it is usually assumed that the degree of belief and the degree of disbelief sum up to 1—see also equation (6)
  - $red = \langle 0.2, 0.7 \rangle$ means that *red* holds to the degree 0.2 and does not hold to the degree 0.7; observe that the sum of the degree of belief and disbelief is smaller than 1, which corresponds to a lack of knowledge as, e.g., in intuitionistic fuzzy sets [2]
  - $red = \langle 0.4, 0.7 \rangle$ means that *red* holds to the degree 0.4 and does not hold to the degree 0.7; observe that the sum of the degree of belief and disbelief is greater than 1, which corresponds to inconsistency, as in [28].   ◁

In what follows $[0, 1]^2 \stackrel{\text{def}}{=} [0, 1] \times [0, 1]$.

Definition 4. A *fuzzy interpretation* of $\mathcal{C}(\mathcal{L})$ is given by a mapping $\theta : \mathcal{P} \longrightarrow [0, 1]^2$. The mapping θ is extended to $\vartheta : \mathcal{F} \longrightarrow [0, 1]^2$ as follows, assuming inductively that $\vartheta(A) = \langle \beta_A, \delta_A \rangle$ and $\vartheta(B) = \langle \beta_B, \delta_B \rangle$:

  - $\vartheta(A) \stackrel{\text{def}}{=} \theta(A)$, where $A \in \mathcal{P}$
  - $\vartheta(\neg A) \stackrel{\text{def}}{=} \langle \delta_A, \beta_A \rangle$

- $\vartheta(A \vee B) \overset{\text{def}}{=} \langle \max\{\beta_A, \beta_B\}, \min\{\delta_A, \delta_B\} \rangle$
- $\vartheta(A \wedge B) \overset{\text{def}}{=} \langle \min\{\beta_A, \beta_B\}, \max\{\delta_A, \delta_B\} \rangle$
- $\vartheta(A \to B) \overset{\text{def}}{=} \langle \min\{\beta_A, \beta_B\}, \min\{\beta_A, \delta_B\} \rangle$.   ◁

Observe that in Definition 4 the semantics for implication is somehow arbitrary. In fuzzy-like reasoning various implications and their generalizations have been found suitable for different application domains. Examples of the considered implications are those of Lukasiewicz-Tarski, Kleene-Dienes, Reichenbach, Gödel, Goguen, Mamdani, Willmott, Yager, Zadeh as well as some other (see [11]). Therefore we have chosen a definition, where $\langle A \wedge B, A - B \rangle$ provides the measure for $A \to B$. This choice is not substantial for our considerations. The methodology we propose is independent of a particular definition of fuzzy interpretation. The only requirement is that we are able to calculate the values of complex formulas on the basis of the values of subformulas.

Observe also that rules cited in the introduction (formulas (1) and (2) there) require special definitions of the implication connective. This becomes even more laborious when rules involving vague concepts are used, e.g.,

exertional cough $\underbrace{\text{is suggestive of}}_{\text{implies to a degree}}$ heart failure.   (9)

In (9) the concept of *exertional cough* is vague. The term *is suggestive of* gives rise to a special kind of implication, which is modeled by assigning a pair of numbers to the conclusion indicating the degree to which it holds and the degree to which it does not hold.

In order to define the four-valued semantics, we have to be able to decide how a given pair of reals from $[0, 1]$ attached to a given formula variable is interpreted as one of the truth values $\{\mathsf{f}, \mathsf{u}, \mathsf{i}, \mathsf{t}\}$. Observe that such an interpretation depends on a particular formula. For example, in a particular application, $red = \langle 0.8, 0.1 \rangle$ might be interpreted as t, while $safe = \langle 0.8, 0.1 \rangle$ might be interpreted as u. In such a case we would have

$$(\underbrace{red}_{\mathsf{t}} \wedge \underbrace{safe}_{\mathsf{u}}) = \langle 0.8, 0.1 \rangle.$$

In this case the interpretation of $(red \wedge safe)$ should be given by the value of the underlying four-valued conjunction $\mathsf{t} \wedge \mathsf{u}$ (which in many logics evaluates to u).

We then have the following definition.

Definition 5. By a *four-valued interpretation* we shall mean any mapping

$$\iota : \mathcal{P} \times [0, 1]^2 \longrightarrow \{\mathsf{f}, \mathsf{u}, \mathsf{i}, \mathsf{t}\}$$

providing a four-valued interpretation of values (from $[0, 1]^2$) of propositions.   ◁

Example 6. An example of ι can be given by

$$\iota(red, \beta, \delta) \stackrel{\text{def}}{=} \begin{cases} \mathsf{f} \text{ when } \beta + \delta = 1 \text{ and } \delta \geq 0.8 \\ \mathsf{u} \text{ when } \beta + \delta < 1 \text{ or} \\ \quad\quad \beta + \delta = 1 \text{ and } 0 \leq \beta < 0.8 \text{ and } 0 \leq \delta < 0.8 \\ \mathsf{i} \text{ when } \beta + \delta > 1 \\ \mathsf{t} \text{ when } \beta + \delta = 1 \text{ and } \beta \geq 0.8. \end{cases}$$

The four-valued interpretation ι is illustrated in Figure 1.   ◁

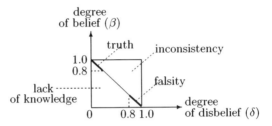

Fig. 1. Degrees of belief and disbelief considered in Example 6.

The following definition extends interpretations to cover all formulas.

Definition 7. Let ι be a four-valued interpretation. By an *interpretation over* ι we shall understand the mapping $\kappa : \mathcal{F} \times [0,1]^2 \longrightarrow \{\mathsf{f}, \mathsf{u}, \mathsf{i}, \mathsf{t}\}$ satisfying

- $\kappa(A, \beta, \delta) \stackrel{\text{def}}{=} \iota(A, \beta, \delta)$ for $A \in \mathcal{P}$
- $\kappa(\neg A, \beta, \delta) \stackrel{\text{def}}{=} \neg\kappa(A, \beta, \delta)$
- $\kappa(A \vee B, \beta, \delta) \stackrel{\text{def}}{=} \kappa(A, \delta, \beta) \vee \kappa(B, \delta, \beta)$
- $\kappa(A \wedge B, \beta, \delta) \stackrel{\text{def}}{=} \kappa(A, \delta, \beta) \wedge \kappa(B, \delta, \beta)$
- $\kappa(A \rightarrow B, \beta, \delta) \stackrel{\text{def}}{=} \kappa(A, \delta, \beta) \rightarrow \kappa(B, \delta, \beta)$,

where $\neg, \vee, \wedge, \rightarrow$ at the righthand side of the equalities are respectively the negation, disjunction, conjunction and implication of the underlying four-valued logic $\mathcal{L}$.   ◁

Let $\mathcal{D}$ be the set of designated values in logic $\mathcal{L}$.[5] In paraconsistent reasoning most frequently $\mathcal{D}$ is $\{\mathsf{t}\}$, but it is also reasonable to assume that it is $\{\mathsf{t}, \mathsf{i}\}$ or, e.g., in some forms of non-monotonic reasoning, $\{\mathsf{t}, \mathsf{i}, \mathsf{u}\}$.

Observe that in a particular application not all possible four-valued interpretations ι (and thus κ) make sense. Usually one would fix one such interpretation or a particular class of such interpretations. Example 6 provides a typical

---

[5] Recall that in many-valued logics the set of designated values consists of truth values that act as t—see [5, 25, 30].

interpretation, but the choice of threshold 0.8 is somewhat arbitrary there. Frequently, one would consider a more flexible definition of ι:

$$\iota(red, \beta, \delta) \overset{\text{def}}{=} \begin{cases} \mathbf{f} \text{ when } \beta + \delta \in [1 - \epsilon_u, 1 + \epsilon_i] \text{ and } \delta \geq \epsilon_f \\ \mathbf{u} \text{ when } \beta + \delta < 1 - \epsilon_u \\ \quad \beta + \delta = 1 \text{ and } 0 \leq \beta < \epsilon_t \text{ and } 0 \leq \delta < \epsilon_f \\ \mathbf{i} \text{ when } \beta + \delta > 1 + \epsilon_i \\ \mathbf{t} \text{ when } \beta + \delta \in [1 - \epsilon_u, 1 + \epsilon_i] \text{ and } \beta \geq \epsilon_t. \end{cases} \quad (10)$$

where $\epsilon_f, \epsilon_u, \epsilon_i, \epsilon_t \in [0, 1]$ are thresholds suitably chosen for deciding whether a given pair $\langle \beta, \delta \rangle$ is to be interpreted as $\mathbf{f}$, $\mathbf{u}$, $\mathbf{i}$ or $\mathbf{t}$.[6]

We are now ready to define the notion of semantic consequence.

Definition 8. Let $\mathbb{F}$ be a set of fuzzy interpretations and $\mathbb{C}$ be a set of four-valued interpretations. Let $F \subseteq \mathcal{F}$ be an arbitrary set of formulas of $\mathcal{C}(\mathcal{L})$ and $A \in \mathcal{F}$ be a formula of $\mathcal{C}(\mathcal{L})$. We say that A is a *semantic consequence of F* w.r.t. $\mathbb{F}$ *and* $\mathbb{C}$, denoted by $F \models_{\mathbb{F},\mathbb{C}} A$, provided that for all fuzzy interpretations $\vartheta \in \mathbb{F}$ and all four-valued interpretations $\iota \in \mathbb{C}$,

if for all $B \in F$ we have that $\kappa(B, \vartheta(B)) \in \mathcal{D}$ then also $\kappa(A, \vartheta(A)) \in \mathcal{D}$,

where κ is the interpretation over ι, as defined in Definition 7 and $\mathcal{D}$ is the set of designated values in $\mathcal{L}$. $\triangleleft$

4   Example: Logic $\mathcal{C}(L_t)$

In this section we show an example of logic belonging to the family of logics introduced in Section 3.1, and examples of its applications.

4.1   Logic $\mathcal{C}(L_t)$

The logic we apply is L_t introduced in [22]. To construct it we use two orderings on truth values, namely the truth ordering and the knowledge ordering. Truth ordering is used for calculations within a single information source while knowledge ordering is used for gathering knowledge from different sources. This approach has been initiated in [4,3] and, in the framework of bilattices, in [16,17].

The *knowledge ordering* $\leq_k$ and the *truth ordering* $\leq_t$ on $\mathcal{B}$ are shown in Figure 2. For example, $\mathbf{u} \leq_k \mathbf{t} \leq_k \mathbf{i}$, $\mathbf{u} \leq_k \mathbf{f} \leq_k \mathbf{i}$ and $\mathbf{f} \leq_t \mathbf{u} \leq_t \mathbf{i} \leq_t \mathbf{t}$. The knowledge ordering coincides with Belnap's knowledge ordering. Since Belnap's truth ordering can give counterintuitive results when used in the types of reasoning we are interested in (see [21]), the truth ordering coincides with the truth ordering of [1,31,22].

Table 2 provides semantics for connectives of L_t. Observe that the implication $\rightarrow$, introduced in [31], is a four-valued extension of the usual logical

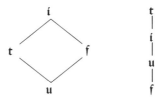

Fig. 2. Knowledge ordering and truth ordering.

Table 2. Truth tables for connectives of L_t.

$\wedge$	f	u	i	t	$\vee$	f	u	i	t	$\rightarrow$	f	u	i	t	$\neg$	
f	f	f	f	f	f	f	u	i	t	f	t	t	t	t		t
u	f	u	u	u	u	u	u	i	t	u	u	u	i	t		u
i	f	u	i	i	i	i	i	i	t	i	i	i	i	t		i
t	f	u	i	t	t	t	t	t	t	t	f	u	i	t		f

implication, suitable for determining set containment and Pawlak-like approximations [23] in the case of four-valued sets.

The reasoning within a single information source is carried out according to Definition 4. In order to gather results concerning the same proposition from different sources we use an additional fuzzy operator $\oplus$:

$$\langle \beta_1, \delta_1 \rangle \oplus \langle \beta_2, \delta_2 \rangle \overset{\text{def}}{=} \langle \max\{\beta_1, \beta_2\}, \max\{\delta_1, \delta_2\} \rangle. \tag{11}$$

In L_t we gather results concerning the same head from different sources by using the disjunction $\vee_k$ w.r.t. knowledge ordering, defined in Table 3.

Table 3. Truth table for $\vee_k$.

$\vee_k$	f	u	i	t
f	f	f	i	i
u	f	u	i	t
i	i	i	i	i
t	i	t	i	t

Example 9. As a simple example of reasoning consider the following rule:

$$\left(severe \ sore \ throat \wedge painful \ swallowing \wedge headache \wedge fever \wedge chills \right) \rightarrow tonsillitis.$$
$$\tag{12}$$

Assume that we ask two doctors for a diagnosis based on (12). The results of examination of symptoms are provided in Table 4, where for all variables we assume the four-valued interpretation given in Example 6. Based on the

---

[6] In particular the chosen thresholds should make clauses for **t** and **f** mutually exclusive.

Table 4. An example of evaluation of symptoms for rule (12).

	sore throat	painful swallowing	headache	fever	chills
doctor 1	$\langle 0.8, 0.1 \rangle \rightsquigarrow \mathbf{u}$	$\langle 0.9, 0.1 \rangle \rightsquigarrow \mathbf{t}$	$\langle 0.2, 0.8 \rangle \rightsquigarrow \mathbf{f}$	$\langle 0.7, 0.3 \rangle \rightsquigarrow \mathbf{u}$	$\langle 0.8, 0.1 \rangle \rightsquigarrow \mathbf{u}$
doctor 2	$\langle 0.9, 0.1 \rangle \rightsquigarrow \mathbf{t}$	$\langle 1.0, 0.0 \rangle \rightsquigarrow \mathbf{t}$	$\langle 0.8, 0.2 \rangle \rightsquigarrow \mathbf{t}$	$\langle 0.9, 1.0 \rangle \rightsquigarrow \mathbf{t}$	$\langle 1.0, 0.0 \rangle \rightsquigarrow \mathbf{t}$

examination of symptoms and Definition 4, the first doctor decides that belief and disbelief for tonsillitis are $\langle 0.7, 0.3 \rangle$ and the second doctor decides that these values are $\langle 0.9, 1.0 \rangle$. Gathering those results gives $\langle 0.7, 0.3 \rangle \oplus \langle 0.9, 1.0 \rangle = \langle 0.9, 0.3 \rangle$, which is interpreted as $\mathbf{i}$ by paraconsistent fuzzy reasoning. Note that according to the first doctor, the conjunction of symptoms in rule (12) results in $\mathbf{f}$. Therefore the first doctor decides that the conclusion (tonsillitis) is $\mathbf{f}$, too. According to the second doctor, the conjunction of symptoms, thus also the value of tonsillitis, is $\mathbf{t}$. Gathering those results by $\vee_k$ results in $\mathbf{i}$, as in the case of paraconsistent fuzzy reasoning. ◁

4.2 Rule-Based Reasoning: A Case Study

Let us now illustrate a more advanced rule-based reasoning, where the conclusions of rules are assigned belief and disbelief degrees on the basis of premises and a medical knowledge base.

Medical knowledge is defined in [8] as a fuzzy relation R, linking the set of symptoms with the set of diagnoses "which reveals the degree of association and the degree of non-association" between symptoms and diagnoses. The methodology proposed in [8] involves three steps:

1. determination of symptoms
2. formulation of medical knowledge based on fuzzy relations
3. determination of diagnosis on the basis of composition of fuzzy relations.

In the approach of [8] and also [29] the diagnosis is evaluated on the basis of a certain distance from symptoms to a given disease. The reasoning in [8] is based on the following rule:

> IF: the state of a given patient P is described in terms
> of a description of symptoms A
> THEN: P is assumed to be assigned diagnosis in terms   (13)
> of a description of diagnoses B, through
> a medical knowledge database.

Medical knowledge is given as a paraconsistent fuzzy relation relating symptoms to diagnoses. For example, indications for and against chosen diseases are provided in Table 5, based on [8].

The degree of belief and disbelief of a diagnosis $d \in D$ is calculated as:

$$\left\langle \max_{s \in S} \left\{ \min\{\beta_A(s), \beta_R(s, d)\} \right\}, \min_{s \in S} \left\{ \max\{\delta_A(s), \delta_R(s, d)\} \right\} \right\rangle, \tag{14}$$

Table 5. The relation between symptoms and disease.

	viral fever	malaria	typhoid
temperature	$\langle 0.4, 0.0 \rangle$	$\langle 0.7, 0.0 \rangle$	$\langle 0.3, 0.3 \rangle$
headache	$\langle 0.3, 0.5 \rangle$	$\langle 0.2, 0.6 \rangle$	$\langle 0.6, 0.1 \rangle$
cough	$\langle 0.4, 0.3 \rangle$	$\langle 0.7, 0.0 \rangle$	$\langle 0.2, 0.6 \rangle$

where S is the set of symptoms, D is the set of diagnoses, A is a (paraconsistent intuitionistic) fuzzy set describing symptoms of a patient, R is the (paraconsistent intuitionistic) fuzzy relation relating symptoms to diagnoses, and $\beta_X(\bar{x})$ $(\delta_X(\bar{x}))$ stand for the degree of belief (disbelief) that $\bar{x}$ satisfies X.

Assuming the set of symptoms $\{temperature, headache, cough\}$ and the set of diseases $\{viral\ fever, malaria, typhoid\}$, (14) gives rise to the following rules:

$$viral\ fever \leftarrow temperature, headache, cough. \tag{15}$$

$$malaria \leftarrow temperature, headache, cough. \tag{16}$$

$$typhoid \leftarrow temperature, headache, cough. \tag{17}$$

Evaluation of conclusions of rules (15)–(16) is given by instantiating formula (14). For example, the degree of belief and disbelief for *malaria* is given by

$$\Big\langle \max\Big\{ \min\{\beta_A(temperature), \beta_R(temperature, malaria)\},$$
$$\min\{\beta_A(headache), \beta_R(headache, malaria)\},$$
$$\min\{\beta_A(cough), \beta_R(cough, malaria)\}\Big\},$$
$$\min\Big\{ \max\{\delta_A(temperature), \delta_R(temperature, malaria)\},$$
$$\max\{\delta_A(headache), \delta_R(headache, malaria)\},$$
$$\max\{\delta_A(cough), \delta_R(cough, malaria)\}\Big\}\Big\rangle,$$

where A is the (paraconsistent intuitionistic) fuzzy set describing symptoms of a patient and R is the relation provided in Table 5.
Table 6 provides examples of symptoms for three patients as well as the evaluated degrees of belief and disbelief for the considered diagnoses and their four-valued interpretation, as considered in Example 6.

Table 6. Symptoms and diagnoses for patients.

	patient 1	patient 2	patient 3
temperature	$\langle 0.8, 0.1 \rangle \rightsquigarrow \mathbf{u}$	$\langle 0.1, 0.9 \rangle \rightsquigarrow \mathbf{f}$	$\langle 0.5, 0.6 \rangle \rightsquigarrow \mathbf{i}$
headache	$\langle 0.6, 0.5 \rangle \rightsquigarrow \mathbf{i}$	$\langle 0.1, 0.9 \rangle \rightsquigarrow \mathbf{f}$	$\langle 0.4, 0.2 \rangle \rightsquigarrow \mathbf{u}$
cough	$\langle 0.7, 0.2 \rangle \rightsquigarrow \mathbf{u}$	$\langle 0.0, 0.9 \rangle \rightsquigarrow \mathbf{f}$	$\langle 0.5, 0.4 \rangle \rightsquigarrow \mathbf{u}$
viral fever	$\langle 0.4, 0.1 \rangle \rightsquigarrow \mathbf{u}$	$\langle 0.1, 0.9 \rangle \rightsquigarrow \mathbf{f}$	$\langle 0.4, 0.4 \rangle \rightsquigarrow \mathbf{u}$
malaria	$\langle 0.7, 0.1 \rangle \rightsquigarrow \mathbf{u}$	$\langle 0.1, 0.9 \rangle \rightsquigarrow \mathbf{f}$	$\langle 0.5, 0.4 \rangle \rightsquigarrow \mathbf{u}$
typhoid	$\langle 0.6, 0.3 \rangle \rightsquigarrow \mathbf{u}$	$\langle 0.1, 0.9 \rangle \rightsquigarrow \mathbf{f}$	$\langle 0.4, 0.2 \rangle \rightsquigarrow \mathbf{u}$

Analyzing the results one can note that:

- malaria is the most plausible indication for the first patient
- none of the considered diseases fits symptoms of the second patient
- there is a weak suggestion of malaria as well as a weak suggestion for typhoid for the third patient.

The four-valued analysis of the above results is based on rules (15)–(16). These rules are interpreted as implications:

$$(temperature \land headache \land cough) \rightarrow viral\ fever \qquad (18)$$

$$(temperature \land headache \land cough) \rightarrow malaria \qquad (19)$$

$$(temperature \land headache \land cough) \rightarrow typhoid. \qquad (20)$$

Based on truth values of L_t given in Table 6 we can conclude that assuming the interpretation provided in Example 6, all conclusions are, in fact, unknown.

In the case of the first patient the implications (18)–(20) reduce to $(\mathsf{u} \land \mathsf{i} \land \mathsf{u}) \rightarrow \mathsf{u}$, which, according to Table 2, is $\mathsf{u} \rightarrow \mathsf{u}$, i.e., u. This means that conclusions and the validity of implications are unknown.

In the case of the second patient the implications (18)–(20) reduce to $(\mathsf{f} \land \mathsf{f} \land \mathsf{f}) \rightarrow \mathsf{f}$, which is t. Here all the conclusions are f meaning that none of the considered diseases fits the symptoms. Moreover, the implications (18)–(20) are t which indicates that the reasoning is sound.

The qualitative interpretation of the results for the third patient are similar to the interpretation for the first patient.

Altogether, the results indicate that the interpretation of Example 6 might be too restrictive when used in this reasoning. However, once assumed, it allows us to interpret the results on the basis of a solid background.

Of course, the choice of the fuzzy interpretation is flexible so the previous constraints can be relaxed, e.g., along the lines of interpretation defined by formula (10) with suitably chosen thresholds. For example, taking $\epsilon_f = \epsilon_t \overset{\text{def}}{=} 0.6$ and $\epsilon_u = \epsilon_i \overset{\text{def}}{=} 0.2$ in formula (10), one can conclude that the diagnosis for the first patient is malaria, since now $\langle 0.7, 0.1 \rangle$ is interpreted as t. In this case the implication (19) reduces to $(\mathsf{t} \land \mathsf{t} \land \mathsf{t}) \rightarrow \mathsf{t}$, i.e., to t, which shows soundness of the reasoning.

Remark 10. Observe that the semantics of implication based on Table 2 makes the implication t also when t is derived from u or i. In rule-based reasoning one would frequently prefer to make such implications f since deriving conclusions from unknown or inconsistent knowledge leads to forms of non-monotonicity. To "block" derivations based on such implications, one can use other implications, like $\Rightarrow$ shown in Table 7. Other implications are also considered in paraconsistent rule-based reasoning (see [22]). ◁

Table 7. Truth table for the implication $\Rightarrow$.

$\Rightarrow$	f	u	i	t
f	t	t	t	t
u	u	t	u	f
i	i	i	t	f
t	f	f	f	t

5   Conclusions

In the current paper we addressed the problem of qualitative interpretation of fuzzy-like paraconsistent reasoning. We proposed a methodology in which suitably chosen qualitative four-valued logics are used to serve this purpose. The interpretation is applied in all steps where fuzzy-like reasoning is carried out. Therefore the resulting interpretation provides the logical value which would be obtained by applying the four-valued reasoning.

In this paper we did not consider first-order $\mathcal{C}(L_t)$. To obtain such logics one has to extend the language as in the case of the classical first-order logic. In the case of $\mathcal{C}(L_t)$ it is reasonable to define the semantics of quantifier $\forall$ by generalizing the conjunction $\wedge$ and the semantics of $\exists$ by means of generalizing the disjunction $\vee$:

$$\forall x[P(x)] \stackrel{\text{def}}{=} \underset{x \in U}{\text{GLB}}^t\{P(x)\} \quad \text{and} \quad \exists x[P(x)] \stackrel{\text{def}}{=} \underset{x \in U}{\text{LUB}}^t\{P(x)\} \,,$$

where the superscript ^{t} indicates that the greatest lower bound (GLB) and least upper bound (LUB) are computed w.r.t. truth ordering.

In future we plan the research on employing approximations in the spirit of Pawlak [23, 12] but also taking into account the approach of [31].

Similar methodology can be applied to interpret the traditional fuzzy reasoning by interpreting fuzzy values qualitatively and tracking the reasoning using a suitably chosen qualitative logic.

Acknowledgments. This paper has been supported in part by the MNiSW grant N N206 399134.

References

1. Amo, S., Pais, M.S.: A paraconsistent logic approach for querying inconsistent databases. International Journal of Approximate Reasoning 46, 366–386 (2007)
2. Atanassov, K.: Intuitionistic fuzzy sets. Fuzzy Sets and Systems 20, 87–96 (1986)
3. Belnap, N.D.: How a computer should think. In: Ryle, G. (ed.) Contemporary Aspects of Philosophy, pp. 30–55. Oriel Press, Stocksfield (1977)
4. Belnap, N.D.: A useful four-valued logic. In: Eptein, G., Dunn, J.M. (eds.) Modern Uses of Many Valued Logic, pp. 8–37. Reidel, Dordrecht (1977)
5. Bolc, L., Borowik, P.: Many-Valued Logics, 1. Theoretical Foundations. Springer, Berlin (1992)

6. Bolc, L., Coombs, M.J. (eds.): Expert System Applications. Springer, Heidelberg (1988)
7. Brandon, M., Rescher, N.: The Logic of Inconsistency. Basil Blackwell, Oxford (1978)
8. De, S.K., Biswas, R., Roy, A.R.: An application of intuitionistic fuzzy sets in medical diagnosis. Fuzzy Sets and Systems 117, 209–213 (2001)
9. Dubois, D., Hadj-Ali, A., Prade, H.: Fuzzy qualitative reasoning with words. In: Wang, P.P. (ed.) Computing with Words, pp. 347–366 (2001)
10. Dubois, D., Lang, J., Prade, H.: Fuzzy sets in approximate reasoning, part 2: logical approaches. Fuzzy Sets and Systems 40(1), 203–244 (1991)
11. Dubois, D., Prade, H.: Fuzzy sets in approximate reasoning, part 1: inference with possibility distributions. Fuzzy Sets and Systems 40(1), 143–202 (1991)
12. Dubois, D., Prade, H.: Putting rough sets and fuzzy sets together. In: Słowiński, R. (ed.) Intelligent Decision Support: Handbook of Applications and Advances of the Rough Sets Theory, pp. 203–232. Kluwer Academic Publishers, Dordrecht (1992)
13. Dubois, D., Prade, H.: Fuzzy Sets and Systems. In: Fuzzy Logic CDROM Library, Academic Press, London (1996)
14. Dubois, D., Prade, H.: What are fuzzy rules and how to use them. Fuzzy Sets and Systems 84, 169–185 (1996)
15. Ebrahim, R.: Fuzzy logic programming. Fuzzy Sets and Systems 117, 215–230 (2001)
16. Fitting, M.C.: Bilattices in logic programming. In: Epstein, G. (ed.) 20th International Symposium on Multiple-Valued Logic, pp. 238–247. IEEE CS Press, Los Alamitos (1990)
17. Ginsberg, M.: Multivalued logics: a uniform approach to reasoning in AI. Computational Intelligence 4, 256–316 (1988)
18. Guglielmann, R., Ironi, L.: The need for qualitative reasoning in fuzzy modeling: robustness and interpretability issues. In: Proc. of 18th International Workshop on Qualitative Reasoning, pp. 113–120 (2004)
19. Innocent, P.R., John, R.I.: Computer aidded medical diagnosis. Fuzzy Sets and Systems 162, 81–104 (2004)
20. Kruse, R., Schwecke, E., Heinsohn, J.: Uncertainty and Vagueness in Knowledge Based Systems. Numerical Methods. Springer, Heidelberg (1991)
21. Małuszyński, J., Szałas, A., Vitória, A.: A four-valued logic for rough set-like approximate reasoning. In: Peters, J.F., Skowron, A., Düntsch, I., Grzymała-Busse, J.W., Orłowska, E., Polkowski, L. (eds.) Transactions on Rough Sets VI. LNCS, vol. 4374, pp. 176–190. Springer, Heidelberg (2007)
22. Małuszyński, J., Szałas, A., Vitória, A.: Paraconsistent logic programs with four-valued rough sets. In: Chan, C.-C., Grzymala-Busse, J.W., Ziarko, W.P. (eds.) RSCTC 2008. LNCS, vol. 5306, pp. 41–51. Springer, Heidelberg (2008)
23. Pawlak, Z.: Rough Sets. Theoretical Aspects of Reasoning about Data. Kluwer Academic Publishers, Dordrecht (1991)
24. Prade, H.: A quantitative approach to approximate reasoning in rule-based expert systems. In: [6], pp. 199–256 (1988)
25. Rescher, N.: Many-Valued Logic. McGraw-Hill, New York (1969)
26. Shortliffe, E.H.: Computer-Based Medical Consutations: MYCIN. Elsevier, Amsterdam (1976)

27. Shwe, M.A., Middleton, B., Middleton, D.E., Henrion, M., Horvitz, E.J., Lehmann, H.P., Cooper, G.F.: Probabilistic diagnosis using a reformulation of the INTERNIST-1/QMR knowledge base. Methods of Information in Medicine 30(4), 241–255 (1991)

28. Sunderraman, R., Wang, H.: Paraconsistent intuitionistic fuzzy relational data model (2004),
    http://www.citebase.org/abstract?id=oai:arXiv.org:cs/0410054

29. Szmidt, E., Kacprzyk, J.: Intuitionistic fuzzy sets in some medical applications. In: Reusch, B. (ed.) Fuzzy Days 2001. LNCS, vol. 2206, pp. 148–151. Springer, Heidelberg (2001)

30. Urquhart, A.: Many-valued logic. In: Gabbay, D.M., Guenthner, F. (eds.) Handbook of Philosophical Logic, vol. 3, pp. 71–116. Reidel, Dordrecht (1986)

31. Vitória, A., Szałas, A., Małuszyński, J.: Four-valued extension of rough sets. In: Wang, G., Li, T., Grzymala-Busse, J.W., Miao, D., Skowron, A., Yao, Y. (eds.) RSKT 2008. LNCS, vol. 5009, pp. 106–114. Springer, Heidelberg (2008)

32. Vojtáš, P.: Fuzzy logic programming. Fuzzy Sets and Systems 124, 361–370 (2001)

33. Zadeh, L.: From computing with numbers to computing with words – from manipulation of measurements to manipulation of perceptions. Int. J. Appl. Math. Comput. Sci. 12(3), 307–324 (2002)

34. Zadeh, L.A.: Fuzzy sets. Information and Control 8, 333–353 (1965)

35. Zadeh, L.A.: Outline of a new approach to the analysis of complex system and decision processes. IEEE Trans. Syst. Man. Cybern. SMC-3, 28–44 (1973)

36. Zadeh, L.A.: Fuzzy logic = computing with words. IEEE Trans. on Fuzzy Systems 4, 103–111 (1996)

Part II

Language

On the Root-Based Lexicon for Polish

Joanna Rabiega-Wiśniewska

Institute of Computer Science, Polish Academy of Sciences,
J.K. Ordona 21, 01-237 Poland
Joanna.Rabiega@ipipan.waw.pl

Abstract. In this paper we present the concept of an electronic lexicon based on morphological roots. The idea of the root-based lexicon returns to traditional linguistic division of a word into a stem and an inflectional suffix. The only difference to the pure linguistic description is that an electronic resource must adapt to the analyzed text. We assume that the lexicon will be used in written text analysis (or synthesis), therefore we operate on grapheme objects.
We used the lexicon of the inflectional analyzer AMOR as the empirical foundation for the root-based lexicon. In the second part of the paper we describe the process of the automatic conversion of the data from the analyzer into the assumed format. The conversion concerns the major inflecting parts of speech: nouns, adjectives and verbs. The results are two-level morphology based entries which bear the whole package of morphological information about lexemes. In the presented form, however, any generalization about Polish inflection or inner root alternations is not available. Thus, we rebuilt the lexicon of roots. As a result we obtained the compressed lexicon which can serve not only for inflection analysis but also applications of word-formation descriptions.

Key words: root-based lexicon, two-level morphology, automatic morphological analysis, lexicon compression

1 Introduction

One of the most important resources for NLP are lexicons that can be easily adapted to various applications. At present, there are several methods of carrying out inflectional analysis automatically and a few different standards of lexical entry description for Polish. Those methods abandon the traditional approach to morphological analysis according to which word-forms are mostly segmented into two morphologically motivated parts, the stem and the ending. Instead of the morphological word-form division, the longest sequence of characters appearing in the whole paradigm is considered as a *semi-stem*. The rest of the characters form a *semi-ending*. This particular lexicon design results in the fact that any morphological analyzer which uses it delivers only a lemma and an appropriate grammatical characteristic of a given word-form. Information about inside morphemes cannot be obtained. The question arises how additional information for each entry should be provided to allow building a lexicon useful

M. Marciniak and A. Mykowiecka (Eds.): Bolc Festschrift, LNCS 5070, pp. 61–82, 2009.
© Springer-Verlag Berlin Heidelberg 2009

for a morpheme-based application, as e.g. a derivative analysis. The principles of current automatic morphological analysis of Polish are presented in section 2.

The paper aims at presenting a root-based electronic lexicon for Polish in which an entry is described by principles of morphological construction. The root-based lexicon has already been exploited in the formal model of Polish nominal derivation [12], since it gives access to all inflection stems of a given lemma and all endings separately. It can be easily adapted for automatic inflectional analysis and/or synthesis.

In section 3 we describe the process of building the lexicon. The two-level morphology model by Koskenniemi [10] served as a theoretical background for the study (subsection 3.1). To prepare a lexicon of roots we used the grammatical dictionary of the AMOR analyzer [14] as the empirical basis (subsection 3.2). The algorithm of the AMOR data conversion into the lexicon of roots is presented in the following steps in the subsection 3.3:

a. synthesis of all word-forms for a given lexeme,
b. segmentation of all synthesized word-forms into the stems and the corresponding endings,
c. collection of all stems of a given lemma and their internal alternations description,
d. the final root choice.

As a result, we have obtained entries that contain the whole inflectional characteristics and details about internal alternations. Let us consider the example (1). A lexicon entry WYŻEŁ 'pointer' is represented by a lemma, wyżeł, i.e. a superficial representative (and a code of grammatical gender, r:2), a root wyżEŁ and two sets. The first set contains textual representations of all internal alternations within the lemma and grammatical codes of their distribution. For example the alternation E:e(N) means that the root alternant E has the text realization e in nominative singular (N). The last component comprises a full package of endings. Discussion of description details is presented in subsection 3.3

(1) wyżeł r:2
     wyżEŁ {E:e(N),O(G,D,A,B,L,V,n,g,d,a,b,l,v);
            Ł:ł(N,G,D,A,B,n,g,d,a,b,l,v),l(L,V)}
     k(O,a,owi,a,em,e,e,y,ów,om,y,ami,ach,y)

Data compression is carried out through the automatic merging the lexicon entries into inflectional subgroups of specific properties. The current set up for the entry WYŻEŁ 'pointer' among other entries of the same paradigm and gender is presented in example (2). First, lexemes of one inflectional group were automatically arranged in paradigm patterns (k:9), then they are bound in nests defined by their alternations (a:35). This results in a consistent form of the data in the lexicon. Details of this process will be provided in section 4.

(2) `a:35 {E:e(N),O(G,D,A,B,L,V,n,g,d,a,b,l,v);`
     `Ł:ł(N,G,D,A,B,n,g,d,a,b,l,v),l(L,V)}`

   `k:9 k(O,a,owi,a,em,e,e,y,ów,om,y,ami,ach,y)`

   `suseł susEŁ r:2 a:35 k:9`
   `wyżeł wyżEŁ r:2 a:35 k:9`
   `. . .`

The root-based lexicon may be a part of an automatic inflectional as well as a derivational tool. The morphological entry description seems to show a new perspective in creating linguistic resources for Polish.

2   Aims of Morphological Analysis

In this section we present current solutions applied to Polish inflection and word-formation in the field of automatic morphological analysis (and synthesis).

The methods of inflectional analysis that were elaborated in the past thirty years give proper results in text processing. However, they do not support linguistic approaches to morphological description. The discussion on this is presented in subsection 2.1.

The dissonance between linguistic theories and practical applications causes a problem with formal description of word-formation. Lexicon resources that fit inflectional analyzers can barely be adapted to serve rules of derivation. We prove that returning to traditional word-form analysis can bring profits at both inflectional and word-formation levels.

2.1   Inflection

The main task of automatic inflectional analysis is to find a lexeme for a given word-form. It is expected that all possible and correct grammatical characteristics are assigned to it. Another task is automatic inflectional synthesis which provides generation of the whole paradigm for a given lexeme. Concerning inflection it is crucial to maintain an appropriate description of word-forms, here a word-form is understood as an interpretation of a separate text unit. As the grammatical (syntactic) interpretation concerns the whole text, the morphological structure of the word-form is of little importance while creating a lexicon for an inflectional analyzer.

For Polish there are several methods of carrying out automatic analysis and synthesis [18,11,22,21,16]. On the basis of such methods, several programs have been implemented [7,4]. It has turned out that the traditional approach to morphological analysis is not sufficient for technical applications. In the linguistic approach to inflection a word-form is divided into two parts: a stem and an inflectional suffix, see example (3).

(3)   wariat-$\emptyset_{nom.sg}$, wariat-a$_{gen.sg}$, ..., wariaci-e$_{dat.sg}$, ... 'madman'
      ćm-a$_{nom.sg}$, ćm-ą$_{intr.sg}$, ..., ciem-$\emptyset_{gen.pl}$, ... 'moth'
      zwięzł-y$_{nom.masc}$, zwięzł-a$_{nom.fem}$, ..., zwięźl-i$_{nom.h-masc}$[1], ... 'brief'
      buja-m$_{1.sg.pres.}$, buja-sz$_{2.sg.pres.}$, buja-$\emptyset_{3.sg.pres.}$, ...,
      bujał-$\emptyset_{3.sg.masc.past}$, bujal-i$_{3.pl.h-masc.past}$, ... 'to rock'

Without a description that copes with stem alternations a traditional segmentation into two parts has to be abandoned. In the seventies Tokarski [19] assumed that an algorithm of inflectional analysis could use only a well-defined part of a word in the processing. His proposal was to bind the last part of word-forms in which they differ within one lexeme with a mechanism converting them into an appropriate entry form and inflectional description. That final sub-string of letters can be called *semi-ending*. For the part common to all word-forms within a lexeme or initial substring of letters the term *semi-stem* was proposed. The illustration of technical word-form division is in example (4). The distribution of semi-endings and semi-stems became a crucial problem to be solved in the domain of computational linguistics [20].

(4)   waria-t$_{nom.sg}$, waria-ta$_{gen.sg}$, ..., waria-cie$_{dat.sg}$, ... 'madman'
      $\emptyset$-ćma$_{nom.sg}$, $\emptyset$-ćmą$_{instr.sg}$, ..., $\emptyset$-ciem$_{gen.pl}$, ... 'moth'
      zwię-zły$_{nom.masc}$, zwię-zła$_{nom.fem}$, ..., zwię-źli$_{nom.h-masc}$, ... 'brief'
      buja-m$_{1.sg.pres.}$, buja-sz$_{2.sg.pres.}$, buja-$\emptyset_{3.sg.pres.}$, ...,
      buja-ł$\emptyset_{3.sg.masc.past}$, buja-li$_{3.pl.h-masc.past}$, ... 'to rock'

The inflectional analyzers of Polish are inspired by Tokarski's approach. Regardless of the direction of the process (*a tergo* or *a fronte*) all programs leave out the description of alternations which is not necessary for sufficient automatic inflectional analysis or synthesis. The lexicon set-up designed especially for the AMOR analyzer [11] will be presented later in subsection 3.2. This lexicon was the basis for the lexicon of root conversion, which is described in details in subsection 3.3.

2.2   Word-Formation

The account of automatic word-formation analysis poses the question of pure formal relationships between derivatives and their basis lexemes. We believe that a formal approach to word-formation description processes in Polish can be based on combining word-forms of certain inflectional properties with a set of well-defined derivational affixes. It means that there must be a clear choice of a basis lexeme for a given derivative; in other words, an unequivocal motivation of the word formation process must be determined. That is why the analysis shown in example (5a) is more desirable than the one shown in example (5b) in a formal application.

(5)   a.   bibliotekar-k(a) ← bibliotek-arz($\emptyset$) ← bibliotek(a)
           'female librarian' ← 'male librarian' ← 'library'

---
[1] The abreviation **h-masc** stands for *human-masculine gender*

b.   bibliotek-ar-k(a) ← bibliotek-arz(ø)
       'female librarian' ← 'male librarian'

       bibliotek-ar-k(a) ← bibliotek(a)
       'female librarian' ← 'library'

The description of word-formation in the framework of automatic text analysis (or synthesis) must fulfill more requirements. Derivational affixes need direct access to an inflectional root of a given basis word-form. Moreover, this should be a specific root of one defined word-form of a basis-lexeme. The importance of choosing basis word-forms in the derivation process was described first by Booij [2,3] and is known as the *inflection feeding derivation* phenomenon. The existence of this phenomenon in Polish has been proven by Cetnarowska [5] and a discussion of morphology division is presented in [13]. Therefore, we need a lexicon which allows us to easily remove inflectional suffixes from inflectional roots. The use of word-forms in the process of derivation helps to limit the number of alternation rules. A great part of alternations is solved at the inflection level of analysis. This is shown in example (6). Example (6a) presents a traditional way of forming derivatives at the level of lexemes. The other one, (6b), demonstrates the proposal of the analysis at the level of word-forms, which is given in reference [12].

(6)   a.   KRÓWKA   ← KROWA, [o:ó]
             'little cow' ← 'cow',   *alternation*

       b.   krów-k(a)   ← krów(ø), gen.pl
             'little cow' ← 'cow',   *inflectional label*

It is now clear that the method of describing Polish morphology for NLP purposes mentioned in the previous subsection cannot serve word-formation. The task of formal derivation description requires designing a new kind of lexicon, one that is based on inflectional roots and suffixes. The concept of the root lexicon which is presented in this article arose within the work on Polish derivation and conversion. In reference [15] the lexicon entries were the basis of formal rules for both word-formation and alternation processes. That description was intended for the future analysis of written text. That is why the formalism operates on a grapheme level of lexical entries[2]. The construction of the lexicon was also meant for textual applications.

3   The Concept of a Root-Based Lexicon

The main task of this work is to describe a lexicon entry by principles of morphological structure. The idea is to obtain access to inflectional stems and suffixes separately. We have adopted a two-level morphology approach developed by Koskenniemi [10] to construct a morphological lexicon of Polish lexemes which inflect. In the following subsection 3.1 we present the two-level model

---

[2] We do not analyse the morphophonological or phonological structure of word-forms.

of morphology, then we describe the structure of a lexical entry in the AMOR lexicon in subsection 3.2. The process of automatic conversion of the AMOR entries into the root-based entries follows in subsection 3.3.

3.1 Two-Level Morphology Approach

At the beginning of 80's, researchers working on computational morphology focused on methodologies using finite-state automata. During the annual meeting of the Linguistic Society of America in 1981 Kaplan and Kay [8] proposed to reformulate the problem of automatic morphology. The first paper on this matter [9] was published at the same time as the dissertation by Kimmo Koskenniemi „Two-Level Morphology: A General Computational Model for Word-Form Recognition and Production" [10], in which he presented a new computationally implemented linguistic model for morphological analysis and synthesis.

Koskenniemi's model incorporates a general formalism for making morphological descriptions of particular languages, and a language-independent program implementing the model. The two-level model is based on a lexicon system and a set of two-level rules. The description of Finnish inflectional morphology is presented in order to validate the model.

The two-level model consists of two representations: the lexicon and the phonemic alphabet. The lexicon contains morphonological representations of word entries and endings. The phonemic surface level consists of phonemes, or letters of the phonemic alphabet. The two levels are considered to be written above each other as two tiers. The upper row consists of lexical representations: word stems and morphotactically permitted endings. The lower row consists of the phonemes of the surface form. The upper and the lower characters correspond to each other pairwise.The relation is formulated with parallel rules which may refer to both of these two representations. The rules "do" nothing as such, they only test whether the correspondence is correct. The actual production or analysis is a separate driver mechanism guided by the rules.

At the lexical level functional morphemes, root lexemes and alternation patterns are linguistically defined as a lexicon. The author gives appropriate restrictions to the lexicon entries, e. g. codes of acceptable morphemes or morpheme groups which can be attached to them. The example below presents an entry for the Finnish word-form kielemme (the root kielE 'language', +n 'genitive singular', /mme 'possessive').

(7)   k i e l E + n / m m e
      k i e l e 0 0 0 m m e

Koskenniemi's description is purely synchronic, comprehensive and concentrates on Standard Finnish. It includes the full inflection of nouns, adjectives and verbs, including affixation of possessive suffixes and clitics, as well as compounding.

The idea of keeping each lexical entry at two levels was adopted to create a lexicon of Polish roots. According to our approach each lexeme will be defined at two levels:

- level of its textual realization, mainly regarded as a label that helps recognizing a lexeme;
- level of graphemic representation, which is understood as a separated inflectional stem and encoded stem alternations, and an additional set of inflectional endings.

We assume that we need three kinds of information to achieve this task. First, a description of POS classes and inflectional groups for Polish is needed, as in example (8a). Using this information we can gather sets of appropriate inflection endings that serve lexemes which belong to them. Second, all word-forms of a given lexeme have to be at our disposal (8b), as well as their inflectional description. After deleting ending candidates, the rest will be the basis stems which will be used to establish the shape of a root and root alternations. The last piece of information crucial for the lexicon structure is a linkage between found stem alternations and mentioned inflectional description (8c). Through this, generating a stem of given inflectional properties will be possible. This process will be presented in detail in subsection 3.3.

(8) a. POS, inflectional group ($ending_1$, $ending_2$, $ending_3$, ...)

  b. LEXEME ($form_1$, $form_2$, $form_3$, ...)

  c. root ($stem_1$, $stem_2$, $stem_3$, ...),
      alternation ($alternation_1$, $alternation_2$, $alternation_3$, ...)

Prospective rules of inflection would operate externally on entries and would not be a part of the lexicon itself.

3.2   The AMOR Lexicon Design

We used lexical entries of the AMOR lexicon in the process of automatic conversion. Construction of the AMOR lexicon is presented in detail in references [11] and [14]. Here, we describe the set-up of entries important for our work.

The lexicon consists of text files and each of them represents one inflectional subgroup according to Tokarski's description [19]. We followed the grammatical classes (parts of speech) distinction according to the typology by Zygmunt Saloni [17]. Every file contains a list of lemmata and patterns which encode full information about inflection.

The inflectional pattern consists of a list of pairs of symbols representing inflection information and parts of word-forms labeled as *semi-endings*.

```
grammatical code = semi-ending
```

The number of pairs depends on the grammatical class, e.g. for nouns there are at least 14 pairs which represent 7 cases and 2 numbers. Semi-endings are parts in which word-forms differ (semi-stems represent parts common to all word-forms within lexemes), that means that they comprise inflectional suffixes and very often parts of morphological roots. Compare 4 patterns for masculine nouns of the fourth declension group in examples (9)-(12).

(9) N1= G1=a D1=owi A1=a B1=em L1=ie V1=ie
     n1=i g1=ów d1=om a1=ów b1=ami l1=ach v1=i
     (aborygen|)
     'aborigine'

(10) N1= G1=a D1=owi A1=a B1=em L1=zie V1=zie
     n1=zi n1=owie g1=ów d1=om a1=ów b1=ami l1=ach v1=zi
     (kloszard|)
     'clochard'

In example (9) the inflectional pattern for the lexeme ABORYGEN meets the requirements of morphological division into a stem and an ending. The pattern presented in (10) differs from the previous one in three semi-endings (codes V1, n1 and v1), although morphologically there is no difference in inflectional endings. Those semi-endings are extended with the character [z], which appears here as a latter part of the digraph [dz]. The digraph [dz] is the result of a stem alternation [z:dz]. In both examples, however, the entry labels aborygen and kloszard overlap with the traditional lexical entries for these lexemes.

Next two examples clearly show the division into a semi-stem and a semi-ending. The entry labels waria and oszu are parts of the traditional lexical entries wariat and oszust. The inflectional endings (equal in the respective case pairs) occur together with the stem alternation variants in the semi-ending position ([t:ci] in example (11), [s:ś] and [t:ci] in example (12)), what results in two separate patterns in the lexicon:

(11) N1=t G1=ta D1=towi A1=ta B1=tem L1=cie V1=cie
     n1=ci g1=tów d1=tom a1=tów b1=tami l1=tach v1=ci
     (waria|t)
     'madman'

(12) N1=st G1=sta D1=stowi A1=sta B1=stem L1=ście V1=ście
     n1=ści g1=stów d1=stom a1=stów b1=stami l1=stach v1=ści
     (oszu|st)
     'swindler/cheat'

There are several patterns which include full word-forms because of alternations at their beginnings as in the feminine noun in example (13). The alternation variants [ć] and [ci] ground the difference between the word-forms, therefore they must be attached to the respective semi-endings in the pattern.

(13) Nz=ćma Gz=ćmy Dz=ćmie Az=ćmę Bz=ćmą Lz=ćmie Vz=ćmo
     nz=ćmy gz=ciem dz=ćmom bz=ćmami lz=ćmach vz=ćmy
     (|ćma)
     'moth'

The inflectional rule binds the pattern with an appropriate subset of lexemes, those whose inflectional characteristics and semi-endings fit in with that pattern, see examples (14)-(16). Many rules represent a small number of lexemes, every root alternation requires preparing a new pattern and therefore a new rule. For example, in the lexicon file containing masculine nouns of the fourth declension group there are currently 146 inflectional rules.

(14)　11: N1= G1=a D1=owi A1=a B1=em L1=zie V1=zie
　　　n1=zi n1=owie g1=ów d1=om a1=ów b1=ami l1=ach v1=zi
　　　kloszard:11

　　　'clochard'

(15)　67: N1=t G1=ta D1=towi A1=ta B1=tem L1=cie V1=cie
　　　n1=ci g1=tów d1=tom a1=tów b1=tami l1=tach v1=ci
　　　waria:67

　　　'madman'

(16)　22: Nz=ćma Gz=ćmy Dz=ćmie Az=ćmę Bz=ćmą Lz=ćmie Vz=ćmo
　　　　　　nz=ćmy gz=ciem dz=ćmom bz=ćmami lz=ćmach vz=ćmy
　　　:22

　　　'moth'

The problem of rules overproduction also concerns the description of verbs. The influence of alternations on the number of inflectional rules is visible in the majority of conjugation groups. A good example is the group 7a (see [19]), represented by about 100 verbs and 16 rules, which cover 1 to 18 lexemes, examples (17) and (18) show two of them.

(17)　0: 0=eć 1=ę 2=isz 3=i 4=imy 5=icie 6=ą r1=my r= r2=cie
　　　1m=ałem 1z=ałam 2m=ałeś 2z=ałaś 3m=ał 3z=ała 3n=ało
　　　4m=eliśmy 4z=ałyśmy 5m=eliście 5z=ałyście 6m=eli 6z=ały
　　　1pm=ałbym 1pz=ałabym 2pm=ałbyś 2pz=ałabyś 3pm=ałby
　　　3pz=ałaby 3pn=ałoby 4pm=elibyśmy 4pz=ałybyśmy
　　　5pm=elibyście 5pz=ałybyście 6pm=eliby 6pz=ałyby t=ąc
　　　b=ano
　　　myśl:0

　　　'to think'

(18)　11: 0d=zieć 1d=zę 2d=zisz 3d=zi 4d=zimy 5d=zicie 6d=zą
　　　r1d=żmy rd=ź r2d=źcie 1md=ziałem 1zd=ziałam 2md=ziałeś
　　　2zd=ziałaś 3md=ział 3zd=ziała 3nd=ziało 4md=zieliśmy
　　　4zd=ziałyśmy 5md=zieliście 5zd=ziałyście 6md=zieli
　　　6zd=ziały 1pmd=ziałbym 1pzd=ziałabym 2pmd=ziałbyś
　　　2pzd=ziałabyś 3pmd=ziałby 3pzd=ziałaby 3pnd=ziałoby
　　　4pmd=zielibyśmy 4pzd=ziałybyśmy 5pmd=zielibyście
　　　5pzd=ziałybyście 6pmd=zieliby 6pzd=ziałyby u=ziawszy
　　　bd=ziano
　　　przesied:11

　　　'to sit for some time'

In the lexicon we distinguished four inflectional groups of adjectives which also contain contemporary and past adjectival participles. In those groups lexemes are subordinated to 29 inflectional rules, although there are only a few sets of inflectional endings. Compare two examples of rules, the regular one presented in example (19), to which over 2200 lexemes are bound, and the rule with the pattern representing an alternation in only 7 lexemes showed in example (20).

(19) 0: Npm=y Npn=e Npz=a Gpmn=ego Gpz=ej Dpmn=emu Dpz=ej
      Apa=ego Apd=y Apn=e Apz=ą Bpmn=ym Bpz=ą Lpmn=ym Lpz=ej
      npo=i npb=e gp=ych dp=ym apo=ych apb=e bp=ymi lp=ych
      efektywn:0

      'effective'

(20) 13: Npm=zły Npn=złe Npz=zła Gpmn=złego Gpz=złej Dpmn=złemu
      Dpz=złej Apa=złego Apd=zły Apn=złe Apz=złą Bpmn=złym
      Bpz=złą Lpmn=złym Lpz=złej npo=źli npb=złe gp=złych
      dp=złym apo=złych apb=złe bp=złymi lp=złych
      zwię:13

      'brief'

The lexicon constructed in the way presented here was invented especially for the program of inflection analysis and synthesis AMOR. Although morphological division of word-forms is not accessible in its original form, we found procedures rebuilding a lexical entry into a two-level-based one. The process of the lexicon conversion is discussed next.

3.3   The Conversion of the AMOR Lexicon

Converting the AMOR lexicon into a lexicon of morphological roots was planned specifically for inflecting classes: nouns, adjectives and verbs. We prepared an algorithm of automatic division of a word-form into a stem and an ending, as well as choosing a unified root for a given lexeme. This resulted in a new lexicon entry, which consists of a root, a set of alternations and a set of endings. Both sets encode enough inflectional information to form appropriate word-forms.

The conversion process runs in three parallel parts, each one was separately defined for one grammatical class. Here, we present the details of steps for nouns transformation. We also give examples for adjective and verb transformation, putting emphasis on differences in processing rules.

We present the steps of the lexicon rebuilding process on the example of one subgroup of nouns. The lexeme WARIAT 'madman' serves as their illustration in examples (22)-(26). The description of the entry structure for adjectives and verbs is given further in the paper.

In order to convert the lexicon automatically a few steps had to be considered. The following subsequent stages of analysis will be described further in this subsection.

Collection of Sets Containing Inflectional Suffixes. At the beginning we manually gathered inflectional endings in sets. As the basis we followed the description of Polish inflection proposed by [19]. Separate sets are needed for each noun declension group in respect of the gender. We built three sets for nouns, and each of them represents inflectional suffixes of nouns belonging to one general grammatical gender: masculine, feminine and neuter. Conjugation and adjectival groups require only one set of endings each. Thus, two other sets were prepared for adjectives and verbs respectively.

The set of inflectional suffixes corresponding to masculine nouns consists of all possible alternate endings separated by a point (.). One of possible morphological endings is a zero-morph (0). The position (separated by commas) refers to a chain of cases, which means that the first chain of alternate endings represents nominative suffixes singular, the second—genitive suffixes singular, etc. The set of suffixes for masculine nouns is presented in example (21). It is worth mentioning that for nouns the value of gender is kept while an entry conversion.

(21) (o.0,a.u,owi.u,a.u.0,em,e.u,e.u.0,
       i.y.owie.e.a,ów.i.y.0,om,ów.y.e.i.a.0,
       ami.mi,ach,i.y.owie.e.a)

Synthesis of All Word-Forms for One Lexeme of the Chosen Class in the Lexicon. The next step consists in synthesis of all word-forms of the lexicon entries for the chosen grammatical classes. Using the inflectional rules of the AMOR, we automatically bound the semi-stem with the semi-ending to produce a list of word-forms for every lexeme (the example concerns masculine nouns of one inflectional group). The grammatical description of word-forms is preserved.

(22)
N = wariat			n = wariaci	
G = wariata			g = wariatów	
D = wariatowi			d = wariatom	
A = wariata			a = wariatów	
B = wariatem			b = wariatami	
L = wariacie			l = wariatach	
V = wariacie			v = wariaci	

Segmentation of One Lexeme Word-Forms into the Stem and the Ending. After comparing the ends of word-forms one after another with the inflectional suffixes in the proper position in the previously presented set (21), the endings are separated and they build a unique set of appropriate suffixes for a given lexeme. The set of the inflectional endings is marked by k, the initial of the word *końcówki* 'endings'.

(23)
N = wariat|			n = wariac|i	
G = wariat|a			g = wariat|ów	
D = wariat|owi			d = wariat|om	

```
A = wariat|a              a = wariat|ów
B = wariat|em             b = wariat|ami
L = wariaci|e             l = wariat|ach
V = wariaci|e             v = wariac|i
```

(24) k(0,a,owi,a,em,e,e,i,ów,om,ów,ami,ach,i)

Collection of All Stems of a Given Lemma. The part which remains after a suffix is separated is a morphological root. All different roots are listed together with the information about the position in which they are realized in texts. The earlier preserved codes of inflectional description are now used to delimit a position of each root in an inflection paradigm and label appropriate inflectional characteristics of a word-form that can be generated from every root.

(25) wariat(N,G,D,A,B,g,d,a,b,l)
      wariaci(L,V)
      wariac(n,v)

Internal Alternations Description and the Final Root Choice. The example (23) illustrates the difference between grapheme and morphonological representation of word-forms [6], [1]. After the inflectional suffixes are separated from the root, characters at the root-suffix border cannot always refer to appropriate phoneme. The most significant problem is with the character [i] and its double denotation in Polish. It is used in writing to denote palatal consonants and often at the same time it refers to a close front unrounded vowel *i*, as in *wariaci*. In our approach we abandoned such mixed text-phonetic interpretations. Therefore, some roots not expected in view of phonology appear in specific contexts during [i] separation, e.g. *wariac-*. Although this is a series phenomenon and it results in root alternation which is explained below, it is unavoidable at the grapheme level of word-form morphological analysis.

The next step of an entry conversion is inside-root alternations description and a label (final, entry) root choice. To process inner alternations automatically we defined pairs of possible character changes first. We also had Polish diacritics in mind. We distinguished pairs of characters as: [sz], [rz], [ch], and changes like [r:rz], [g:dz], but also [e:0], [p:pi] and [d:dzi]. To find an alternation change in nominal groups we compared roots of one lexeme in pairs. Each time a pair of roots consisted of the root in nominative case and the other roots of the given lexeme one after another. Then, considering the defined character changes, the single character which appeared in the nominative root was chosen to be the head grapheme of the alternation. The head grapheme is marked by capitalization. Its variants found in the following roots are the representatives and are gathered in sets together with the grammatical code connecting alternations with a root and a suffix. In the example below (26) the alternation of the head grapheme [T] is shown. [T] is represented by three alternative graphemes [t], [ci], [c]. Each of them proceeds a set of encoded cases exactly the same as the sets presented in example (25) which accompany earlier extracted roots.

(26)   wariaT {T:t(N,G,D,A,B,g,d,a,b,l),ci(L,V),c(n,v)}

Below is the example of a root, with more inner alternations, which comes from the same inflectional group:

(27)   kaREŁ {R:rz(N),r(G,D,A,B,L,V,n,g,d,a,b,l,v);
               E:e(N),0(G,D,A,B,L,V,n,g,d,a,b,l,v);
               Ł:ł(N,G,D,A,B,n,g,d,a,b,l,v),l(L,V)}
       'dwarf'

The conversion results in a lexicon entry which consists of two parts. The first part is a lemma in a traditional form, that is e.g. a noun in nominative singular, a verb in infinitive or an adjective in nominative singular masculine form. In the case of nouns the value of gender follows the lemma. The second part comprises a morphological root and two sets of ordered alternations and inflectional suffixes. Example (28) brings together the description of the lexeme WARIAT 'madman'.

(28)   wariat r:m1
       wariaT {T:t(N,G,D,A,B,g,d,a,b,l),ci(L,V),c(n,v)}
               k(,a,owi,a,em,e,e,i,ów,om,ów,ami,ach,i)

In the subsection 3.2 we presented the example of the AMOR inflection schema for the lexeme ĆMA 'moth' (13). All the word-forms filled the schema because the alternations start already at the beginning of each word-form. In the conversion process those alternations were solved and now a lexicon entry presents those changes in the root:

(29)   ćma r:z
       ĆOM {Ć:ć(N,G,D,A,B,L,V,n,d,a,b,l,v),ci(g);
             0:0(N,G,D,A,B,L,V,n,d,a,b,l,v),e(g);
             M:m(N,G,A,B,V,n,g,d,a,b,l,v),mi(D,L)}
             k(a,y,ę,ę,ą,e,o,y,0,om,y,ami,ach,y)

The description of adjectives in the root lexicon is similar to the noun set-up except for the gender, which is a selective category for nouns but an inflectional category for adjectives in Polish. Therefore it is not separately given inside an entry. In order to distinguish word-forms of different genders all inflectional suffixes appear in the fixed order inside each set. Generally they are placed in a sequence of cases and for a given case in the following gender order: *masculine, feminine, neuter* in singular and *masculine, non-masculine* in plural. The endings are separated by a point (.). In four plural cases the gender cannot be determined (*genitive, dative, instrumental* and *locative*). Below, we present two entries for adjectives. The first one for the lexeme EFEKTYWNY 'effective' is given in example (30). The second one, ZWIĘZŁY 'brief', in example (31) shows an entry with two inner root alternations. Both examples come from the same subgroup of adjectives in the AMOR lexicon.

(30) efektywny
      efektywn k(y.e.a,ego.ej,emu.ej,ego.y.e.ą,ym.ą,ym.ej,y.e.a,
               i.e,ych,ym,ych.e,ymi,ych,i.e)

(31) zwięzły
      zwięZŁ {Z:z(Npm,Npn,Npz,Gpmn,Gpz,Dpz,Apa,Apd,Apn,Apz,Bpmn,
              Lpz,npb,gp,dp,apo,apb,bp,lp),ź(npo);Ł:ł(Npm,Npn,Npz,
              Gpmn,Gpz,Dpz,Apa,Apd,Apn,Apz,Bpmn,Lpz,npb,gp,dp,apo,
              apb,bp,lp),l(npo)}
              k(y.e.a,ego.ej,emu.ej,ego.y.e.ą,ym.ą,ym.ej,y.e.a,
               i.e,ych,ym,ych.e,ymi,ych,i.e)

While describing verbs in means of morphological roots we benefit from Tokarski's account [19]. According to Tokarski verbal word-forms are built of three elements: a root, a thematic infix (or alternating infixes) and personal (tense) endings. Tokarski prepared a classification of Polish verbs by grouping together verbs of the same thematic infix or the same alternation pattern for infix change through an inflectional paradigm. There are eleven verbal groups numbered 1-11, the higher the group number the more complex are the infix changes. The eleventh group comprises irregular verbs.

The AMOR lexicon of verbs consists of text files corresponding to Tokarski's classification. Verbs representing one group or one subgroup are described in separate files which also contain their inflectional schemas. Currently, there are 19 files used by AMOR, including two irregular groups, see examples in subsection (3.2). The automatic conversion concerned only those groups that included any thematic infix patterns. Conversion of irregular verbs is possible, of course, but was done manually.

We decided to keep a separate set of thematic infixes for each verb group. That is why, in contrary to nouns or adjectives, verbs required one more step of conversion. After the inflectional suffixes were separated from the root candidates, thematic infixes were set apart. Earlier, we defined sets of infixes for every inflectional group of verbs. The result of verb conversion, a lexicon entry for the verb BUJAĆ 'to rock' is presented in example (32).

(32)   bujać
       buj   s(a:0,1,2,3,4,5,1m,1z,2m,2z,3m,3z,3n,
             4m,4z,5m,5z,6m,6z,1pm,1pz,2pm,2pz,3pm,
             3pz,3pn,4pm,4pz,5pm,5pz,6pm,6pz,b;
             aj:6,r,r1,r2,t)
             k(ć,m,sz,0,my,cie,ą,0,my,cie,łem,łam,
             łeś,łaś,ł,ła,ło,liśmy,łyśmy,liście,łyście,
             li,ły,łbym,łabym, łbyś,łabyś,łby,łaby,łoby,
             libyśmy,łybyśmy,libyście,łybyście,liby,łyby,
             ąc,no)

The second part of the entry consists of the set of thematic infixes connected to the appropriate grammatical codes (marked by s) and the set of inflectional suffixes, ordered in the same way as the infixes (marked by k).

In the case of inner root alternations within verb entries, an additional set of alternate thematic infixes is made. And again the head alternate gets all text realizations with their inflectional descriptions. The example (33) presents such a verbal entry.

```
(33)  szeleścić
      szeleŚC {Ś:sz(1,6,t,b),ś(0,2,3,4,5,r,r1,r2,
              1m,1z,2m,2z,3m,3z,3n,4m,4z,5m,5z,6m,
              6z,1pm,1pz,2pm,2pz,3pm,3pz,3pn,4pz,
              4pm,5pm,5pz,6pm,6pz);
              C:c(0,2,3,4,5,1m,1z,2m,2z,3m,3z,3n,
              4m,4z,5m,5z,6m,6z,1pm,1pz,2pm,2pz,3pm,
              3pz,3pn,4pz,4pm,5pm,5pz,6pm,6pz),
              cz(1,6,t,b),ć(r,r1,r2)}
      s(:1,6,r,r1,r2,t,b;i:0,2,3,4,5,1m,1z,
        2m,2z,3m,3z,3n,4m,4z,5m,5z,6m,6z,1pm,
        1pz,2pm,2pz,3pm,3pz,3pn,4pz,4pm,5pm,
        5pz,6pm,6pz)
      k(ć,ę,sz,0,my,cie,ą,0,my,cie,łem,łam,
        łeś,łaś,ł,ła,ło,liśmy,łyśmy,liście,
        łyście,li,ły,łbym,łabym,łbyś,łabyś,
        łby,łaby,łoby,łybyśmy,libyśmy,
        libyście,łybyście,liby,łyby,ąc,ono)
```
'to rustle'

The presented procedure of automatic lexicon conversion was applied to almost every lexicon entry located within inflectional subgroups of nouns, adjectives and verbs. However, a small subset of lexemes of each grammatical class remained due to their complicated inner structure. This was caused by several inner alternations that change one root in a significant way. Such lexemes would require too many rules to achieve an appropriate description in the lexicon. The subsets containing lexemes problematic for automatic conversion were elaborated manually. The lexical entries TYDZIEŃ 'week' and CIĄĆ 'to cut' represent such noun and verb conversion results.

```
(34)  tydzień r:m3
      ty00DON  {0:0(N,A),g(G,D,L,V,B,n,g,d,a,b,l,v);
                0:0(N,A),o(G,D,L,V,B,n,g,d,a,b,l,v);
                D:dzi(N,A),d(G,D,L,V,B,n,g,d,a,b,l,v);
                0:e(N,A),0(G,D,L,V,B,n,g,d,a,b,l,v);
                N:ń(N,A),ni(G,D,L,V,B,n,g,d,a,b,l,v)}
                k(,a,owi,a,em,u,u,e,i,om,i,ami,ach,e)
```

(35) ciąć C
      {C:t(1,2,3,4,5,6,r,r1,r2,t),ci(0,
              1m,1z,2m,2z,3m,3z,3n,4m,4z,5m,5z,6m,
              6z,1pm,1pz,2pm,2pz,3pm,3pz,3pn,4pz,
              4pm,5pm,5pz,6pm,6pz,b);
          s(n:1,6,t;nie:2,3,4,5;nij:r,r1,r2;
              ą:0,1m,2m,3m,1pm,2pm,3pm;ę:1z,2z,3z,
              3n,4m,4z,5m,5z,6m,6z,1pz,2pz,3pz,3pn,
              4pz,4pm,5pm,5pz,6pm,6pz,b)
      k(ć,ę,sz,0,my,cie,ą,0,my,cie,łem,łam,
          łeś,łaś,ł,ła,ło,liśmy,łyśmy,liście,
          łyście,li,ły,łbym,łabym,łbyś,łabyś,
          łby,łaby,łoby,łybyśmy,libyśmy,
          libyście,łybyście,liby,łyby,ąc,to)

4   Compression of the Root-Based Lexicon

Although the automatic conversion of the AMOR data for three large inflecting classes delimited grapheme boundaries of morphological roots and inflection suffixes, and the lexicon of roots gives access to inner root alternations, it is not easy to compare inflectional or alternation patterns. This is because information about alternation changes and inflectional suffixes is repeated in every entry. This influences the size of the lexicon. To make our lexicon more compact and to get access to pattern distribution we grouped together lexemes fitting the same patterns of both inflectional and alternation changes.

The reorganization of lexicon data was done automatically by a script designed especially for the purpose. The objective was first to compare the sets of inflectional suffixes within every inflectional group, define a number of inflectional patterns and then gather together all lexemes which inflect according to those patterns respectively. After the inflection patterns were obtained, the sets of alternation changes were investigated. Again, they were all compared within one alternation pattern. This resulted in a list of alternation patterns linked to sets of lexemes which fit in. We did not consider entries elaborated manually at this moment. They should be added to the final version of the lexicon.

After compression the representation of a lexeme in the root lexicon changes. Lexemes and patterns (alternation and inflection) are kept separately. A lexeme is bound to the alternation pattern (if there is any alternation at all) and the pattern of inflectional suffixes by their numbers. Examples below present the entries of the previously described lexemes from the current set up of the lexicon of roots (36)–(39).

(36) wariat wariaT r:1 a:60 k:1
      karzeł kaRZEŁ r:2 a:42 k:9

```
a:42 {R:rz(N),r(G,D,A,B,L,V,n,g,d,a,b,l,v);
      E:e(N),0(G,D,A,B,L,V,n,g,d,a,b,l,v);
      Ł:ł(N,G,D,A,B,n,g,d,a,b,l,v),l(L,V)}
a:60 {T:t(N,G,D,A,B,g,d,a,b,l),ci(L,V),c(n,v)}

k:1 k(0,a,owi,a,em,e,e,i,ów,om,ów,ami,ach,i)
k:9 k(0,a,owi,a,em,e,e,y,ów,om,y,ami,ach,y)
```

(37) efektywny efektywn a:1 k:1
      zwięzły zwięZŁ a:19 k:1

```
a:1 -
a:19 {Z:z(Npm,Npn,Npz,Gpmn,Gpz,Dpz,Apa,Apd,Apn,Apz,
      Bpmn,Lpz,npb,gp,dp,apo,apb,bp,lp),ź(npo);
      Ł:ł(Npm,Npn,Npz,Gpmn,Gpz,Dpz,Apa,Apd,Apn,Apz,
      Bpmn,Lpz,npb,gp,dp,apo,apb,bp,lp),l(npo)}

k:1 k(y.e.a,ego.ej,emu.ej,ego.y.e.ą,ym.ą,ym.ej,y.e.a,
      i.e,ych,ym,ych.e,ymi,ych,i.e)
```

(38) bujać buj s:1 k:1

```
s:1 s(a:0,1,2,3,4,5,1m,1z,2m,2z,3m,3z,3n,4m,4z,5m,5z,
      6m,6z,1pm,1pz,2pm,2pz,3pm,3pz,3pn,4pm,4pz,5pm,5pz,
      6pm,6pz,b;aj:6,r,r1,r2,t)

k:1 k(ć,m,sz,0,my,cie,ą,0,my,cie,łem,łam,łeś,łaś,ł,ła,ło,
      liśmy,łyśmy,liście,łyście,li,ły,łbym,łabym,łbyś,
      łabyś,łby,łaby,łoby,libyśmy,łybyśmy,libyście,
      łybyście,liby,łyby,ąc,no)
```

(39) szeleścić szeleŚC a:1 k:1

```
a:1 {Ś:sz(1,6,t,b),ś(0,2,3,4,5,r,r1,r2,1m,1z,2m,2z,
      3m,3z,3n,4m,4z,5m,5z,6m,6z,1pm,1pz,2pm,2pz,3pm,
      3pz,3pn,4pz,4pm,5pm,5pz,6pm,6pz);
      C:c(0,2,3,4,5,1m,1z,2m,2z,3m,3z,3n,4m,4z,5m,5z,
      6m,6z,1pm,1pz,2pm,2pz,3pm,3pz,3pn,4pz,4pm,5pm,
      5pz,6pm,6pz),cz(1,6,t,b),ć(r,r1,r2)}
s:1 s(:1,6,r,r1,r2,t,b;i:0,2,3,4,5,1m,1z,2m,2z,
      3m,3z,3n,4m,4z,5m,5z,6m,6z,1pm,1pz,2pm,2pz,3pm,
      3pz,3pn,4pz,4pm,5pm,5pz,6pm,6pz)
k:1 k(ć,ę,sz,0,my,cie,ą,0,my,cie,łem,łam,łeś,łaś,ł,ła,
      ło,liśmy,łyśmy,liście,łyście,li,ły,łbym,łabym,
      łbyś,łabyś,łby,łaby,łoby,łybyśmy,libyśmy,libyście,
      łybyście,liby,łyby,ąc,ono)
```

At present, various statistics on the root lexicon are at our disposal. Once the patterns are numbered and ordered it is visible how many lexemes use them in inflection. In Table 1 we put together some statistics for the analyzed classes. We present a summed up number of alternation and inflection numbers for separate inflection groups of nouns, adjectives and verbs. It is important to make two comments at this point. It is possible that in two or more different groups there are exactly the same sets of inflectional suffixes or alternations. The numbers in the table do not refer to unique sets of either alternation or inflection patterns. Concerning the manner of counting patterns we must add that the "empty" alternation patterns were also taken into consideration. By "empty" we understand here the case of no alternation at all in an inflectional paradigm. However, it is marked, for technical reasons, as one of the patterns with an appropriate number (consider the example of the adjective EFEKTYWNY 'effective').

Table 1. Results of the lexicon conversion

Class	N^o of Lexemes	N^o of Alternation Patterns	N^o of Inflection Patterns
Nouns	51528	446	189
Adjectives	36297	36	5
Verbs	12617	40	34

We can also tell, for example, how many lexemes undergo a specified alternation change or how many lexemes use a given inflectional pattern. A comparison of those data gives us some insight into the whole inflection system formation.

Table 2 presents the frequency of different alternation patterns in one subgroup of feminine nouns (z1). The pattern a:6 describes lexemes without root alternation.

Some alternation pairs repeat in the presented subgroup, they constitute, however, different patterns. Examples (40) and (41) present two alternation patterns of the nominal alternation [ni:n]. The distribution of the alternation variants clearly differs. The variant [n] occur in four cases in the former and only in the *genitive plural* in the latter example.

(40)   a:1 {N:ni(N,A,B,V,n,d,a,b,l,v),n(G,D,L,g)}

(41)   a:7 {N:ni(N,G,D,A,B,L,V,n,d,a,b,l,v),n(g)}

In the same subgroup of feminine nouns (z1) we analyzed the distribution of lexemes in different inflectional patterns. Table 3 shows the frequency of the lexicon entries for each inflectional pattern.

5   Results and Application

The experiment of converting the AMOR lexicon into the lexicon of roots was done primarily while working on the formal description of Polish derivation [15].

Table 2. Number of lexemes undergoing alternation patterns in z1

Pattern N^o	Alternation	Example of Entry	N^o of Entries
a:1	ni:n	dźwignia 'lever'	514
a:2	zi:z	bazia 'catkin'	1
a:3	bi:b	skrobia 'starch'	4
a:4	mi:m	ziemia 'soil'	2
a:5	ni:n	Bośnia 'Bosnia'	2
a:6	-	fala 'wave'	2630
a:7	ni:n	brzoskwinia 'peach'	2
a:8	ci:c	babcia 'grandmother'	11
a:9	ni:n	niania 'nanny'	4
a:10	ni:n:ń	pieczarkarnia 'mushroom house'	95
a:11	d:dź, 0:e, ni:n:ń	studnia 'well'	1
a:12	si:s:ś	gosposia 'domestic help'	13
a:13	ś:si, 0:e, ni:n	wiśnia 'cherry (tree)'	1
a:14	zi:z:ź	nózia 'little leg'	8
a:15	0:e	kropla 'drop'	3
a:16	o:ó	topola 'poplar'	2
a:17	ni:n:ń	odtwórczyni 'female performer'	96
			3389

In the mentioned dissertation we presented empirical material: the conversion covered over 35 thousand nominal lexemes, 27 thousand adjectival lexemes and 7 thousand verbal lexemes. The design of the lexicon entries significantly helped in composing formal rules of derivation and rules of alternation which accompany them.

Table 3. Frequency of lexemes inside the inflection patterns

Inflectional Pattern	Example of Entry	N^o of Entries
k:1	dźwignia 'lever'	2724
k:2	empatia 'empathy'	241
k:3	fala 'wave'	188
k:4	babcia 'grandmother'	30
k:5	pieczarkarnia 'mushroom house'	106
k:6	studnia 'well'	2
k:7	odtwórczyni 'female performer'	95
k:8	pani 'madam'	1
k:9	Genua 'Genoa'	1
k:10	genua 'genoa jib'	1
		3389

The data presented in this paper underwent the process of conversion once again especially for the purpose of this paper. The AMOR lexicon has been enriched in the last few years and in effect there were over 51 thousand nominal lexemes, 36 adjectival lexemes and 12 thousand verbal lexemes at our disposal. By carrying out the conversion twice we confirmed the effectiveness of the described method.

The consequence of the entry structure design in the root lexicon is that it is possible to synthesize not only chosen word-forms (e.g. the form of the locative singular of the noun PALMA 'palm', that is *palmie*) but also specified inflectional roots (e.g. the root of the locative singular of the noun PALMA 'palm', that is *palmi-*). The second method of synthesis is predicted in derivation rules (and alternation rules) presented in the model of nominal derivation [12]. This approach to derivation assumes operations on morphological roots and the rules solve morphological alternation at the boundary of the stem and the derivational affix if necessary. Example (42) presents one rule which generates nouns with the suffix -NI*(a)*. The input for the rule is the specified root, namely *palmi*, then the text realization *arnia* of the suffix -NI*(a)* is attached (other rules control the choice of a grapheme form of the suffix). The output is a nominative form of the derivative PALMIARNIA 'palm house'.

$$(42) \quad \text{-NI}(a) : \text{F}_{N.loc.sing.fem} + \text{-NI}(a) \rightarrow \text{F}_{N.nom.sing.fem} \text{ (PALMIARNIA)}$$

The lexicon of roots shows the description of lexical units in a new way, taking into account their inflection as well as inner-root alternations. This approach can slightly deviate from the traditional linguistic description of Polish inflectional stems because it is not based on phonetic properties of morphemes. It is assumed that the lexicon will serve grammatical identification of a word in a written text. It is a.practical solution to base such analysis on its grapheme representation. However, the inflectional description is independent of the method of data storage. Therefore, the lexicon can be used for different tools of text analysis.

Although the lexicon was built beside the formal description of derivation, its entries served as empirical illustration of derivation and alternation rules, and the main idea of its application in the future was described as a morphological analyzer implementation. An analyzer which would operate on the lexicon would have access to much more information. It would be able to synthesize e.g.:

- a set of inflectional roots (stems) for a given lexeme,
- a set of inflectional suffixes for a given lexeme,
- sets of inflectional suffixes for one inflectional group or subgroup,
- a set of root alternations for a given group of lexemes, or
- sets of root alternations for one inflectional group.

According to our knowledge, the existing inflectional analyzers for Polish have not provided such a detailed morphological review so far. The root lexicon allows one to extend the aims of inflectional analysis (synthesis).

The second natural application for the root lexicon would be to implement derivational rules [15]. The lexicon can obviously help to operate on word-forms or pure inflectional roots, that is why the word-formation analyzer could use the formal description to segment word units automatically. It should be stressed that the program of word-formation analysis (and synthesis) should perform inflection at the same time.

Finally, the lexicon of roots could be the basis for formulating automatically hypotheses concerning inflection or derivation of new words found in a text. A tool for new word inflection investigation would take advantage of the property of the lexicon thanks to which sets of inflectional suffixes are separated from roots and the inflectional characteristics are bound to them. After the suffix candidates are separated, the remaining part, the root, could be transformed according to alternation classification. As a result, a hypothetical representation of a lexeme (a list of possible lexemes) is achieved. Applying derivational rules first, a hypothesis for a basis lexeme could be generated.

6 Summary

Building the lexicon of roots was influenced by the two-level morphology approach. The idea of lexical data description without a complicated formula seems to be appropriate for Polish inflection. Any algorithm developed to operate on the lexicon will not interfere in the entries. Access to the inner structure of word-forms is already granted by the sophisticated design of the entries containing them.

We assume that the conversion of an electronic lexicon for Polish, not specifically that of AMOR, into a root lexicon could be repeated. The procedure will work if there are given sets of word-forms grouped together with sets of inflectional suffixes corresponding to them. In the paper we presented the lexicon which was generated twice.

Possible applications were presented. One that is obvious is the implementation of a morphological analyzer that, aside from lemma recognition or word-form inflectional characteristic identification, could provide information on e.g. inner root alternations or lexemes of the same inflectional pattern. Until now, the lexicon of roots served the formal description of Polish derivation. The review of root-based entries helped to create the rules which may also be automated. Combining the lexicon and derivational description would allow one to formulate hypotheses of morphological structure of words in texts.

Acknowledgments. We would like to express our appreciation to Dariusz Wiśniewski. Work on the root lexicon would not be possible without his help with programming the procedures presented in this paper.

References

1. Bień, J.S.: Koncepcja słownikowej informacji morfologicznej i jej komputerowej weryfikacji. Dissertationes Universitatis Varsoviensis, vol. 383. Wydawnictwa Uniwersytetu, Warszawskiego (1991)
2. Booij, G.: Against Split Morphology. Yearbook of Morphology 1993, 27–50 (1994)

82   J. Rabiega-Wiśniewska

3. Booij, G.: Inherent versus Contextual Inflection and the Split Morphology Hypothesis. Yearbook of Morphology 1995, 1–16 (1996)
4. Broda, B., Piasecki, M., Radziszewski, A.: Towards a Set of General Purpose Morphosyntactic Tools for Polish. In: Intelligent Information Systems XVI. Challenging Problems of Science. Computer Science, pp. 441–450 (2008)
5. Cetnarowska, B.: On Inherent Inflection Feeding Derivation in Polish. Yearbook of Morphology 1999, 153–183 (2001)
6. Gruszczyński, W.: Fleksja rzeczowników pospolitych we współczesnej polszczyźnie pisanej. Prace Językoznawcze 122. Zakład Narodowy im. Ossolińskich (1989)
7. Hajnicz, E., Kupść, A.: Przegląd analizatorów morfologicznych dla języka polskiego. Technical Report 937, Instytut Podstaw Informatyki Polskiej Akademii Nauk, Warszawa (2001)
8. Kaplan, R.M., Kay, M.: Phonological rules and finite-state transducers. In: Linguistic Society of America Meeting Handbook, Fifty-Sixth Annual Meeting, New York, December 27–30 (1981)
9. Kay, M.: When Meta-Rules are not Meta-Rules. In: Sparck Jones, K., Wilks, Y. (eds.) Automatic Natural Language Parsing, pp. 94–116. Ellis Horwood, Chichester (1983)
10. Koskenniemi, K.: Two-Level Morphology: A General Computational Model for Word-Form Recognition and Production. PhD thesis, Helsinki University (1983)
11. Rabiega-Wiśniewska, J.: Podstawy lingwistyczne automatycznego analizatora morfologicznego AMOR. Poradnik Językowy 10 (619), 59–78 (2004)
12. Rabiega-Wiśniewska, J.: A formal model of Polish nominal derivation. In: Human Language Technologies as a Challenge for Computer Science and Linguistics, Proceedings of 2nd Language & Technology Conference, Poznań, April 21-23, 2005, pp. 323–327. Wydawnictwo Poznańskie Sp. z o.o, Poznań (2005)
13. Rabiega-Wiśniewska, J.: Wpływ fleksji na derywację – dyskusja podziału morfologii. LingVaria 2(6), 41–57 (2008)
14. Rabiega-Wiśniewska, J., Rudolf, M.: Towards a Bi-Modular Automatic Analyzer of Large Polish Corpora. In: Kosta, R., Błaszczak, J., Frasek, J., Geist, L., Żygis, M. (eds.) Investigations into Formal Slavic Linguistics. Contributions of the Fourth European Conference on Formal Description of Slavic Languages – FDSL IV, held at Potsdam University, November 28-30, 2001, pp. 363–372 (2003)
15. Rabiega-Wiśniewska, J.: Formalny opis derywacji w języku polskim. Rzeczowniki i przymiotniki. PhD thesis, Uniwersytet Warszawski (2006)
16. Saloni, Z., Gruszczyński, W., Woliński, M., Wołosz, R.: Słownik gramatyczny języka polskiego. Wiedza Powszechna (2007)
17. Saloni, Z., Świdziński, M.: Składnia współczesnego języka polskigo. Wydawnictwo Naukowe PWN, Warszawa (1998)
18. Szafran, K.: Automatyczna analiza fleksyjna tekstu polskiego (na podstawie Schematycznego indeksu a tergo Jana Tokarskiego). PhD thesis, Uniwersytet Warszawski, Warszawa (1994)
19. Tokarski, J.: Fleksja polska. PWN, Warszawa (1973)
20. Tokarski, J.: Schematyczny indeks a tergo polskich form wyrazowych. PWN, Warszawa (1993)
21. Woliński, M.: Morfeusz — a Practical Tool for the Morphological Analysis of Polish. In: Kłopotek, M., Wierzchoń, S., Trojanowski, K. (eds.) Intelligent Information Processing and Web Mining, IIS:IIPWM'06, pp. 503–512. Springer, Heidelberg (2006)
22. Wołosz, R.: Efektywna metoda analizy i syntezy morfologicznej w języku polskim. Problemy Współczesnej Nauki. Teoria i Zastosowania. Inżynieria Lingwistyczna. Akademicka Oficyna Wydawnicza EXIT, Warszawa (2005)

Representation of Uzbek Morphology in Prolog

Gayrat Matlatipov[1,2] and Zygmunt Vetulani[2]

[1] Department of Information Technology, Faculty of Mathematics
Urgench State University
H. Olimjon Street, 14, Urgench, Uzbekistan
gayrat22@yahoo.com
[2] Department of Computer Linguistics and Artificial Intelligence
Faculty of Mathematics and Computer Science
Adam Mickiewicz University
ul. Umultowska 87, 61-614 Poznań, Poland
vetulani@amu.edu.pl

Abstract. In the paper we address issues related to the morphology of the Uzbek language. In Uzbek, as in many other agglutinative languages, some single text-words correspond to sentences in non-agglutinative languages. Morphological processing is therefore a crucial operation in the automatic processing of Uzbek. We approach the theory of Uzbek morphology in terms of morphotactic and morphophonemic rules. We present the UZMORPP system of automatic morphological parsing for the Uzbek language. The Prolog implementation of this system is provided.

Key words: natural language processing (NLP), agglutination, morphology, Uzbek language, lexicon, Prolog tools design, morphophonemics, morphotactic

Introduction

Nowadays, the problem of the automated processing of a natural language has a special urgency. The research presented in this paper is a contribution to the development of natural language processing tools for the Uzbek language. Morphological parsing is the process of breaking a word into its smallest meaningful components called morphemes [7,2]. In this paper we describe a morphological parsing tool implemented in Prolog which performs segmentation of Uzbek words into roots and suffixes. We consider written Uzbek language which conforms to the spelling principles laid out in the document "Principal Orthographic Rules for the Uzbek Language. The Uzbekistan Cabinet of Minister's Resolution No. 339" [1].

The Uzbek language is agglutinative [6], i.e. it is a language whose words are generated by adding affixes to the root forms. In such languages, given a word in its root form we can derive a new word by adding an affix to this root form, and then derive another one by adding another affix, and so on. Thus in many cases, a single Uzbek word may correspond to a many-word sentence or

M. Marciniak and A. Mykowiecka (Eds.): Bolc Festschrift, LNCS 5070, pp. 83–110, 2009.
© Springer-Verlag Berlin Heidelberg 2009

phrase in a non-agglutinative language. In agglutinative languages the syntactic analysis of a text (considered as composed of sentences) is relatively simple once the morphological analysis is completed, but the number of words we have to process is huge. The morphological processing is therefore a crucial operation in the automatic processing of the Uzbek language. In particular its role is much more important for designing systems with natural language competence then in case of non-agglutinative languages (as e.g. Polish and other Slavonic languages) [8].

EXAMPLE: *kelolmaganlardanmisiz?*   Are you one of those who couldn't come?

The large number of suffixes and the combination of these suffixes in different orders make morphological analysis non trivial. This above example is a complex one, but even relatively simple words may cause problems. For example, the word "olma" (don't take) may be considered as being composed of two morphemes, "ol" and "–ma". Given the input string:

olma

the parsing program would output annotated morphemes:

[ol:{root, verb, "to take"}; -ma:{negation}]

Morphological parsing for Uzbek is not always that simple, as in many cases it is possible to segment a given word in various ways. If there is no additional disambiguating information, the morphological algorithm cannot determine which of these is the correct one. The algorithm must therefore have the ability to explore alternative solutions, i.e. to backtrack to the next alternative. In the case considered here it will return the following solution:

[olma:{root, common_noun, "apple"}]

where "olma" is recognised just as a root. After having found the last solution the algoritm should terminate.

The above example illustrates the non-deterministic parsing scheme as applied in the **Uzbek MORPhological Parser / UZMORPP**[3], presented in this paper. Of course it may also be called in a deterministic manner in order to have all possible morphological parsing results returned together.

In the next sections we will describe the program design and the user configurable options.

1   Input and Output

The tool accepts as input a list of words[4] from an external input text file (e.g. uzb.tst):

---

[3] The UZMORPP morphological tool for the Uzbek languages was designed and implemented by Gayrat Matlatipov as a result of the discussions and deliberations of both co-authors.

[4] In the Uzbek language, as in many other agglutinative languages, some single text words correspond to sentences in non-agglutinative languages.

```
olma.    ('apple', 'do not take')
kelmadilar.    ('They did not come')
bormadingmi?    ('Didn't you go?'
... ... ...
```

It will return a list of all possible morphological segmentations of a word with corresponding descriptions, e.g.:

```
Word: olma-[    (1)olma:{root, common_noun, "apple"};
(2)ol:{root, verb, "to take"}; -ma: {negation}]
```

In this example "olma" has two possible segmentations: as a noun (1) and as a verb (2).

```
Word: kelmadilar - [kel:{root, verb, "to come"}; -ma:
       {negation}; -dilar: { past simple tense, 3-person,
       plural}]
```

```
Word: bormadingmi -[bor:{root, verb, "to go"}; -ma:{negation};
       -ding:{ past simple tense, 2-person, singular}; -mi:
       {question suffix}]
```

2   Lexicon

The main data structure to encode the linguistic knowledge necessary for the morphological analysis is the lexicon. We will use DCG (Definite Clause Grammar) notation to encode it. One of the most common advantages for adding Prolog goals to a DCG is the simplification of the encoding of the lexicon. Rather than encoding the lexicon using separate DCG rules for each lexical item, it is much simpler and less redundant to have a single DCG rule for each category (part of speech). The following example is for *nouns* (n).

$$n \; \text{-->} \; [\text{Word}], \; \{n(\text{Word})\}.$$

The above rule says that the nonterminal n can cover any terminal symbol that is a noun. (The Prolog fact *n(Word)* is to be read *'Word' is a noun*. Together with this single rule, we need a dictionary in the form of Prolog facts as follows:

```
...
n(kursi).    %    chair
n(ayol).     %    woman
n(olma).     %    apple
...
```

The utility of this technique is magnified in the context of lexical items associated with additional arguments. If an argument is directly computable from the word itself, the lexical entry can perform the computation and the dictionary entry need not give a value for the argument. For arguments that are idiosyncratically related to the word such as, for example, grammatical number, the dictionary entry will contain this information in tabular form.

```
p_pr(number,person) -->
        [Word], {p_pr(Word, Number, person)}.
        %p_pr stands for ''personal pronoun''
p_pr(men, [singular, 1-person]).
p_pr(sen, [singular, 2-person]).
...
p_pr(ular, [plural, 3-person]).
```

(More examples for various parts of speech are listed in the Appendix A)

2.1 Irregular Words

Irregular words are those that cannot be parsed by regular spelling rules. We consider a list of entries each containing an irregular word and the correct parsing result for that word.

For example:

```
irregular_pronoun_case( mening,X,["men", possessive_case| X ]).
```

When the program tries to parse a word, it first looks at the irregular word list.

3 User-Callable Predicates for the UZMORPP (Uzbek MORPhological Parser) Program

The morphological parser can be executed in a deterministic or nondeterministic way [5]. If the program is required to return just one result at a time, it will backtrack to find some still untried alternatives. The only exception to this rule is when the word being parsed is in the list of irregular words. After all, if a word is irregular, there is no point in applying general spelling rules.

In this section we present the main callable predicates of the morphological parser UZMORPP.

The predicate do/0 is used to initiate the UZMORPP system. This predicate reads from the text file (here 'uzb.tst'):

```
do :-       prolog_to_os_filename(OsFNameIn, 'uzb.tst'),
        (exists_file(OsFNameIn),!;
        pr('File not found: ',OsFNameIn), fail),
        open(OsFNameIn, read, HandleIn),
        (read_sentence(HandleIn, SentAtr),
        SentAtr \= [],nl, nl,
        pr('Original sentence: ', SentAtr), nl,
        prepare_sentence(SentAtr, _), nl, fail;
        close(HandleIn)).
```

The predicate `read_sentence(+stream, -list)` – read sentences from the input file and convert them to lists of words. Every sentence should be separated with '.', '!', '?':

```
eos(46,46):-!. % '.'
eos(33,33):-!. % '!'
eos(63,63):-!. % '?'
eos(-1,46):-!.
% If there are no other sentences (EOF) the predicate will fail.
```

The `read_sentence/2` predicate is non-deterministic and backtracks to find the next sentence.

Every word should be separated with a comma or space and must contain only Uzbek characters. The Uzbek alphabet has two compound characters (o' and g') where a latin letter is followed by the apostrophe (ASCII code 39). In order to correctly process Uzbek words using these characters we first change the apostrophe to "_" (ASCII code 95). This is controlled by the following predicate:

```
is_uzb_char(39). % ' (apostroph)
is_uzb_char(C):-between(65,122,C).
uzb_char(C, NewC):-
        between(65,90,C), !,
        NewC is C+32.
uzb_char(C, C) :-
        between(97,122,C), !.
uzb_char(39,95).   % ' -> _
```

The predicate `prepare_sentence(+Atoms, -List)` converts an `atom` to a list of morphemes. The variable `Atoms` should be instantiated to an atom or a list of atoms. The predicate will unify `List` with a list of lists where each inner list contains the parsing result. This predicate is non-deterministic and will backtrack to some alternative morphological parsing.

It is usual to divide the morphological analysis of the Uzbek language into two interrelated parts: morphotactic and morphophonemic analysis [3].

1. Morphotactic rules. These rules are about the application and the order of suffixes, i.e. they decide which suffixes can be attached to a given word (according to the part of speech : noun, verb, etc.) and in which order. Words are grouped in different categories according to their functions, and a suffix that can be attached to a word of one category may not be well fitted to words of another category.
2. Morphophonemic rules. These rules are about the forms of suffixes. According to some properties of a word, the form of a suffix may change. For example, the suffix indicating the *dative case* (typically -*ga*) may take one of the three possible forms -*ka*, -*qa*, -*ga*, according to the last consonant of the word which can be -*k*, -*q* and other, respectively.

3.1 Morphotactics of Uzbek

In the Uzbek language there are constraints on the attachment of some suffixes to some words. In order to articulate these constraints we use the following categories: N (Noun), V (Verb), A (Adjective), C (Conjunction), D (Adverb), Pp (Preposition), I (Interjection), Abr (Abbreviation), L (Letter), P (Pronoun), Pn (Proper noun) and Num (Number).

We can subdivide the class of suffixes into two parts: conjugational suffixes and derivational suffixes. A conjugational suffix that is defined for some category can be attached to all of the words in that category. A conjugational suffix does not change the meaning of the word that it is attached to; it only expresses some functional properties of the word (such as the possessive or tense).

A derivational suffix changes the meaning of the word that it is attached to, i.e. it forms a new word. The suffix can also change category of the word. For example, a noun may become a verb after a derivational suffix is attached. Also, the number of words that a derivational suffix can be attached to varies from a single word to nearly all words in the category. The Uzbek language has a large number of derivational suffixes. In this study we present only the most widely used ones. They are listed in the Table 1. The source category indicates the category of the words allowing the suffix. The destination category indicates the category of the new word obtained after the suffix is attached.

Table 1. Derivational suffixes for word categories

SOURCE CATEGORY	SUFFIXES	DESTINATION CATEGORY	EXAMPLE
A (Adjective)	*-ni, - ning,* *-ga,-da,* *-dan*	N *(Noun)*	ya shi+ni, ya shi+ning, ya shi+ga, ya shi+da, ya shi+dan /good, of the good one, to the good one, at the good one, from the good one/
A (Adjective)	*-(i)si,-lari,* *-niki*	N (Noun)	ya shi+si, ya shi+lari, ya shi+larniki /one of good, some of good/
A (Adjective)	*-man,-san,-* *, (i)miz,* *-siz, -lar*	V (Verb)	ya shi+man, ya shi+san, ya shi, ya shi+miz, ya shi+siz, ya shi+lar /{I am, you are, he/she/it is, we are,you are,they are} good/
Num (Number)	*-ni,-ning, -* *ga,-da,-dan*	N *(Noun)*	***quantitative:*** besh- beshni, beshning, beshga, beshda, beshdan /five – five of, to the five of, from the five of/ ***collective:*** beshov+affix belongings(i) –beshovining, beshovini, beshovidan, beshoviga, beshovida; /five – five of, to the five of,

Table 1. Derivational suffixes for word categories—cont.

SOURCE CATEGORY	SUFFIXES	DESTINA- TION CATEGORY	EXAMPLE
			from the five of/ beshala+affix belongings (-si)- beshalasining, beshalasini, beshalasiga, beshalasida, beshalasidan; /five – five of, to the five of, from the five of/ ***single-piece:*** beshta + affix belongings (si)- beshtasining, beshtasini, beshtasiga, beshtasida, beshtasidan; /five – five of, to the five of, from the five of/ ***ordinal:*** beshinchi-beshinchining, beshinchini, beshinchiga, beshinchida, beshinchidan; /fifth – of fifth, to the fifth, from the fifth/
Num (Number)	*-m, -ng,* *-(i)si,* *(i)miz,* *-(i)ngiz,* *lari, niki*	N (Noun)	***collective:*** beshov – beshoving, beshovi, beshovimiz, beshovingiz, beshovlari; beshala - beshalang, beshalasi, beshalamiz, beshalangiz, beshalalari; /you are the five of, five of, we are five of, you are five of, they are five of/ ***single-piece:*** beshta - beshtang, beshtasi, beshtamiz, beshtangiz, beshtalari; /you are the five of, five of, we are five of, you are five of, they are five of/ ***indefinite:*** beshtacha - beshtachang, beshtachasi, beshtachamiz, beshtachangiz, beshtachalari; /you are the five of, five of, we are five of, you are five of, they are five of/ ***ordinal:*** beshinchi - beshim, beshinching, beshinchisi, beshinchimiz, beshinchingiz; beshinchilari; /the fifth – of the fifth, to the fifth, from the fifth/

Table 1. Derivational suffixes for word categories—cont.

SOURCE CATEGORY	SUFFIXES	DESTINATION CATEGORY	EXAMPLE
Num (Number)	-man,-san, ∅, -(i)mniz, -siz, -lar	V (Verb)	***single-piece:*** beshta – beshtamiz, beshtasiz, beshtalar, /we/you/they are five of/ ***indefinite:*** beshtacha- beshtachamiz, beshtachasiz /about five of, we/you are about five of/ ***ordinal:*** beshinchi-beshinchiman, beshinchisan, beshinchi, beshinchimiz, beshinchisiz; beshinchilar; /the fifth – {I am, you are,he/she/it is, we/you/they are} the fifth/
P (Pronoun)	-dagi	N (Noun)	**personal** : men- meniki; /I, mine/ **reflexive:** o'zim-o'zimniki; /myself, mine/ **demonstrative:** bu-bu+n+dagi; /this, in it/ **question:** kim- kimniki;/who, whose/ **definite:** hamma-hammaniki; /all, everybody's/
N (Noun)	-la,-a,-ay -lan, -lash,-ir, -ik,-iq	V (Verb)	kuy+la (song, sing), o'yn+a /game, play/, kuch+ay /forse, be strong/, ot+lan /horse, be ready/, gap+lash /sentence, talk/, gap+ir /sentence, speak/, yo'l+iq /a road, meet/, kech+ik /a night, be late/
A (Adjective)	-la,-(a)r	V (Verb)	toza+la /clean/, qisqa+r /short,be shortly/, ko'k+ar /blue, grow/

In the Uzbek language, the use of a suffix may limit the choice of further suffixes. For example, the interrogative suffix -*mi* cannot directly follow the negation suffix -*ma*. The possesive suffix must follow the plural suffix. Below, we specify order of the conjugational suffixes for nouns and verbs according to Uzbek morphotactic rules.

NOUN:

1. Case suffixes (-*ning*: genitive, -*ga*, -*ka*, -*qa*: dative, -*ni*: accusative, -*da*: locative, -*dan*: ablative)

2. Plural suffix (*-lar*)
3. Possessive suffixes (*-(i)m, -(i)ng, -(s)i, -(i)miz, -(i)ngiz, -(s)i*).
4. Diminutive-endearment suffixes (*-cha*)
5. Noun formation suffixes
   (people: *-chi, -dosh, -kash, -bon, -paz, -dor, -shunos, -xon, -soz, -kor;*
   places: *-zor, -iston, -xona* ;
   abstract things: *-lik, -chilik*)

VERB:

1. Factitive suffix (*-dir*)
2. Passive voice suffix (*-(i)l, -(i)n*)
3. Reflexive voice suffix (-(i)n; -(i)l; -lan)
4. Reciprocal voice suffix (-(i)sh, -lash)
5. Causative voice suffix (-(i)t, -tir, -ir, -ar, -iz)
6. Negation (bo'lishsizlik) suffix (*-ma*)
7. Question (so'roq) suffix (*-mi*)
8. Tense suffixes:
   (a) Present-future tense (Hozirgi-kelasi zamon fe'li): kel+a+man
   (b) Present continuous tense (Hozirgi zamon davom fe'li): kel+ayap+man,
       kel+moqda+man
   (c) Indefinite/simple past tense (Aniq o'tgan zamon fe'li): kel+di+m
   (d) Recent past/present perfect tense (Yaqin o'tgan zamon fe'li):
       kel+gan+man
   (e) Distant past/past perfect tense (Uzoq o'tgan zamon fe'li): kelgan edi+m
   (f) Past continuous tense (O'tgan zamon davom fe'li): kelar edi+m
   (g) Future tense (Kelasi zamon fe'li): kel+ar+man
   (h) Future intentional tense (Kelasi zamon maqsad fe'li):
       kel+moqchi+man

3.2   Morphophonemics of Uzbek

In this section, we will define morphophonemic rules used in Uzbek. These rules
are used, in general, to determine the form of a suffix that will be attached to
a word. In addition to the suffix formation, some of the rules may operate on the
word itself instead of the suffix; i.e. the rules change the form of the word. Such
phenomena are rare, but in order to obtain a complete morphological model of
the Uzbek language we must consider these exceptional cases as well.

In what follows, we present all rules that are used in our morphological model.
These rules include some well-known rules such as the *consonant harmony rule*,
and some rules which are used for a very limited number of cases such as *vowel
deletion rule 1*.

Before presenting these rules, we must first define the Uzbek alphabet and
the categorization of the letters in the Latin alphabet for Uzbek language:

Uzbek alphabet = {a,b,d,e,f,g,h,i,j,k,l,m,n,o,p,q,r,s,t,u,v,x,y,z,o',g'}

Vowels = {a, e, i, o, u, y, o'}
Consonants = {b,d,f,g,h,j,k,l,m,n,p,q,r,s,t,v,x,z,g'}
Harsh consonants = {f,h,k,p,s,t}
Soft consonants = {b,c,d,g,g',j,l,m,n,r,v,z}

What follows is a list of some morphophonemic rules for Uzbek language. Some of the rules apply to each of the suffixes that are attached to a word, while the others apply only to the first suffix that is attached to a word. To make the rules easy to read, we have used abbreviations: x denotes the first letter of the suffix, y denotes the last letter of the current word, v denotes the last vowel of the current word.

RULE 1. *Consonant harmony rule: If x is a vowel and y is in {q, k}, then y is replaced by {g', g}, respectively.*

EXAMPLE: *chiroq* (a lamp) + *-ing* ↦ *cirog'ing* (your lamp) *kuylak* (a dress) + *-im* ↦ *kuylagim* (my dress)

RULE 2. *Vowel deletion rule: If x is a vowel, then delete v.*

EXAMPLE: *og'iz* (a mouth) + *-im* ↦ *og'zim* (my mouth)
ayir (to separate) + *-il* ↦ *ayril* (to be separated)

RULE 3. *Double consonant rule: If x is a vowel, then y doubles.*

EXAMPLE: *rab* (the lord) + *-im* ↦ *rabbim* (my lord)

RULE 4. *Possessive suffix rule:If the suffix is -si, then x (which is 's') may or may not drop.*

EXAMPLE: *mavqe* (a position) + *-si* ↦ *mavqesi* (his/her position)
mavqe (a position) + *-si* ↦ *mavqei* (his/her position)

Most of the rules listed above have some exceptions, i.e. they are not valid for all words. Only rules 1 and 4 are generally valid. We do not discuss these exceptional cases here.

4   Using Prolog to Represent Uzbek Morphology

For each word category (part of speech) we have defined appropriate (Prolog) predicates encoding morphotactic relations. Tables 2 and 3 list these predicates for various categories used in the UZMORPP implementation. In both tables the variable X stands for the *input word,*, Y for the residual, i.e., *the part of the word* left over at the end of the process of suffix identification, Z is a suffix (with description) ($X = Y + Z$).

EXAMPLE: kelmadi (didn't come) – has two suffixes (*-ma:* for *negation* and *-di:* for *past simple, 3rd person, singular*) and the root of verb: *-kel* (to come). The following two rules (predicates) are used in order to analyse this word:

1. `tenses(X,Y,Z)`—X—*kelmadi*, Y—*kelma* (this word is used for the next rule as input), Z—*di* (past simple,3rd person, singular)
2. `negation(X,Y,Z)`—X—*kelma*, Y—*kelma*, Y—*kel*, Z—*ma* (negation)

Table 2. Predicates for identification of conjugational suffixes.

NAME OF PREDICATE	DESCRIPTION
np(X,P)	**For the noun: X returns the parsing result P in form of a list of morphemes**
cases(X,Y,Z)	The case of a noun indicates its grammatical function in a word: -ning, possessive case; -ga, dative case; -ni, accusative case; -da, locative case; -dan, ablative case
possesive(X,Y,Z)	Possessive suffixes: -(i)m,singular,1-person; -(i)ng,singular,2-person; -(s)i,singular,3-person; -(i)miz,plural,1-person; -(i)ngiz,plural,2-person; -(s)i,plural,3-person
belongings_s(X,Y,Z)	Suffix of belongingship −niki (eshikniki − of the door)
in_limit_s(X,Y,Z)	Suffix of indication of limitation -gacha (eshikgacha − to the door).
plur_sing(X,Y,Z)	Grammatical number of a word. To obtain the plural form add the suffix −lar
diminutive_endearment(X,Y,Z)	Diminutive-endearment suffixe −cha (quyoncha − a little rabbit)
verb_morph(X,P)	**For the verb: X returns the parsing result P in form of a list of morphemes**
question(X,Y,Z)	Interrogation suffix: −mi (bordi+mi? − did he/she go?)
imperative(X,Y,Z)	Imperative suffix: Come!, Go!
negation(X,Y,Z)	Negation suffix for verbs −ma (kel+ma, don't come)
reciprocal_voice(X,Y,Z)	Reciprocal suffix: kel+ish+di (they have came)
reflexive_voice(X,Y,Z)	Reflexive suffix(i)n,(i)l (sev+in, enjoy)
tenses(X,Y,Z)	The verb tenses in Uzbek may be categorized according to the time frame: past tenses, present tenses, and future tenses

EXAMPLE:

```
case(onamning, onam, [-ning, possessive_case])
(my mother's, my mother, [-'s, possessive_case])
```

Table 3. Predicates for derivational suffixes.

NAME OF PREDICATE	DESCRIPTION
noun_verb(X,Y,Z)	noun+suffix → verb
noun_adj(X,Y,Z)	noun+suffix → adjective
number_noun(X,Y,Z)	number+suffix → noun
number_verb(X,Y,Z)	number+suffix → verb
pronoun_noun(X,Y,Z)	pronoun+suffix → noun
adj_verb(X,Y,Z)	adjective+suffix → verb

EXAMPLE:

```
noun_verb(otlan, ot, [-lan, derivational suffixes])
(otlan - be ready, ot - horse)
```

5   Concluding Remarks

In this paper, we have examined the morphological structure of Uzbek. We also presented UZMORPP (Uzbek MORPhological Parser) a morphological parser implemented in Prolog. The program uses lexicons for roots (1000 entries) (cf. Appendix C) and suffixes (108) (Appendix D). The morphotactic and morphophonemic rules are implemented in the form of Prolog predicates. The morphotactic rules are included in the source code of the program (cf. Appendix B). An operational prototype of UZMORPP is installed for free inspection at: http://main3.amu.edu.pl/~zlisi/projects/uzmorpp/uzmorpp.rar (access 1.01.2009).

Acknowledgements. This research was partially supported by the postdoc grant of the first author within his Erasmus Mundus Fellowship at the Adam Mickiewicz University (from 2008 to 2009).

References

1. — (1995): Principal Orthographic Rules for the Uzbek Language. The Uzbekistan Cabinet of Minister's Resolution No. 339, of August 24, 1995.
2. Gazdar, G., Mellish, C. (1989): Natural Language Processing in Prolog, Addison Wesley.
3. Güngör, T. and Kuru, S. (1993): Representation of Turkish morphology in ATN. Bo°aziçi University, Istanbul.

4. Pulatov, A., Juraeva, N. (2002): Working out the formal model of the grammar of Uzbek language. Uzbek Mathematical Journal #1, Tashkent, Uzbekistan, 2002.
5. Schlachter, J. G. (2003): ProNTo Morph: Morphological Analysis Tool for use with ProNTo (Prolog Natural Language Toolkit), University of Georgia. (http://www.ai.uga.edu/mc/pronto, access 2008/11/01).
6. Usmonova, M., Azlarov, E., Sharipov, Gh. (1991): O'zbek Tili. Toshkent: O'qituvchi.
7. Vetulani, Z., Obrębski, T. (1997): Morphological tagging of texts using the lemmatizer of the 'POLEX' electronic dictionary, in: B. Lewandowska-Tomaszczyk, P.J. Melia (ed.) Practical Applications in Language Corpora, Proceedings, Łódź University Press, Łódź, 1997, 496–505.
8. Vetulani, Z. (2004): Komunikacja człowieka z maszyną. Komputerowe modelowanie kompetencji językowej, Akademicka Oficyna Wydawnicza EXIT, Warszawa, 2004.

Appendices

A   Example of Lexicon for Various Parts of Speech

```
n_c(gul).                        % Common noun, "flower"
n_c(odil).                       % Proper noun
p_pr(men,[singular,1-person]).   % Personal pronoun , "I"
r_pr(o_zim).                     % Reflexive pronoun, "myself"
d_pr(bu).                        % Demonstrative pronoun, "this"
i_pr(kim).                       % Interrogative pronouns , "who"
df_pr(hamma).                    % Definite pronouns, "everyone"
n_pr(hechkim).                   % Negative pronouns, "nobody"
id_pr(kimdir).                   % Indefinite pronouns, "someone"
v_t(ol).                         % Transitive verb, "to take"
v_i(kel).                        % Intransitive verb, "to come"
con(va).                         % Conjunction, "and"
adj(qizil).                      % Adjective, "red"
```

B   The Source Code of the UZMORPP (Uzbek MORPhological Parser) Program

The UZMORPP program uses as input a list of words from an external input text file. It returns all possible segmentations of a word with descriptions.

To run the program it is necessary to have SWI PROLOG (or any other compatible) installed.

Operations:

1. Put the input text file uzb.tst in the same directory as UZMORPP
2. Consult the UZMORPP source file (dictionaries are integrated into the source code)
3. Run SWI Prolog
4. Run UZMORPP from the prolog command line with "?- do."

(The source code below is complete except for the grammatical resource were
only selected examples are supplied for demo purposes).
  Parsing results will be returned to the file result.txt.

```
do:-
    prolog_to_os_filename(OsFNameIn, 'uzb.tst'),
    ( exists_file(OsFNameIn),!;
      pr('File not found: ', OsFNameIn), fail),
    open(OsFNameIn, read, HandleIn),
    (
    read_sentence(HandleIn, SentAtr),
    SentAtr \= [],
    nl, nl, pr('Original sentence: ', SentAtr), nl,
    prepare_sentence(SentAtr, _),
    % pr('Prepared sentence: ', L), nl,
    fail;
    close(HandleIn)
    ).

read_sentence(HandleIn, Result):-
    \+ at_end_of_stream(HandleIn),
    get0(HandleIn, C),
    r_sentence(HandleIn, C, Result).
 read_sentence(HandleIn, Result):-
    \+ at_end_of_stream(HandleIn),
    read_sentence(HandleIn, Result).

r_sentence(HandleIn, CIn, [AWord|RestWords]):-
    remove_space(HandleIn, CIn, C),
    \+ eos(C, _),
    read_token(HandleIn, C, Word, COut), !,
    name(AWord, Word),
    r_sentence(HandleIn, COut, RestWords).
r_sentence(_, _, []).

read_token(HandleIn, CIn, Token, COut):-
    is_uzb_char(CIn), !,
    read_uzb_word(HandleIn, CIn, Token, COut).
read_token(HandleIn, 44, [99, 111, 109, 109, 97], C):- %% 'comma'
    \+ at_end_of_stream(HandleIn),
    get0(HandleIn, C).

read_uzb_word(HandleIn, CIn, [NewC|RestC], COut):-
    uzb_char(CIn, NewC), !,
    get0(HandleIn, C),
    read_uzb_word(HandleIn, C, RestC, COut).
read_uzb_word(_, CIn, [], CIn).

is_uzb_char(39). % '
is_uzb_char(C) :-
    between(65, 122, C).
```

```prolog
uzb_char(C, NewC):-
   between(65, 90, C), !,
   NewC is C + 32.
uzb_char(C, C):-
        between(97, 122, C), !.
uzb_char(39, 95).    % ' -> _

eos(46, 46):-!. %% '.'
eos(33, 33):-!. %% '!'
eos(63, 63):-!. %% '?'
eos(-1, 46):-!. %% EOF

sign(44, [99, 111, 109, 109, 97]). %% 'comma'

remove_space(_, CIn, CIn):-
   is_uzb_char(CIn), !.
remove_space(_, CIn, COut):-
   eos(CIn, COut), !.
remove_space(_, 44, 44):- !. %% 'comma'
remove_space(HandleIn, _, COut):-
   get0(HandleIn, C),
   remove_space(HandleIn, C, COut).

prepare_sentence([], []):-!.
prepare_sentence([comma|T], [comma|ResL]):- !,
   prepare_sentence(T, ResL).

% Morphological dictionary
prepare_sentence([H|T], [wd(H, AtrL)|ResL]):-
   ( morpho_uzb(H, AtrL, PartofSpeech), writeq(PartofSpeech),
     pr(' word: ',H),write(' - '),pisz(AtrL),!;
     (pr(' Word: ',H),write(' - not found'),nl /*, fail*/)
   ),
   prepare_sentence(T, ResL).

:-consult(noun).
morpho_uzb(H, AtrL,noun) :- np(H, AtrL).

:-consult(verb).
morpho_uzb(H, AtrL,verb) :- verb_morph(H, AtrL).

:-consult(conjunction).
morpho_uzb(H, [AtrL],conjunction) :- con(H, AtrL).

banner:-write_ln('Uzbek morphological analyse'), nl.

pr(Str,Term):-
   write(Str), writeq(Term).

pisz([]):- nl.
pisz([X|Y]) :- write(X),write(', '),pisz(Y).

remove_empty([],[]):-!.
remove_empty([[]|Y],Z) :-  remove_empty(Y,Z),!.
remove_empty([X|Y],[X|Z]) :-  remove_empty(Y,Z).
```

```
np(X,R) :-
    cases(X,X0,R1),
    belongings_s(X0,X2,R2),              % +niki
    in_limit_s(X2,X3,R4),                % +gacha
    possesive(X3,X4,R5),                 % +im,ing,i,imiz,ingiz,lari
    plur_sing(X4,X5,R6),                 % +lar
    diminutive_endearment(X5,Y,R7),      % +cha
    n(Y,R8),
    remove_empty([Y,R8,R7,R6,R5,R4,R2,R1],R).

cases(X,Y,R) :-
    name(X,X0),
    case_s(Suf,R),
    append(Front, Suf, X0),
    name(Y,Front).

% case endings(suffixes)
case_s("ning",[-ning,possessive_case]).
case_s("ga",[-ga,dative_case]).
case_s("ni",[-ni,accusative_case]).
case_s("da",[-da,locative_case]).
case_s("dan",[-dan,ablative_case]).
case_s("",[]).

possesive(X,Y,R) :-
    name(X,X0),
    poss_s(Suf,R),
    append(Front, Suf, X0),
    name(Y,Front).

% Possessive suffixes
poss_s("m",[-m,singular,1-person]).
poss_s("im",[-im,singular,1-person]).
poss_s("ng",[-ng,singular,2-person]).
poss_s("ing",[-ing,singular,2-person]).
poss_s("si",[-si,singular/plural,3-person]).
poss_s("i",[-i,singular/plural,3-person]).

poss_s("miz",[-miz,plu,1-person]).
poss_s("imiz",[-imiz,plural,1-person]).
poss_s("ngiz",[-ngiz,plural,2-person]).
poss_s("ingiz",[-ingiz,plural,2-person]).

poss_s("",[]).

% Affix belongings

belongings_s(X,Y,R) :-
    name(X,X0),
    bs(Suf,R),
    append(Front, Suf, X0),
    name(Y,Front).

bs("niki",[-niki,belongings_suffix]).    % uka+m+niki
```

```
bs("",[]).
```

% Affix indication of limitation / restriction

```
in_limit_s(X,Y,R) :-
   name(X,X0),
   lr(Suf,R),
   append(Front, Suf, X0),
   name(Y,Front).
```

```
lr("gacha",[-gacha,indication_limitation]).     % maktab+gacha
lr("",[]).
```

% Plural suffix
```
plur_sing(X,Y,R) :-
   name(X,X0),
   plur_s(Suf,R),
   append(Front, Suf, X0),
   name(Y,Front).
```

```
plur_s("lar",plural).
plur_s("",singular).
```

% Diminutive-endearment suffixes

```
diminutive_endearment(X,Y,R) :-
   name(X,X0),
   dim_end_suf(Suf,R),
   append(Front, Suf, X0),
   name(Y,Front).
```

```
dim_end_suf("cha",[diminutive/endearment_suffixes]).
dim_end_suf("",[]).
```

```
n(X,R) :-
   n_c(X),string_to_atom('common_noun', R);
   n_p(X),string_to_atom('proper_noun', R);
   p_pr(X,R1),append([personal_pronoun],R1,R);
   r_pr(X),string_to_atom('reflexive_pronoun', R);
   d_pr(X),string_to_atom('demonstrative_pronoun', R);
   i_pr(X),string_to_atom('interrogative_pronoun', R);
   df_pr(X),string_to_atom('definite_pronoun', R);
   n_pr(X),string_to_atom('negative_pronoun', R);
   id_pr(X),string_to_atom('indefinite_pronoun', R);
   adj(X),string_to_atom('adjective', R).
```

% Common nouns
```
n_c(kampir).   % an old woman      n_c(echki).   % a goat
n_c(kaptar).   % a pigeon          n_c(ayol).    % a woman
n_c(mushuk).   % a cat             n_c(bola).    % a child
n_c(sichqon).  % a mouse           n_c(ona).     % a mother
n_c(uka).      % a little brother  n_c(maktab).  % a school
n_c(sumka).    % a bag             n_c(stol).    % a table
n_c(bog_).     % a garden          n_c(yoz).     % summer
```

```
n_c(o_g_il).    % a son              n_c(kuylak).  % a shirt
n_c(ko_ngl).    % soul               n_c(odam).    % a human
n_c(singil).    % a little sister    n_c(qish).    % winter
n_c(dev)        % a giant            n_c(og_iz).   % a mouth
n_c(burun).     % a nose

% Proper nouns
n_p(karim).                          n_p(salim).
n_p(g_ayrat).                        n_p(odil).
n_p(toshkent).

% Personal pronouns    (kishilik olmoshlari)
p_pr(men,[singular,1-pers]).              % I
p_pr(sen,[singular,2-pers]).              % you
p_pr(u,[singular,3-pers]).                % he/she/it
p_pr(biz,[plural,1-pers]).                % we
p_pr(siz,[plural/respect,2-pers]).        % you
p_pr(siz,[plural/respect,2-pers]).        % they

% Reflexive pronouns    (o'zlik olmoshlari)
r_pr(o_zim).        % myself
r_pr(o_zing).       % yourself
r_pr(o_zi).         % himself, herself, itself
r_pr(o_zimiz).      % ourselves
r_pr(o_zlaringiz).  % yourselves
r_pr(o_zlari).      % themselves

% Demonstrative pronouns    (ko'rsatkich olmoshlari)
d_pr(u).        % that             d_pr(bu).     % this
d_pr(shu).      % this             d_pr(o_sha).  % that
d_pr(mana bu).  % this             d_pr(ana u).  % that

% Interrogative pronouns    (so'roq olmoshlari)
i_pr(kim).      % who              i_pr(nima).    % what
i_pr(qanday).   % how              i_pr(qanaqa).  % what kind of
i_pr(qaysi).    % which            i_pr(qachon).   % when
i_pr(qancha).   % how much         i_pr(nechta).   % how many
i_pr(nechanchi). % which one

% Definite pronouns    (belgilash olmoshlari)
df_pr(hamma).   % everybody        df_pr(barcha).    % all
df_pr(harkim).  % everyone         df_pr(harqaysi).  % each
df_pr(harnima). % anything         df_pr(harnarsa).  % anything

% Negative pronouns    (bo'lishsizlik olmoshlari)
n_pr(hechkim).      % nobody       n_pr(hechnima).   % nothing
n_pr(hechqachon).   % never        n_pr(hechqaysi).  % none of
n_pr(hechqayerda).  % nowhere      n_pr(hechqanday). % never

% Indefinite pronouns    (gumon olmoshlari)
id_pr(kimdir).      % someone      id_pr(nimadir).    % something
id_pr(qandaydir).   % any          id_pr(qanchadir).  % some
id_pr(allakim).     % somebody     id_pr(allanima).   % something
```

```
id_pr(allaqanday).  % a/an        id_pr(allaqancha).  % some
id_pr(allaqaysi).   % a/an        id_pr(allanechuk).  % a/an
id_pr(birnecha).    % a few       id_pr(birqancha).   % several
id_pr(birnima).     % something   id_pr(birnarsa).    % something
id_pr(ba_zi).       % some        id_pr(birov).       % somebody
id_pr(biron).       % a/an

verb_morph(X,R) :-
    imperative(X,Y,R0),
    verb_root(Y,R1),
    remove_empty([R1,R0],R).

verb_morph(X,R) :-
    question(X,X0,R0),
    tenses(X0,X1,R1),
    \+ ( \+ R0=[], R1=[]),   % ber+mi? - incorrect
    negative(X1,X2,R2),
    reciprocal_voice(X2,X3,R3),
    reflexiv_voice(X3,Y,R4),
    verb_root(Y,R5),
    (X = Y, append(R5,[imperative_mood],R),!;
    remove_empty([R5,R4,R3,R2,R1,R0],R) ).

verb_root(X,R) :-
    (v_t(X),R=[X,transitive_verb],!);
    (v_i(X),R=[X,intransitive_verb]).

imperative(X,Y,R) :-
    name(X,X0),
    imper_suf(Suf,R),
    append(Front, Suf, X0),
    name(Y,Front).

imper_suf("ing",[-ing,imperative_mood,respect]).
imper_suf("inglar",[-inglar,imperative_mood,plural]).

question(X,Y,R) :-
    name(X,X0),
    question_suf(Suf,R),
    end(Front, Suf, X0),
    name(Y,Front).

question_suf("mi",[-mi,question]).
question_suf("",[]).

negative(X,Y,R) :-
    name(X,X0),
    verb_negation(Suf,R),
    append(Front, Suf, X0),
    name(Y,Front).

verb_negation("ma",[-ma,negation]).
verb_negation("",[]).
```

```prolog
reciprocal_voice(X,Y,R) :-
   name(X,X0),
   reciprocal_suf(Suf,R),
   append(Front, Suf, X0),
   name(Y,Front).

reciprocal_suf("ish",[-ish,reciprocal_voices]).   % after consonant
reciprocal_suf("sh",[-sh,reciprocal_voices]).     % after vowel
reciprocal_suf("",[]).

reflexiv_voice(X,Y,R) :-
   name(X,X0),
   reflexiv_suf(Suf,R),
   append(Front, Suf, X0),
   name(Y,Front).

reflexiv_suf("in",[-in,reflexiv_voice]).   % after consonant
reflexiv_suf("n",[-n,reflexiv_voice]).     % after vowel
reflexiv_suf("il",[-il,reflexiv_voice]).   % after consonant
reflexiv_suf("l",[-l,reflexiv_voice]).     % after vowel
reflexiv_suf("",[]).

tenses(X,Y,R) :-
   name(X,X0),
   verb_tenses(Suf,R),
   append(Front, Suf, X0),
   name(Y,Front).
% Present/future Simple (Xozirgi/kelasi zamon darak) suffixes
% after consonant (undoshdan keyin)
verb_tenses("aman",[-aman,present/future_simple,singular,1-person]).
verb_tenses("asan",[-asan,present/future_simple,singular,2-person]).
verb_tenses("asiz",[-asiz,present/future_simple,singular,2-person,
      respect]).
verb_tenses("adi", [-adi, present/future_simple,singular,3-person]).
verb_tenses("amiz",[-amiz,present/future_simple,plural,1-person]).
verb_tenses("asizlar",[-asizlar,present/future_simple,plural,
      2-person]).
verb_tenses("adilar",[-adilar,present/future_simple,plural,3-person]).
% after vowel (unlidan keyin)
verb_tenses("yman",[-yman,present/future_simple,singular,1-person]).
verb_tenses("ysan",[-ysan,present/future_simple,singular,2-person]).
verb_tenses("ysiz",[-ysiz,present/future_simple,singular,2-person,
      respect]).
verb_tenses("ydi", [-ydi, present/future_simple,singular,3-person]).
verb_tenses("ymiz",[-ymiz,present/future_simple,plural,1-person]).
verb_tenses("ysizlar",[-ysizlar,present/future_simple,plural,
      2-person]).
verb_tenses("ydilar",[-ydilar,present/future_simple,plural,3-person]).

% Past Simple (O'tgan zamon darak) suffixes
verb_tenses("dim",[-dim, past_simple,singular,1-person]).
verb_tenses("ding",[-ding, past_simple,singular,2-person]).
```

```prolog
verb_tenses("dingiz",[-dingiz,
      past_simple,singular,2-person,respect]).
verb_tenses("di",[-di, past_simple,singular,3-person]).
verb_tenses("dik",[-dik, past_simple,plural,1-person]).
verb_tenses("dingizlar",[-dingizlar,past_simple,plural,2-person]).
verb_tenses("dilar",[-dilar,past_simple,plural/respect,3-person]).

% Present Continuous (Xozirgi davom) suffixes
% after consonant (undoshdan keyin)
verb_tenses("ayapman",[-ayapman,present_continuous,singular,
      1-person]).
verb_tenses("ayapsan",[-ayapsan,present_continuous,singular,
      2-person]).
verb_tenses("ayapsiz",[-ayapsiz,present_continuous,singular,
      2-person,respect]).
verb_tenses("ayapdi",[-ayapdi,present_continuous,singular,3-person]).
verb_tenses("ayapmiz",[-ayapmiz,present_continuous,plural,1-person]).
verb_tenses("ayapsizlar",[-ayapsizlar,present_continuous,plural,
      2-person]).
verb_tenses("ayapdilar",[-ayapdilar,present_continuous,
      plural/respect,3-person]).
% after vowel/consonant (unlidan keyin)
verb_tenses("yapman",[-yapman,present_continuous,singular,1-person]).
verb_tenses("yapsan",[-yapsan,present_continuous,singular,2-person]).
verb_tenses("yapsiz",[-yapsiz,present_continuous,singular,2-person,
      respect]).
verb_tenses("yapdi",[-yapdi,present_continuous,singular,3-person]).
verb_tenses("yapmiz",[-yapmiz,present_continuous,plural,1-person]).
verb_tenses("yapsizlar",[-yapsizlar,present_continuous,plural,
      2-person]).
verb_tenses("yapdilar",[-yapdilar,present_continuous,
      plural/respect,3-person]).

% Past Continuous (O'tgan davom fe'li) verb suffixes

% after consonant (undoshdan keyin)
verb_tenses("ayotgandim",[-ayotgandim,past_continuous,singular,
      1-person]).
verb_tenses("ayotganding",[-ayotganding,past_continuous,singular,
      2-person]).
verb_tenses("ayotgandingiz",[-ayotgandingiz,past_continuous,singular,
      2-person,respect]).
verb_tenses("ayotgandi",[-ayotgandi,past_continuous,singular,
      3-person]).
verb_tenses("ayotgandik",[-ayotgandik,past_continuous,plural,
      1-person]).
verb_tenses("ayotgandingizlar",[-ayotgandingizlar,past_continuous,
      plural,2-person]).
verb_tenses("ayotgandilar",[-ayotgandilar,past_continuous,
      plural/respect,3-person]).

% after vowel (unlidan keyin)
```

```
verb_tenses("yotgandim",[-yotgandim,past_continuous,singular,
    1-person]).
verb_tenses("yotganding",[-yotganding,past_continuous,singular,
    2-person]).
verb_tenses("yotgandingiz",[-yotgandingiz,past_continuous,singular,
    2-person,respect]).
verb_tenses("yotgandi",[-yotgandi,past_continuous,singular,3-person]).
verb_tenses("yotgandik",[-yotgandik,past_continuous,plural,1-person]).
verb_tenses("yotgandingizlar",[-yotgandingizlar,past_continuous,
    plural,2-person]).
verb_tenses("yotgandilar",[-yotgandilar,past_continuous,
    plural/respect,3-person]).

% Present perfect (Xozirgi tugal fe'li) verb suffixes
verb_tenses("ganman",[-ganman,present_perfect,singular,1-person]).
verb_tenses("gansan",[-gansan,present_perfect,singular,2-person]).
verb_tenses("gansiz",[-gansiz,present_perfect,singular,2-person,
    respect]).
verb_tenses("gan",[-gan,present_perfect,singular,3-person]).
verb_tenses("ganmiz",[-ganmiz,present_perfect,plural,1-person]).
verb_tenses("gansizlar",[-gansizlar,present_perfect,plural,
    2-person]).
verb_tenses("ganlar",[-ganlar,present_perfect,plural/respect,
    3-person]).
verb_tenses("",[]).

% Vocabulary of Transitive verbs
v_t(ol).    % take          v_t(ber).    % give
v_t(och).   % open          v_t(ko_r).   % watch
v_t(yoz).   % write         v_t(ich).    % drink
v_t(qo_y).  % put           v_t(so_ra).  % ask
v_t(yuv).   % wash

% Vocabulary of Intransitive verbs
v_i(bor).    % go           v_i(kel).     % come
v_i(o_tir).  % sit          v_i(uxla).    % sleep
v_i(yur).    % walk         v_i(zavqlan). % enjoy

adjp(X,R) :-
    (
     comparative(X,Y,R1),     % +roq
     (adj(Y,Ind),adjgrp(Ind,R2),!;
      adj(Y)
     ),
     remove_empty([Y,adjective,R2,R1],R)
    );
    (
     similar_to(X,X1,R1),                % odam+simon, dev+sifat
     locative_temperal(X1,X2,R2),        % +ki,qi,gi,dagi
     possesive(X2,X3,R3),                % +im,ing,i,imiz,ingiz,lari
     plur_sing(X3,X4,R4),                % +lar
     orientr(X4,X5,R5),                  % toshkent+lik
```

```prolog
    diminutive_endearment(X5,Y,R6),        % +cha
    n(Y,R7),
    remove_empty([Y,R7,R6,R5,R4,R3,R2,R1],R)
    ).

% Making adjective from noun
adjp(X,R) :-
   adj_s(X,Y,R1),
   (n_c(Y),!;
   n_p(Y)),
   remove_empty([Y,noun,R1],R).

adj_s(X,Y,R):-
   name(X,X0),
   adj_suf(Suf,R),
   append(Front, Suf, X0),
   name(Y,Front).

adj_suf("li",[-li,noun_to_adj]).
adj_suf("dor",[-dor,noun_to_adj]).
adj_suf("mand",[-mand,noun_to_adj]).
adj_suf("siz",[-siz,noun_to_adj]).
adj_suf("lik",[-lik,noun_to_adj]).
adj_suf("gi",[-gi,noun_to_adj]).
adj_suf("qi",[-qi,noun_to_adj]).
adj_suf("ki",[-ki,noun_to_adj]).
adj_suf("chan",[-chan,noun_to_adj]).
adj_suf("simon",[-simon,noun_to_adj,similar_to]).
adj_suf("sifat",[-sifat,noun_to_adj,similar_to]).
adj_suf("kor",[-kor,noun_to_adj]).
adj_suf("cha",[-cha,noun_to_adj]).
adj_suf("namo",[-namo,noun_to_adj]).
adj_suf("parvar",[-parvar,noun_to_adj]).
adj_suf("aki",[-aki,noun_to_adj]).
adj_suf("shumul",[-shumul,noun_to_adj]).
adj_suf("",[]).

comparative(X,Y,R) :-
   name(X,X0),
   comp_s(Suf,R),
   append(Front, Suf, X0),
   name(Y,Front).

comp_s("roq",[-roq,comparative]).
comp_s("",[]).

% Locative-Temporal adjective

locative_temporal(X,Y,R) :-
   name(X,X0),
   loc_temp_s(Suf,R),
   append(Front, Suf, X0),
   name(Y,Front).
```

```
loc_temp_s("gi",[-gi,locative_temporal]).
loc_temp_s("qi",[-qi,locative_temporal]).
loc_temp_s("ki",[-ki,locative_temporal]).
loc_temp_s("dagi",[-dagi,locative_temporal]).
loc_temp_s("",[]).

% Relative
similar_to(X,Y,R) :-
    name(X,X0),
    sim_s(Suf,R),
    append(Front, Suf, X0),
    name(Y,Front).

sim_s("simon",[-simon,similar_to]).
sim_s("sifat",[-sifat,similar_to]).
sim_s("",[]).

% Relative
orientr(X,Y,R) :-
    name(X,X0),
    or_s(Suf,R),
    append(Front, Suf, X0),
    name(Y,Front).

or_s("lik",[-lik,orient_belongings]).
or_s("",[]).

% Dictionary of adjectives
adj(qizil,1).         % red          adj(oq,1).          % white
adj(keng,2). ).       % wide         adj(tor,2).         % narrow
adj(baland,2).        % high         adj(past,2).        % low
adj(uzun,2).          % long         adj(chuqur,2).      % deep
adj(yirik,2).         % thick        adj(mayda,2).       % petty
adj(dumaloq,2).       % round        adj(egri,2).        % crooked
adj(yassi,2).         % even         adj(qiyshiq,2).     % bumpy
adj(do_ng,2).         % bumpy        adj(shirin,3).      % sweet
adj(achchiq,3).       % bitter       adj(nordon,3).      % sour
adj(sho_r,3).         % salty        adj(taxir,3).       % sour
adj(bemaza,3).        % tasteless    adj(chuchmal,3).    % vague
adj(muloyim,4).       % kind         adj(odobli,4).      % well-behaved
adj(shirinso_z,4).    % eloquent     adj(chaqqon,4).     % adroit
adj(sho_x,4).         % jolly        adj(yalqov,4).      % lazy
adj(dangasa,4).       % lazy         adj(chechan,4).     % talkative
adj(qo_rs,4).         % rude         adj(qaysar,4).      % stubborn
adj(sassiq,5).        % rotten       adj(qo_lansa,5).    % stinking
adj(badbo_y,5).       % fetid        adj(xushbo_y,5).    % fragrant
adj(muattar,5).       % fragrant     adj(xursand,6).     % glad
adj(xafa,6).          % sad          adj(shod,6).        % happy
adj(kasal,6).         % sick         adj(injiq,6).       % capricious
adj(yosh,7).          % young        adj(qari,7).        % old
adj(keksa,7).         % old          adj(zamonaviy,8)    % modern
adj(oilaviy,8).       % family       adj(ma_naviy,8).    % spiritual
```

```
                                   adj(siyosiy,8).   % political
adjgrp(1,colour).
adjgrp(2,form_shape).    % hajmi,shakl va ko'rinish ma'nosidagi
     sifatlar
                    % guruhi
adjgrp(3,taste). % maza-ta'm  ma'nosidagi sifatlar guruhi
adjgrp(4,character). % xarakter-xususiyat  ma'nosidagi sifatlar
     guruhi
adjgrp(5,odour). % hid  ma'nosidagi sifatlar guruhi
adjgrp(6,psychic_state).    % odamning psixik xolat  ma'nosidagi
                    % sifatlar guruhi
adjgrp(7, human_qualities).  % odamning tabiiy xolat  ma'nosidagi
                    %sifatlar guruhi
adjgrp(8,possessive). % xoslik ma'nosidagi sifatlar guruhi, aslida
     otga
                    % -viy,-iy qushish blan yasaladi

con(va,combining_conjunction).         % and
con(hamda,combining_conjunction).      % and
con(bilan,combining_conjunction).      % with
con(ammo,contrastin_conjunction).      % but
con(lekin,contrastin_conjunction).     % but
con(biroq,contrastin_conjunction).     % but, however
con(yo,disjunctive_conjunction).       % or
con(yoki,disjunctive_conjunction).     % or
con(yoxud,disjunctive_conjunction).    % or else
```

C Roots

Conjuctions:
ham, va, chunki, lekin, hatto, biroq, agar, balki, faqat, yoki, yo

Adjectives:
adabli, agressiv, aldoqchi, a'lo, al'ternativ, aniq, arzon, asosiy,
bahaybat, baland, band, baxramand, beadab, bebosh, bedor, behayo,
bejirim, beso'naqay, betartib, bezor, bitmas, bo'sh, butun, chigal,
chiroyli, chuqur, dadil, dag'al, dilkash, dumaloq, durust, erkin, faol,
g'alati, garang, g'oyib, go'zal, iflos, ikkinchi, intizomsiz, issiq,
ixcham, jasoratli, jasur, jonkuyar, kam, katta, keng, kichkina, ko'p,
ko'r, kuchli, lo'nda, manzur, marhum, mashhur, mavhum, maxsus, mayin,
muhtaram, murakkab, musaffo, mutlaq, nizoli, noto'g'ri, og'ir, ogoh,
o'rtacha, oson, oxirgi, o'xshash, past, qadoq, qaqshatqich, qari, qarshi,
qat'iy, qattiq, qimmat, qisqa, qiyin, qiziq, qodir, qora, qorong'i,
qo'rqoq, qulay, quruq, rost, rozi, salqin, sekin, sharqiy, shaxdam,
shinam, soda, sovuq, soxta, taniqli, tashqi, tasodifiy, tasodifiy, taxir,
tayinli, tayyor, tekis, telba, teng, ters, tez, tinch, tirik, to'g'ri,
turli, tushunarli, uchinchi, umumiy, uyg'oq, uzoq, uzun, vahshiy,
xursand, yangi, yaqin, yaqqol, yarim, yaxshi, yetarli, yirik, yolg'iz,
yomon, yopiq, yorug', yosh, yovvoyi, yumshoq, yuraksiz, zarur, zich, zid,
ziyrak, zo'r

Verbs:

afsuslan, ajrat, alda, almashtir, anglat, aniqla, aralash, aralashtir,
aybla, ayt, badbo'y, baland, bahola, bajar, baqir, bekin, bemaza, ber,
bil, birlashtir, bitir, bo'l, bor, bor, boshla, boshla, boshqar, bos,
bo'ya, buril, buril, buz, chaqir, chaqir, charaqla, chegarala, chekla,
chiq, chiz, cho'k, chop, choz, davola, egalla, ergash, eri, eshit, eshit,
esla, esla, gapir, gaplash, g'alati, hayda, hayratlantir, ich, iflosla,
isbotla, ishla, ishon, itar, izla, izoxla, jaroxatla, jazola, jo'nat,
jo'nat, kamay, kamgap, kara, kara, katnash, kayna, kaytar, kech, kechir,
kech, kel, kel, kengaytir, kes, kes, ket, kichkina, kir, kiy, kiy, komil,
ko'taril, ko'chir, ko'r, ko'rsat, ko'tar, kul, kuchli, kurash, kut, kut,
kuzat, hotirjam, maqta, mashq, maxsus, mayin, mos, min, muhokama, mumkin,
musobaqalash, mustaqil, muvofaqiat, muzla, nafratlan, nishonla, o'ra,
och, o'g'irla, ogohlantir, og'ri, ol, o'lcha, o'ldir, o'qi, oq, o'rgan,
o'rnash, ornat, oshir, os, o'stir, o'stir, o't, o'tir, o'tir, ot, o'yla,
o'yna, o'zgartir, parlan, parxez, pishir, portlat, qadrla, qara, qaytar,
qil, qisqartir, qista, qiyna, qo'sh, qo'y, qol, qo'lla, qol, qoniqtir,
qo'rq, qo'shil, qoy, qoz, qur, qurolsizlantir, qutqar, qutqar, quy,
rivojlantir, rohatlan, sakra, salishtir, sana, sanch, saqla, sarfla,
sayla, sev, sez, sekin, sezgir, shaxsiylashtir, shong'i, shoshil, silkit,
silliqla, sina, sindir, sindir, shirin, silliq, sog'in, so'ra, sot, sur,
suz, syg'ur, tabrikla, ta'minla, tanla, taqiqla, tarqat, tashla,
tasvirla, taxminla, tekshir, tep, tila, tirka, tirmash, tishla, to'la,
to'ldir, top, to'pla, toqnash, to'xta, tozala, tug'ul, tur, tushir,
tushun, tushuntir, tuyul, tuz, uch, uchrash, umidsizlantir, ur, uril, ur,
urul, urush, ushla, uxla, uxla, uylan, uyushtir, vaqtinchalik, xayda,
xidla, ximoyalan, xoxla, yarat, yasha, ye, yetish, yig', yiqil, yon,
yo'qot, yoqtir, yordamlash, yot, yoy, yoz, yugur, yur, yut, yuv, zararla

Nouns:

adabiyot, adyol, aeroport, agentlik, aholi, ahvol, ajdod, aka, aksariyat,
al'bom, aloqa, Amerika, armiya, arqon, artilleriya, arxeologiya, asab,
asbob, uskuna, ashula, askar, asos, asr, astronomiya, atmosfera,
atrof-muxit, avariya, avtobus, avtomashina, axborot, aybdor, aylana,
ayol, a'zo, baho, bahona, bahor, bakteriya, baliq, bank, banka, baxil,
baxt, bayroq, bekat, belgi, bemor, bet, yuz, bilim, bino, biologiya,
birikma, birlashma, biznes, bobo, bog', bojxona, bola, bo'lak, bolalar,
bomba, borliq, bo'ron, bosh, boshqarma, boshliq, boshpana, bosim, boylik,
bozor, bug'doy, bulut, burch, burchak, burun, butilka, buvi, buyrak,
buyruq, buzilish, byudjet, byulleten, chaqaloq, chegara, chiroq, cho'qqi,
choy, dafn, dala, dalil, damba, danak, daqiqa, daraja, daraxt, dars,
daryo, dastur, davlat, davo, davr, daxl, dehqonchilik, delegat,
demokratiya, dengiz, deraza, detal, devor, diktator, din, diplomat,
diydor, do'kon, do'llar, dori, doska, do'st, dunyo, dur, dushman,
dvigatel, ekologiya, ekstremist, elchi, elchixona, elektr, elektron,
enzim, er, erk, erkak, ertalab, eshik, eslatma, fabrika, fan, faol, farq,
federal, ferma, fikr, film, firma, fizika, foiz, fuqaro, fursat, futur,
g'amxo'rlik, g'alaba, g'alayon, gap, g'arb, garov, g'ayrat, gaz, gazlama,
g'ildirak, g'olib, go'sht, g'ov, gul, guruch, guruh, hafta, hajviya,
halqa, harakat, harorat, hasharot, hat, havo, haykal, hayot, hayvon,

hazil, hazina, hikoya, his-hayajon, hisob, hodisa, hokim, hokimiyat,
hosil, hujayra, hujjat, hukumat, ichak, idora, idrok, ildiz, imkoniyat,
inflyatsiya, Ingliz, injeneriya, iqlim, iqtisod, ish, ishbilarmon,
ishonch, isitma, ism, issiqlik, isyonchi, it, jadval, jamiyat, jang,
jangchi, janub, jarayon, jarohat, javon, jigar, jinnixona, jinoyat, jins,
jonzod, jo'xori, joy, jurnal, jyuri, kafillik, kamchilik, kamera, kamyob,
kapitalizm, kapto'k, karta, kasallik, kasalxona, kecha, kechinma,
kechqurun, kelishuv, kema, kengash, kerak, kimyo, kinofilm, kishi, kitob,
kiyim, ko'cha, kofe, ko'kat, ko'l, kollej, komanda, ko'mir, kompaniya,
kompyuter, kongress, ko'prik, koptok, ko'rsatma, ko'z, kredit, krizis,
krovat, kuch, kuchuk, kul, kun, kunduz, Kuz, laboratoriya, lazer, loy,
loyiha, ma'no, madaniyat, magazin, magnit, mahorat, makka, maktab,
ma'lumot, mamlakat, ma'muriyat, manba, manfaatdor, ma'no, manzara,
maqsad, markaz, marosim, mashina, mashinka, masofa, matematika, materik,
maydon, mehmonxona, meva, mikroskop, militsiya, millat, miltiq, mineral,
ministr, minut, miqdor, misol, miya, modda, mol, molekula, mo'ljal, moy,
muammo, muhandis, muhtoj, mukofot, mulla, muqova, mushak, mushuk, musiqa,
mustamlaka, muz, nabira, namuna, narsa, narxi, nashr, natija, nazariya,
nishona, nomzod, non, nurlanish, nutq, o'rmon, ob-havo, ochlik, o'choq,
ofitser, o'g'il, og'iz, og'riq, oila, okean, olomon, oltin, ona, o'pka,
o'q, o'qituvchi, oqsil, organ, organizm, o'rmon, orol, o'rta, orzu,
oshxona, o'simlik, osmon, oson, o'spirin, ot, ota, ota-ona, o'tirish,
ovoz, ovqat, oy, o'yin, oyoq, ozchilik, panjara, parad, parashyut,
parlament, partizan, pasport, paxta, pichoq, poezd, pol, po'lat, poytaxt,
propaganda, pul, qabila, qahramon, qaltislik, qamoqxona, qaror, qarz,
qayiq, qirg'oq, qish, qishloq, qiz, qiziq, qiziqish, qog'oz, qoida, qo'l,
qoldiq, qomat, qon, qo'ng'iroq, qonun, qor, qoratuproq, qorin, qo'rquv,
qo'y, qo'zg'olonchi, qul, qum, qurbon, qurol, qush, quti, quvg'in,
quyosh, radar, radio, rahm, rais, rak, raketa, rang, raqam, rasm,
reaktsiya, reja, rentgen, rezina, rivoj, robot, ruh, sabab, sabzavot,
samara, samolyot, sana, san'at, san'atkor, sanoat, savdo, savol, sayohat,
sel, senat, sezgi, shahar, shakar, shakl, shamol, sharq, shart, shaxs,
she'r, shimol, shlyapa, shovqin, sifat, sigir, sim, sinf, singil, sir,
siyosat, skelet, soat, soch, soqchi, so'qmoq, sovrin, so'z, spirt, sport,
stakan, sud, surat, sut, suv, suyak, tabiat, tajriba, talab, ta'lim,
tana, tank, tarix, tarqoqlik, taslim, tayoq, teatr, telefon, teleskop,
televideniya, temir, teri, terorist, texnologiya, tezlik, til, tilak,
tilla, tinchlik, tish, tizim, tog', tog'a, tog'ora, tomon, to'qnashuv,
tor, tortishish, tosh, tova, tovush, transport, truba, tufli, tumon, tun,
turkum, tuxum, tuz, uchuvchi, umid, universitet, unsur, urf-odat,
urinish, uslub, uy, uzuk, vaqt, vertolyot, viloyat, virus, vodiy, vulqon,
xabar, xalq, xat, xato, xayot, xiyonat, xizmat, xodisa, xona, xonim,
xotin, xromosoma, xudo, xuruj, yarim, yashin, yer, yerlik, yil, yog',
yo'l, yo'lak, yo'ldosh, yo'lovchi, yomg'ir, yo'nalish, yong'in, olov,
yonilg'i, yordam, yosh, yoz, yuk, yuklama, yulduz, yurak, yuz, yuza,
zachyot, test, zahar, zarb, zarra, zaxiradagi, zehin, zilzila, zina

D   Suffixes

-ning, -ga, -ni, -da, -dan, -m, -im, -ng, -ing, -i, -si, -miz, -imiz,
-ngiz, -ingiz, -niki, -gacha, -lar, -cha, -ng, -ing, nglar, -inglar, -mi,
-ma, sh, -ish, -n, -in, -l, -il, -aman, -asan, -asiz, -adi, -amiz,
-asizlar, -adilar, -yman, -ysan, -ysiz, -ydi, -ymiz, -ysizlar, -ydilar,
-dim, -ding, -dingiz, -di, -dik, -dingizlar, -dilar, -yapman,-ayapman,
-yapsan, -ayapsan, -yapdi, -ayapdi, -yapmiz, -ayapmiz, -yapsiz, -ayapsiz,
-yapsizlar, -ayapsizlar, -yapdilar, -ayapdilar, -yotgandim, -ayotgandim,
-yotganding, -ayotganding, -yotgandi, -ayotgandi, -yotgandik,
-ayotgandik, -yotgandingiz, -ayotgandingiz, -yotgandingizlar,
-ayotgandingizlar, -yotgandilar, -ayotgandilar, -ganman, -gansan, gansiz,
-gan, -ganmiz, -gansizlar, -ganlar, -li, -dor, -mand, -lik, -gi, -qi,
-ki, -chan, -simon, -sifat, -kor, -cha, -namo, -parvar, -aki, -shumul,
-roq, -gi,-qi, -ki, -dagi,

Inflection of Polish Multi-Word Proper Names with Morfeusz and Multiflex

Agata Savary[1], Joanna Rabiega-Wiśniewska[2], and Marcin Woliński[2]

[1] Université François Rabelais de Tours, France,
agata.savary@univ-tours.fr,
[2] Institute of Computer Science, Polish Academy of Sciences, Poland,
{joanna.rabiega,marcin.wolinski}@ipipan.waw.pl

Abstract. We discuss morphological properties of Polish multi-word proper names. We present a cooperating framework of two morphological tools: *Morfeusz*, a morphological analyser and generator for Polish simple words, and *Multiflex*, a cross-language morpho-syntactic generator of multi-word units. We discuss interface constraints required for the interoperability of these tools, and we show how the resulting platform allows one to describe the morpho-syntactic behaviour of some interesting examples of Warsaw multi-word toponyms.

Key words: computational morphology, Polish proper names, multi-word units, inflection and variability of compounds

1 Introduction

Proper names and other named entities are of crucial quantitative and qualitative importance for natural language processing (NLP) due to their frequent occurrence in corpora and their rich semantic content. Despite continual efforts in the NLP community aiming at named entity extraction and modelling, the formal linguistic description of proper names remains a challenge.

In many application domains, such as technical sublanguages, most of the proper names are multi-word units. As in the case of single words they are subject to inflection. In highly inflected languages, such as Polish, each multi-word nominal typically yields seven or fourteen inflected forms (depending on whether it does or does not inflect for number), whose behaviour may be idiosyncratic. Moreover, even if compound proper names have the reputation of institutionalized phrases, they often show a varying degree of flexibility on the orthographic, lexical, syntactic, and/or semantic level. For example, they may allow for some component inflections, omissions, substitutions, insertions, and embedding. This results in a number of their possible textual realisations which should possibly be recognized as variants of the same base form. For instance, it should be indicated that all of the following sequences, along with their corresponding inflected forms, refer to the same roundabout in Warsaw: *Rondo Zgrupowania AK „Radosław", Rondo Zgrupowania „Radosław", Rondo Radosława,* and *Rondo Babka.*

M. Marciniak and A. Mykowiecka (Eds.): Bolc Festschrift, LNCS 5070, pp. 111–141, 2009.

We are interested in the morphological properties of Polish multi-word proper names in general. Our approach is lexical: we aim at explicit description of all grammatically correct inflected forms of a multi-word proper name and their variants.

In this paper we present a cooperating framework of two morphological tools: *Morfeusz*, a morphological analyser and generator for Polish simple words, and *Multiflex*, a cross-language morpho-syntactic generator of multi-word units. We discuss interface requirements for the interoperability of these tools. The resulting platform is characterized by: a rich flexemic tagset of Polish, a graph-based description of multi-word paradigms, and compact inflection patterns due to unification. We show how the platform allows one to describe the morpho-syntactic behaviour of some interesting examples of Warsaw multi-word toponyms.

Finally, we address some remaining problems such as:

- conflation of derivational and semantic variants,
- the necessity of distinguishing variants possessing the same inflection features,
- "extreme" elliptical variants transforming multi-word units into inflectionally intriguing single words (e.g. *ulica Kazimerza Pułaskiego* > *Pułaskiego* in all cases), and
- morpho-graphical problems caused by abbreviations and acronyms.

The results presented in this paper are a part of a larger project aiming at modelling Warsaw topography (streets, places, institutions, monuments, etc.) and transportation system (bus- and tram-stops, underground stations, etc.), as well as a large-scale lexical description of Warsaw toponyms, in view of information extraction and dialogue systems. We also wish to examine the feasibility of integrating Polish toponyms within a multilingual ontology of proper names, considering machine translation, multilingual information extraction and text alignment.

2   State of the Art

Onomastic studies on the origin, history and regional particularities of Polish proper names have been performed for decades [12,28]. However, relatively few efforts have been made in view of a large-scale formal description of their grammatical, in particular inflectional, behaviour. The printed dictionary [10] is dedicated to spelling, pronunciation, formation and inflection of proper names in Polish. Among its 11,000 entries, numerous foreign names are described with respect to their declension, which copes with inflectional suffixes atypical for Polish inflection paradigms. The lexicon [6] is dedicated to formal and systematic description of inflection paradigms of single and compound proper names in Polish. The study aims to describe almost all paradigms that serve inflection of the most important categories of proper names: person names (Polish and loan-words), geographical names (Polish and loan-words) and names of significant chains of shopping centres and supermarkets in Poland. At the basis of this research, a database of about 4,500 proper names and their inflection patterns was created, which, according to our knowledge, is not freely available at the moment. The Polish electronic inflectional dictionary [33], used by *Morfeusz* in our study, provides full inflectional paradigms for almost 245,000 Polish uni-word lemmas. It contains about 3,000 first names and 2,000 most popular Polish family names, as well as a sufficient number of toponyms and relational adjectives stemming from proper names.

In the context of natural language processing the quantitative and qualitative importance of proper names and other named entities is unquestionable. They appear in up to 83% of user queries on the Web [43]. They are semantically rich and prove good clues for NLP applications such as document classification [13,8] or information retrieval [5,22]. With MUC [9] and CoNLL [35] campaigns, named entity extraction has become one of the crucial challenges in the NLP community.

Polish initially remained a less studied language within these mainstream efforts. However, some interesting issues have been investigated recently within the Polish NLP community. Automatic lemmatization of proper names using a string distance metric is proposed in [25] and [24]. In [20] this method is applied to automatic recognition of proper names extracted from the corpus of real-life dialogues. The same paper also presents a method of automatic recognition of proper names during a rule-based semantic annotation of dialogue texts with concept names. This approach has been proposed within the European project LUNA (http://ist-luna.eu). One of its focal issues is the behaviour of proper names in Polish spoken dialogues. The syntactic behaviour of names, in particular their variability, has been shown in [27]. It constitutes the main obstacle for direct application of the existing official lists of proper names to their identification in texts. This conclusion is one of the main motivations for the work presented below.

One of the interesting particularities of Slavic languages, such as Polish, is their rich inflectional morphology. As argued in [38], taking them into account in designing language models enhances the universality of those models. Thus, for instance, a multilingual lexical model for proper names proposed in [45] is based in particular on studies of the morphological complexity of Serbian, Polish and Bulgarian [1,19].

Compound proper names belong to a larger class of multi-word units (MWUs). As shown in [14] and [39], the linguistic (orthographic, morphological, syntactic and semantic) variability of MWUs is their essential property in natural language corpora. Moreover, their morpho-syntactic non-compositionality and idiosyncrasy [40] calls for lexicalized models in which compound units are listed and explicitly described. In [38], a comparative study of eleven such approaches (e.g. [15,14,21,2] and [41]) in seven languages was performed. We can expect the *Multiflex* formalism assumed in this paper to be sufficiently expressive for a large-scale description of inflection and variation of compounds.

As shown below, the prerequisite for automatic processing of compound proper names is the morphological treatment of their components. These contain both one-word proper names and common words. The existing morphological analyzers and generators for Polish single words use stochastic or rule-based methods [11]. *Morfeusz* belongs to the latter category.

3   Morpho-Syntactic Behaviour of Polish Multi-Word Proper Names

3.1   The Data

We limit the scope of our research to Polish proper names of the transportation system and public places in Warsaw. Thus, we consider the following types of places: Warsaw administrative units, traffic routes, stopping places, parks and gardens, cemeteries, public institutions and facilities, mansions, monuments, commercial centres, business establishments.

Administrative Units. The city of Warsaw is divided into eighteen districts. The names of the districts consist of one word, however there are two exceptions:

   (1)   *Praga Północ* 'North Praga', *Praga Południe* 'South Praga'

Although the district names describe wide-spread areas, they may appear in names of bus- and tram-stops:

   (2)   *przystanek Bielany, przystanek Śródmieście, przystanek Ursynów*
      'Bielany stop, Śródmieście stop, Ursynów stop'

The districts are further subdivided into smaller areas such as housing estates. Those are usually given traditional Polish one-word or multi-word names (3), however recently foreign names have been given to some modern residential quarters as in (4).

   (3)   *Rakowiec* 'Rakowiec quarter', *Za Żelazną Bramą* 'Behind the Iron Gate'

   (4)   *Villa Nova*

Traffic Routes. The major part of the lexicon of Warsaw transportation system refers to traffic routes. There are specific lexemes which belong to these types of names, such as: *ulica* 'street', *Aleja* 'avenue', *Rondo* 'roundabout', *Plac* 'square, plaza', *Skwer* 'square', *Trasa* 'route, line', *Trakt* 'route, way', *Most* 'bridge', and *Wał* 'wall'. Only the first one – *ulica* – is not capitalised in proper nouns and is usually omitted in spoken language:

   (5)   *ulica Bracka, Bracka* 'Bracka Street'

   (6)   *Aleja Zjednoczenia, *Zjednoczenia* 'Union Avenue'

The routes are named after historical and fictional persons, historical events, fauna, flora and other places or names of different types (e.g. the names of military units). These embedded names may each consist of one or more words, as in:

(7)   *ulica ⟨Fryderyka Chopina⟩* 'Frederic Chopin Street'

(8)   *Aleja ⟨Na Skrapie⟩* 'At Scarp Avenue'

(9)   *Rondo ⟨Generała Charlesa de Gaulle'a⟩* 'General Charles de Gaulle Roundabout'

(10) *Plac ⟨J. W. Wilsona⟩* 'J. W. Wilson Square'

(11) *Trasa ⟨Toruńska⟩* 'Toruń Way', Toruń is the name of a city

(12) *Most ⟨Siekierkowski⟩* 'Siekierki Bridge', Siekierki is the name of a historical district

Stopping Places. Stopping places in the public transportation system, such as bus- and tram-stops, underground and railway stations, and airports, are particularly significant from both quantitative and qualitative points of view, in the context of applications such as information extraction and dialogue systems (cf. section 7). The key-lexemes attributed to this type contain: *Dworzec* 'Station', *przystanek* 'stop' and *Port Lotniczy* 'Airport', as for instance in:

(13) *Dworzec Centralny* 'Central Station'

(14) *Dworzec PKS Warszawa Zachodnia* 'West-Warsaw Bus Station'

(15) *Port Lotniczy im. F. Chopina* 'Frederic Chopin Warsaw Airport'

Most names of stops have three kinds of origin. They can:

– be given after the names of the streets they are located at, as in example (16)
– refer to public or private institutions they are close to, as in example (17)
– combine a name of a street and a name of an institution, as in example (18)

(16) *przystanek Nowy Świat* 'New World bus-stop'

(17) *przystanek CH Arkadia* 'Arkadia Shopping Center bus-stop'

(18) *przystanek Płocka-Szpital* 'Płocka Str.-Hospital bus-stop'

Other Places and Objects. Other names which we consider in our study describe places in the city. Our analysis includes urban places such as: names of parks and gardens (19), cemeteries (20), public institutions and facilities (21), mansions (22), monuments (23), commercial centres (24) and business establishments. For the last ten years a lot of new financial and business centers have been built in Warsaw. Some of them were given foreign—almost exclusively English—names as in example (25).

(19) *Park Skaryszewski im. I. J. Paderewskiego* 'Skaryszew Park of I. J. Paderewski'
      Pola Mokotowskie 'Mokotów Fields'

(20) *Komunalny Cmentarz Północny* 'North Municipal Cemetery'
      Muzułmański Cmentarz Tatarski 'Muslim Tatar Cemetery'

(21) *Muzeum Azji i Pacyfiku* 'Museum of Asia and Pacific'
      Mazowiecki Teatr Muzyczny Operetka Mazovia 'Music Theater Operetta'

(22)   *Pałac na Wodzie* 'Palace on the Isle'
       Zamek Królewski 'Royal Castle'

(23)   *Pomnik Bitwy o Monte Cassino* 'Battle of Monte Cassino Monument'
       Pomnik Braterstwa Broni 'Monument to Brotherhood in Arms'
       Kolumna Zygmunta III Wazy 'Column of Sigismund III Vasa'

(24)   *Galeria Mokotów* 'Mokotów Gallery'
       Złote Tarasy 'Golden Terraces'

(25)   *Blue Point* (office building)
       Millenium Plaza (banking center)

3.2   Linguistic Analysis of the Data

Major proper names have been listed and described for decades within traditional dictionaries and encyclopedias meant for human readers. These data implicitly assume the reader's linguistic and extra-linguistic knowledge. Thus, they are rarely applicable as such to automatic processing of corpora. Construction of machine-usable lexicons requires a formal, disciplined approach in which all relevant properties are explicitly described.

In this section we present an overview of the morpho-syntactic features of urban proper names, as well as usage manners of Polish speakers with respect to these names. Our study is based on the spoken language corpus resulting from the LUNA project (cf. section 2), as well as on our own Warsaw native speakers' life-long experience.

Abbreviations. Polish proper names frequently contain abbreviations, as in examples (26) through (31). They are written in upper or lower case. According to Polish spelling dictionaries the letter case may determine the precise meaning of the abbreviation, e.g. in (26) and (27) it marks the singular or the plural number of 'avenue'.

(26)   *al. Lotników* 'Avenue of Pilots', *al.* = *aleja* 'avenue' (sing.)

(27)   *Al. Jerozolimskie* 'Avenue of Jerusalem', *Al.* = *Aleje*, 'avenue' (pl.)

(28)   *ul. I. J. Paderewskiego* 'I. J. Paderewski Street', *ul.* = *ulica* 'street', *I. J.* = initials for *Ignacy Jan*

(29)   *Port Lotniczy im. F. Chopina* 'Frederic Chopin Warsaw Airport', *im.* = *imienia* 'dedicated to'

(30)   *Rondo Gen. Charles'a De Gaulle'a* 'General Charles De Gaulle Roundabout, *Gen.* = *Generała* 'general' (gen.)

(31)   *ul. Prof. Janusza Groszkowskiego* 'Professor Janusz Groszkowski Street', *Prof.* = *Profesora* 'professor' (gen.)

Acronyms and Initialisms. Acronyms and initialisms are abbreviations formed from initials of the given proper name's constituents. To distinguish between the two terms, we restrict acronyms to pronounceable words formed from the initial letters or syllables, see examples (32)–(34). If each letter is pronounced individually, we speak of initialisms, as in examples (35) and (36). Foreign acronyms and initialisms are sometimes pronounced according to rules of the language of their origin, as in example (37). There are also acronym-initialism hybrids (38).

(32) *ZUS* /zus/ = **Z**akład **U**bezpieczeń **S**połecznych 'Social Security Office'

(33) *Polfa* /'polfa/ = **Pol**ska **Fa**rmacja 'Polish Pharmacy'

(34) *Polmos* /'polmos/ = **Pol**ski **Mo**nopol **S**pirytusowy 'Polish Spirits Monopoly'

(35) *PKP* /peka'pe/ = **P**olskie **K**oleje **P**aństwowe 'Polish State Railways'

(36) *ONZ* /oen'zet/ = **O**rganizacja **N**arodów **Z**jednoczonych 'United Nations Organization'

(37) *BBC* /bibi'si/ = **B**ritish **B**roadcasting **C**orporation

(38) *SGPiS* /esgie'pis/ = **S**zkoła **G**łówna **P**lanowania *i* **S**tatystyki 'Main School of Planning and Statistics'

The distinction between these types of abbreviations is relevant to inflection. The majority of acronyms and acronym-initialism hybrids in Polish inflect according to inanimate nominal patterns, see example (39). Initialisms may or may not inflect, depending on how well their ultima fits into the system of nominal morphological endings in Polish. For instance, the ultima of example (35) is /-pe/, which is a rather unusual Polish nominal ending, thus the whole initialism remains uninflected as presented in example (40). On the contrary, the ultima of example (36) is /-et/, which is common for many masculine inanimate nouns, e.g. *bilet* 'ticket'. Therefore, it undergoes inflection according to the same pattern, see example (41).

(39) *ZUS* (nom.), *ZUS-u* (gen.), *ZUS-owi* (dat.), etc.

(40) *PKP* (nom., gen., dat., etc.)

(41) *ONZ* (nom.), *ONZ-u* (gen.), *ONZ-owi* (dat.), etc.
      bilet (nom.), *biletu* (gen.), *biletowi* (dat.), etc.

Numerals. Numbers in Polish proper names, such as shown in examples (42) through (45), are usually represented with Arabic or Roman digits. Thus, a particular description is needed in order to link digital representations of numbers with the numerals they refer to.

Morphologically, numbers are divided into cardinal and ordinal numerals. These two classes show essential differences in morphological and syntactic behaviour in Polish. Cardinal numbers constitute a morphological class on their own, with rather complex morpho-syntactic properties as discussed in [31], [34] and [42]. Ordinal numbers show adjectival inflection, and there are no morphological hints allowing their distinction from regular adjectives.

(42) *ul. II Armii Wojska Polskiego* 'The Second Polish Army Street'

(43) *ul. IX Poprzeczna* 'The Ninth Transversal Street'

(44) *ul. 36 Pułku Piechoty Legii Akademickiej* 'The Thirty Sixth Regiment of Academic Infantry Street'

(45) *ul. Posag 7 Panien* 'Dowry of Seven Maids Str.'

Synonyms and Variations. Similarly to other multi-word units, compound proper names are subject to inflectional, syntactic and semantic variation. In this sense synonyms are defined here as different textual sequences referring to the same name.

Inflectional variants rely most often on inflectional patterns of their constituents, as well as on syntagmatic agreement and government rules. For instance, in order to generate all possible inflected forms of example (15) we need to be able to inflect its two first lexemes *port* and *lotniczy*, and to know that they agree in gender, number and case, while the other constituents do not vary:

(46) *Portu Lotniczego im. F. Chopina, Porcie Lotniczym im. F. Chopina*, etc. (gen., loc., etc.)

Syntactic variants result most often from ellipses. It is due to the fact that full official names may be rather long and impractical in everyday language. Components that are most frequently omitted are:

- the *ulica* 'street' keyword, systematically missing in spoken language, as in example (47).
- "non-essential" complements, as in examples (48) and (49). Proper understanding of such shortnames implies one's extra-linguistic knowledge of the city.
- first names, as in example (50).
- personal titles such as *generał, profesor*, etc., as in example (51).

(47) *ulica Chmielna, Chmielna* 'Chmielna Str.'
      ulica Humanistów, Humanistów 'Humanistów Str.'

(48) *ulica Bitwy Warszawskiej 1920 r.* 'Warsaw Battle 1920 Str.'
      ulica Bitwy Warszawskiej 'Warsaw Battle Str.'
      ulica Bitwy 'Battle Str.'

(49) *Rondo Zgrupowania AK „Radosław"* '*Radosław* Division of the Domestic Army Roundabout'
      Rondo Zgrupowania „Radosław" '*Radosław* Division Roundabout'
      Rondo Radosława 'Radosław's Roundabout'

(50) *ulica Władysława Broniewskiego* 'Władysław Broniewski Str.'
      ulica Broniewskiego
      Broniewskiego

(51) *ulica Gen. Antoniego Józefa Madalińskiego* 'General A. J. Madaliński Str.'
      ulica Antoniego Józefa Madalińskiego
      ulica Antoniego Madalińskiego
      ulica Madalińskiego
      Madalińskiego

The phenomenon of systematic *semantic variation* can be presently observed within Warsaw urban toponyms. In the early 90's many names of places referring to the previous socialist period of the Polish history have been eliminated. Although the old names frequently suffer from negative connotations, they continue to be used, in particular by elderly people who have problems with remembering their recent substitutes. Thus, many diachronic synonyms co-exist, as in:

(52)   *Al. Jana Pawła II* 'John-Paul II Avenue'
        previously: *Al. Juliana Marchlewskiego* 'Julian Marchlewski Avenue'

(53)   *Al. Solidarności* 'Avenue of Solidarity'
        previously: *Al. Karola Świerczewskiego* 'Karol Świerczewski Avenue'

In present days renaming places in Warsaw continues, even if the old names bear no negative connotation. Warsaw residents prove reluctant to integrate such new names into their every day vocabulary. For instance, in example (54) the former—official—name is less frequently used than the later—traditional one. In (55) the former name refers to the street whose part has been recently given the later name.

(54)   *Rondo Zgrupowania AK „Radosław"* 'Radosław Division of the Domestic Army Roundabout'
        previously: *Rondo Babka* 'Babka Roundabout'

(55)   *ul. Jana Rosoła* 'Jan Rosół Str.'
        al. Jana „Anody" Rodowicza 'Jan "Anoda" Rodowicz Avenue'

Warsaw citizens sometimes invent their own popular names for places or buildings. It so happens that those informal names are better known than the official ones. The possible sources of such semantic synonymy include the following phenomena:

- If the official name is a periphrasis, it can be substituted by its base word if no confusion arises, as in example (56).
- If the shape or physical feature of a monument or building brings some associations, it may be given a descriptive synonym. For example, the monument mentioned in example (57) represents four statues of soldiers standing still with their heads facing the ground. Consequently, they are usually called the 'Four Sleeping Ones'.
- Colour can play a role in naming. Example (58) mentions a high-rise building in Warsaw which should have been of the golden colour according to the initial architectural plans. The final effect, however, is of the shiny glass-blue colour. Thus, the former official name was quickly changed but three versions of the name co-exist.
- Irony may be a source of linguistic creativity. A recent language joke concerns the roundabout mentioned in example (30), which was decorated several years ago with an artificial 15-meter-high palm tree, rather unnatural for the Warsaw landscape. Warsaw residents reacted to this search for exotic effects by inventing the ironic name, shown in example (59).

(56) *aleja Prymasa Tysiąclecia* 'Avenue of the Millenium Cardinal'
informally: *aleja Wyszyńskiego* 'Wyszyński Ave.'

(57) *Pomnik Braterstwa Broni* 'Monument to Brotherhood in Arms'
informally: *Pomnik Czterech Śpiących* 'Monument of the Four Sleeping'
rarely: *Pomnik Czterech Smutnych* 'Monument of the Four Sad'

(58) *Błękitny Wieżowiec* 'Blue Building'
initially: *Złoty Wieżowiec* 'Golden Building'
informally: *Srebrny Wieżowiec* 'Silver Building'

(59) *Rondo De Gaulle'a* 'De Gaulle Roundabout'
informally: *Rondo de Palma* 'Palm-tree Roundabout'

4   Formalisms and Tools

High-quality automatic processing of urban multi-word proper names requires
the precise and exhaustive description of their morpho-syntactic properties addressed in the previous sections. Our response to this need is fully lexicalized.
Proper names are listed, and their inflected forms and variants are explicitly
described on a two-layer basis. First, the inflectional morphology of single words
is treated by *Morfeusz*, a general-purpose morphological analyser and generator,
containing a large-scale lexicon of Polish. Second, a cross-language multi-word
morpho-syntactic generator *Multiflex* is used to express the behaviour of compounds in the light of their constituents' properties, as well as of some syntagmatic patterns and idiosyncratic rules. The tools have been developed independently. They have been recently integrated within a common framework, whose
first experimental application is the proper name description project addressed
in this paper.

In this section we present the formalisms implemented in the resulting platform, and show how they can be applied to proper names. We also discuss the
interoperability facilities which were necessary for both tools to cooperate.

4.1   Morfeusz

Morfeusz is a morphological analyser of Polish available at `http://nlp.ipipan.`
`waw.pl/~wolinski/morfeusz`. The program (or to be more precise its current
version called *Morfeusz SIaT*) is based on Jan Tokarski's lemmatization rules
for Polish [44]. The set of lemmatisation rules is large (c.a. 20,000) but the
rules are generic in the sense that they capture all grammatical forms possible
in the language, but do not give information which of the possibilities is actually
realized for any given lexeme [47]. This causes the analyser to overgenerate
interpretations.

A new version of the program, named *Morfeusz SGJP*, is under development.
This new version is based on an inflectional dictionary of Polish [33]. The
dictionary contains exact information on inflection for almost 245,000 Polish
lexemes, including several thousand proper names, which correspond to about
4,000,000 different textual words.

For the present work, a generating module has been added to *Morfeusz SGJP* which is able to provide all inflected forms for any given lexeme. The interface between this module and Multiflex is discussed in section 4.3.

The tagset used by *Morfeusz* was designed at the Institute of Computer Science, Polish Academy of Sciences (pol. IPI PAN) [46,26]. The design and the repertoire of morphosyntactic classes and categories assumed in the IPI PAN tagset is based on the rich body of work on Polish morphosyntax by Zygmunt Saloni and his colleagues (e.g., [29,30,31,4,3,44,32,33]).

The IPI PAN tagset used by *Morfeusz* is based on a homogeneous set of morphological, morphosyntactic and (as secondary) syntactic criteria. The main criteria for delimiting grammatical classes are morphological (how a given class inflects; e.g., nouns inflect for case, but not for gender) and morphosyntactic (in which categories it agrees with other classes; e.g., Polish nouns agree in gender with adjectives and verbs). Such an approach leads to more detailed classes than traditional parts of speech (POS); we call such classes *flexemic classes*. Some of these flexemic classes correspond rather directly to the traditional parts of speech, e.g., noun, adjective, adverb, preposition, conjunction, particle; others are more fine grained, e.g., various verbal classes such as infinitive, four classes for the four participial forms (two adjectival and two adverbial), non-past verb, impersonal *-no/-to* form, imperative, l-participle (word-forms like *przysz-l-i*), gerund. The traditional classes of numerals and pronouns, usually defined on the basis of their semantics, are redefined in purely morphological and morphosyntactic terms: the flexemic class of numerals only consists of cardinal numerals, and three specific classes for nominal pronouns are distinguished: non-3rd person pronoun, 3rd person pronoun, and the pronoun *siebie*. Other flexemic classes are: depreciative noun, ad-adjectival adjective, post-prepositional adjective, future *być*, agglutinative *być* (forms like *-śmy*), *winien*-like verb, predicative, alien nominal/other. Grammatical categories assumed in the tagset include traditional categories such as number, case, person, degree, aspect, gender, as well as more restricted categories. Some of them were first introduced in the work of Jan Tokarski and Zygmunt Saloni [44], such as accentability, post-prepositionality, acommodability, agglutination, vocability and negativity.

The following examples should give the impression of the granularity and the positional nature of the tagset:

- *ulicami*: ulica 'street', subst:pl:inst:f — noun, plural number, instrumental case, feminine gender;
- *przyszedł*: przyjść 'come', praet:sg:m1.m2.m3:perf — l-participle, singular number, one of the three masculine genders (i.e., in fact, this is an abbreviation for three different tags), perfective aspect;
- *czterech*: cztery 'four', num:pl:nom.voc:m1:rec|num:pl:gen.loc:m1.m2. m3.n2.f:congr|num:pl:acc:m1:congr — three separate tags are possible: the first one nominative or vocative in masculine personal; the second one genitive or locative in any of masculine subgenders, neuter impersonal, or feminine; the third one accusative in masculine personal. The first tag bears the value rec of accomodability while the other two have the value congr (see section 5.1 for the meaning of these values).

4.2 Multiflex

Multiflex [36,37] is a cross-language morpho-syntactic generator of multi-word units. It relies on a graph-based formalism in which all inflected forms of a compound and their variants are described within one graph. Compounds having the same morpho-syntactic behaviour are assigned to the same graph. Each path in the graph describes one or more inflected forms and variants. A unification mechanism allows one to account for agreement within constituents, and to represent huge inflection paradigms compactly.

Description on the Language Level. Multiflex formalism adapts to the given natural language by encoding its morphological model, i.e. the list of all inflectional categories (number, gender, case, person, etc.) and their corresponding values (e.g. singular and plural for number, feminine, masculine and neuter for gender, etc.), as well as the existing inflectional classes (noun, verb, adjective, etc.). Each class needs information on the categories it inflects for and those that are relevant to it but have constant values. For instance, Polish nouns usually inflect for number and case, and they have a fixed gender.

Fig. 1 shows the encoding of the Polish morphological model according to the IPI PAN tagset, as explained in section 4.1. As it can be easily seen, this model admits two numbers (*Nb*), seven cases (*Case*), nine genders (*Gen*), three persons (*Pers*), three degrees (*Deg*), etc. Adjectives (*adj*) are supposed to inflect for number, case, gender and degree. Cardinal numerals (*num*) should have a fixed number and inflect for case, gender and accomodability (agreeing as in *dwaj*, or governing as in *dwóch*). The whole model includes 12 inflectional categories, 38 values and 32 classes.

```
Polish
⟨CATEGORIES⟩
Nb:              sg , pl
Case:            nom, gen, dat, acc, inst, loc, voc
Gen:             m1, m2, m3, f, n1, n2, p1, p2, p3
Pers:            pri, sec, ter
Deg:             pos, comp, sup
Asp:             imperf, perf
...
⟨CLASSES⟩
subst:           (Nb,⟨var⟩),(Case,⟨var⟩),(Gen,⟨fixed⟩)
num:             (Nb,⟨fixed⟩),(Case,⟨var⟩),(Gen,⟨var⟩),(Accom,⟨var⟩)
numcol:          (Nb,⟨fixed⟩),(Case,⟨var⟩),(Gen,⟨fixed⟩),(Accom,⟨var⟩)
adj:             (Nb,⟨var⟩),(Case,⟨var⟩),(Gen,⟨var⟩),(Deg,⟨var⟩)
adja:
...
```

Fig. 1. Morphological model of Polish in *Multiflex*

Segmentation of a Compound Lemma. The morpho-syntactic description implemented in *Multiflex* relies on the assumption that it is possible to detect boundaries between individual components within a compound. How a single component (token) is defined depends on the underlying morphological module for the single words *Multiflex* works with. In this study, the *Morfeusz*-like token definition is admitted. A lexical token is one of the following:

- a contiguous sequence of the alphabet characters between two non alphabet characters (e.g. *drzewo* is one token)
- a sub-sequence appearing within a contiguous sequence of the alphabet characters, in a closed list of well defined cases (e.g. *kiedy|śmy* contains two tokens)
- several contiguous sequences of alphabet and non alphabet characters, in a closed list of well defined cases (e.g. *ping-pong, d'Arc*)
- a single non alphabet character such as a digit or a punctuation mark (space, hyphen, apostrophe, dot, etc.)

Components of a compound lemma are numbered, which allows for referring to them while describing the compound inflected forms and their variants. For instance, Fig. 2 shows the division of compound proper names (29) and (49) into single components many of which are spaces, dots and quotes.

Port ⟨blank⟩ Lotniczy ⟨blank⟩ im . ⟨blank⟩ F . ⟨blank⟩ Chopina
$1 $2 $3 $4 $5 $6 $7 $8 $9 $10 $11

Rondo ⟨blank⟩ Zgrupowania ⟨blank⟩ AK ⟨blank⟩ „ Radosław "
 $1 $2 $3 $4 $5 $6 $7 $8 $9

Fig. 2. Numbering single components within compounds in *Multiflex*

Annotating Components within a Compound. Naturally enough, the inflection paradigm of a compound lemma, according to *Multiflex*, consists mostly of pointing at which inflected form of a compound requires which transformations of its single components. For instance, in order to obtain the locative forms of the following compound toponyms in Warsaw:

(60)   *Most Grota-Roweckiego* 'Grot-Rowecki bridge'

(61)   *Aleje Jerozolimskie* 'Avenue of Jerusalem', pl.

we have to inflect their head nouns *Most* 'bridge' and *Aleje* 'avenues' for the locative while keeping their number unchanged (singular and plural, respectively). All remaining constituents in example (60) do not change. In example (61) the adjective *Jerozolimskie* is subject to agreement rules with the head noun *Aleje*, i.e. it gets inflected for the locative with no change in number. The results are:

(62)   *Moście Grota-Roweckiego*

(63)   *Alejach Jerozolimskich*

Thus, in order to inflect compounds we need to be able to analyse single words morphologically, as well as to synthesize desired forms. That is where *Multiflex* depends on a morphological module for simple words, e.g. *Morfeusz*. It is necessary to annotate each single component of a compound that is subject to inflection.[3] Note that components of a compound lemma do not necessarily represent lemmas themselves (e.g. *Aleje* is not a lemma in example (61)), therefore they need to be annotated with their own lemmas, accompanied by their inflectional values as in a *Morfeusz* tag (cf. [20]). Whenever a lemma is morphologically ambiguous, a cardinal number allows the indication of the desired homonym. For instance, the following *Morfeusz* annotation is attributed to the component *Aleje* within the compound (61):

(64)   aleja.0:subst:pl:nom:f
       'lemma *aleja*, first homonym, noun, plural, nominative, feminine'

which can be graphically represented as in Fig. 3.

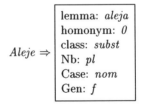

Fig. 3. A possible annotation of the single component *Aleje*

Similarly, processing all possibly inflected components in the compound lemmas (60) and (61) results in annotations shown in figures 4 and 5.

The form and meaning of the inflection codes *NC-CXXXX* and *NC-CXC* attached to the compound entries are explained in the following subsection.

Inflection Graphs for Compounds. The *Multiflex* graph-based formalism is meant for precise and exhaustive description of the morphological behaviour of compounds, i.e. each correct form must be described and no incorrect form can be admitted. Each compound is assigned a graph identifier, as *NC-CXXXX* and *NC-CXC*[4] in figures 4 and 5. Each path in a graph contains zero or more boxes,

---

[3] Such a 'two-layer' description of compounds, first on the level of single words, then on the level of their sequences, is the most commonly assumed in related works, however some approaches see compounds as plain sequences of characters—cf. [38].

[4] Names of the inflection graphs are arbitrary and follow a convention allowing easy navigation in already existing graphs. Here *NC* stands for *Noun-Compound*, *X* stands for an uninflected unit, and *C* for a possibly inflected one.

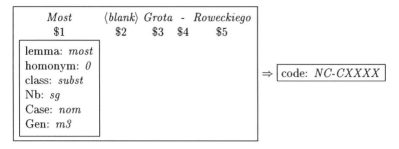

Fig. 4. Lemma annotation for *Most Grota-Roweckiego*

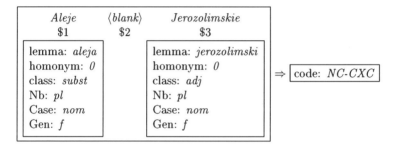

Fig. 5. Lemma annotation for *Aleje Jerozolimskie*

and describes one or more inflected forms or variants of a compound. The path begins with the leftmost arrow and ends with the rightmost encircled box. Morphological descriptions appearing inside the boxes refer to single constituents, while those appearing under the boxes refer to the whole compound.

For instance, Fig. 6 represents the inflection graph *NC-CXXXX* for compound (60). It contains a unique path referring to five constituents of the compound lemma. If a constituent number appears in a box with no morphological description, as in the case of *$2* through *$5*, then it should be left unchanged. However, if it is accompanied by one or more category-value equations, then the constituent (here *$1*) must be inflected for the desired form. ·

The right-hand side of a category-value equation may have one of the three forms:

– A *constant value.* It is taken from the domain of the corresponding category, as specified in the language model in Fig. 1, as in *Nb=pl; Gen=f;*

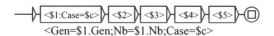

Fig. 6. Inflection graph *NC-CXXXX* for compound inflecting like *Most Grota-Roweckiego*

Case=acc. It indicates that the current component has to be inflected for the corresponding form.

- A *unification variable.* It is introduced by the *$* sign, as in *Nb=$n; Gen=$g; Case=$c.* It may take any value from the corresponding category domain, unless it is limited by unification constraints with respect to other constituents (see Fig. 8). For instance, in *Case=$c* the variable *$c* may potentially take values *nom* through *voc.*
- An *inherited value.* It is introduced by a constituent number followed by a category, e.g. *Nb=$1.Nb; Gen=$2.Gen; Case=$5.Case.* It is fixed and equal to the value which the corresponding constituent has in the compound lemma. For instance, *Gen=$1.Gen* means that the gender must be the same as the gender of the first constituent in the lemma.

For instance, when the graph in Fig. 6, is applied to the annotated lemma in Fig. 4 the first constituent *$1, Most,* inflects freely for case (see *Case=$c* in the first box). The whole compound obtains the same case value as the first constituent takes in each particular form during inflection (see *Case=$c* under the path). Gender and number values of the compound are equal to those possessed by the first constituent in the compound lemma (see *Gen=$1.Gen; Nb=$1.Nb* under the path), here *m3* and *sg,* as in the annotation in Fig. 4.

Complete exploration of a graph consists in following each path as many times as there are possible values for all the unification variables present on the path. Here, the unique path is explored seven times (variable *$c* may take seven values), which results in seven compound inflected forms, annotated with their lemma and *Morfeusz*-like tags, as shown in Fig. 7.

Most Grota-Roweckiego, Most Grota-Roweckiego:subst:m3:sg:nom
Mostu Grota-Roweckiego, Most Grota-Roweckiego:subst:m3:sg:gen
Mostowi Grota-Roweckiego, Most Grota-Roweckiego:subst:m3:sg:dat
Most Grota-Roweckiego, Most Grota-Roweckiego:subst:m3:sg:acc
Mostem Grota-Roweckiego, Most Grota-Roweckiego:subst:m3:sg:inst
Moście Grota-Roweckiego, Most Grota-Roweckiego:subst:m3:sg:loc
Moście Grota-Roweckiego, Most Grota-Roweckiego:subst:m3:sg:voc

Fig. 7. Annotated inflected forms of *Most Grota-Roweckiego*

Unification variables not only allow to freely inflect a component for a certain category, but to express agreement and government rules as well. For instance, the graph in Fig. 8 contains the same unification variable *$c* for both the first and the third constituent. The variable may take any of the seven cases provided that each time it is identical for these two constituents. Thus, instead of $7 \times 7 = 49$ different form combinations we obtain only seven correct forms, one of which has two variants, which results in eight forms, as shown in Fig. 9.

Note that the inherited values allow the graph to be independent of the inflectional features of its head. For instance, in Fig. 8 the equations *Gen=*

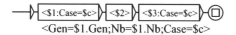

$$\langle Gen=\$1.Gen;Nb=\$1.Nb;Case=\$c\rangle$$

Fig. 8. Inflection graph *NC-CXC* for compounds inflecting like *Aleje Jerozolimskie*

Aleje Jerozolimskie, Aleje Jerozolimskie:subst:f:pl:nom
Alei Jerozolimskich, Aleje Jerozolimskie:subst:f:pl:gen
Alej Jerozolimskich, Aleje Jerozolimskie:subst:f:pl:gen
Alejom Jerozolimskim, Aleje Jerozolimskie:subst:f:pl:dat
Aleje Jerozolimskie, Aleje Jerozolimskie:subst:f:pl:acc
Alejami Jerozolimskimi, Aleje Jerozolimskie:subst:f:pl:inst
Alejach Jerozolimskich, Aleje Jerozolimskie:subst:f:pl:loc
Aleje Jerozolimskie, Aleje Jerozolimskie:subst:f:pl:voc

Fig. 9. Annotated inflected forms of *Aleje Jerozolimskie*

$1.Gen; Nb=$1.Nb allow the graph to be assigned not only to feminine plural compounds as in (61), but also to masculine/neuter and singular ones, as in Fig. 10.

Bypassing Components. As mentioned in section 3.2 urban toponyms show a high degree of variability. They tend in particular to be more frequently used in their abbreviated forms than in their full official forms. Omitting a component is easy to express in a graph by bypassing the appropriate box.[5]

For instance, recall that example (49) admits at least three morpho-syntactic variants, all of which can be inflected for case. The corresponding annotation and inflection graph are shown in Fig. 11. The lowest path describes the official full version of the toponym, *Rondo Zgrupowania AK „Radosław"*, where all nine components are present, and where only the initial head noun *Rondo* inflects for case.[6] The middle path corresponds to the slightly abbreviated variant, *Rondo Zgrupowania „Radosław"*, with the fifth component, i.e. the *AK* acronym, and the following blank space missing. The uppermost path triggers the shortest (and most commonly used) version, *Rondo Radosława*, where *Radosław* shifts into the head's complement position and takes the genitive case.

4.3   Interoperability and Collaborative Framework

As shown in the two previous sections, the morpho-syntactic description of compounds in *Multiflex* is based on a 'two-layer' approach. Single words are described first, then each inflected compound form is seen roughly as a particular

---

[5] Inserting a new component is also natural enough to be expressed within a graph, however we have not run across a toponym example requiring this feature.

[6] Inflection for number is a controversial question for toponyms as they usually name unique objects. However, an exceptional plural usage is acceptable as in: *Twierdzisz, że w Warszawie istnieją dwa Ronda Radosława?* 'Do you claim that there are two Radoslaw Roundabouts in Warsaw?'

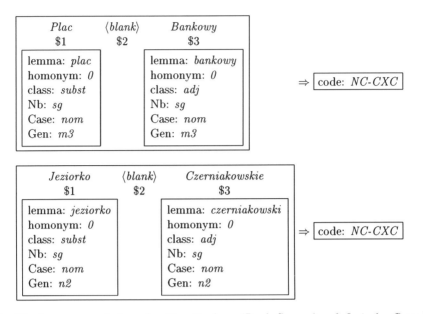

Fig. 10. Lemma annotations for *Plac Bankowy* 'Bank Square' and *Jeziorko Czernia-kowskie* 'Czerniakowskie Lake'

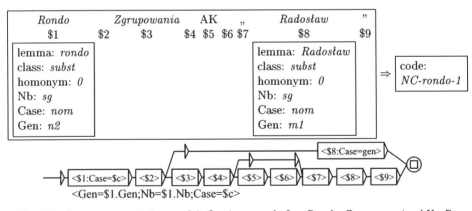

Fig. 11. Lemma annotation and inflection graph for *Rondo Zgrupowania AK „Ra-dosław"*

combination of the inflected forms of its components. If the combination is regular (i.e. respects "typical" syntagm patterns and agreement rules), the corresponding graph is shared by a large class of compounds (as *NC-CXC* in Fig. 5 and Fig. 10. Conversely, if it is idiosyncratic, the graph applies to isolated compounds (as *NC-rondo-1* in Fig. 11).

Multiflex was initially developed within a large-scale multilingual lexicographic project based on the volume [7] and supported by the *Unitex* system [23]. However, numerous other morphological systems for single words have been developed for many languages. Therefore, for the reasons of interoperability, *Multiflex* does not impose its own model but is meant to cooperate with any other external morphological module for single words, hereafter called the *underlying module*, as long as the following interface constraints are observed.

Common Morphological Model. The underlying module and *Multiflex* must share the same morphological model. The present model for Polish, common for *Morfeusz* and *Multiflex*, is described in Fig. 1. It represents the inflectional morphology, both of single and compound words, as a group of categories (number, gender, case, etc.), each of which admits a list of values (singular, plural, feminine, etc.). *Multiflex* interprets this model in the sense of a Cartesian product, i.e. if an inflectional class (noun, adjective, etc.) inflects for a number of categories, then, implicitly, each value of one category combines with each value of another category. For instance, if a noun inflects for number and case, it admits at least $2 \times 7 = 14$ inflected forms. Exceptions to this rule must either be controlled by the underlying module (no form is generated for a particular combination of features), or be explicitly mentioned in the corresponding inflection graphs.

However, some inflectional phenomena, particularly in Slavic languages, are not necessarily of the nature of a Cartesian product. For instance, in the model for Serbian in [16], some particular value combinations are systematically prohibited, others are modelled by 'no-care' values. With *Multiflex* applied to this model in [17], such constraints cannot be expressed at the most general language level, but have to be mentioned for each individual inflection graph concerned.

Recognition of Token Boundaries. The underlying module should provide a clear-cut definition of token boundaries. As shown in [38, sec. 4.1], the distinction between single words and compounds on the purely graphical level is controversial. The definition of an indivisible graphical item may or may not:

- allow token boundaries to occur within contiguous sequences of letters; for instance, *Białystok* 'literally: White Slope = a city name' contains an adjective and a noun with no separating space, however, both components inflect as in a typical *Adjective Noun* syntagm, *Białego/stok̲u̲, Białym/stok̲i̲e̲m̲*, etc.,
- admit separators within a token, e.g. *aujourd'hui* 'today' in French may be seen either as a single word or as two words separated by an apostrophe,
- view sequences of non-alphabet characters as single tokens or as sequences of tokens; e.g. a numeral such as *2008* may be seen as one or four tokens.

Here again, *Multiflex* does not impose its own view on a token boundary but adapts to the underlying module via an interface function. With *Morfeusz*, an indivisible token is defined as in section 4.2 where segmentation of a compound lemma is discussed.

Generation of Particular Inflected Forms on Demand. Given a lemma and a morphological tag, the underlying module should generate all inflected forms of the lemma corresponding to the tag.

For a given lemma, a homonym number and an inflectional class, the generating module of *Morfeusz* produces the set of all inflected forms of the word together with their annotations, as in Fig. 12. Choosing the desired form within this set is done by a matching function which solves possible factorizations and alternatives, as in the case of the form *czterech* explained in section 4.1.

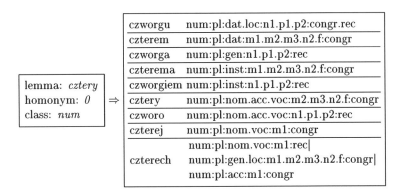

Fig. 12. Annotated inflected forms of *cztery*

As a result, we obtain a process which is roughly opposite to annotation. For instance, for the same lemma and morphological tag as in Fig. 3, the generated form is *aleje*, as shown on the left-hand side of Fig. 13. Note, however, that the form generated for a particular morphological tag is not necessarily unique because of variants, as seen in the middle of the same figure.

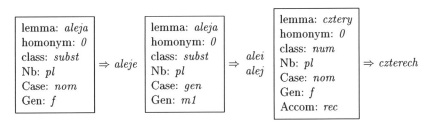

Fig. 13. Generating inflected forms of *aleja* and *cztery*

5   Some Interesting Urban Toponyms

5.1   Numerals Within Polish Compounds

Polish ordinal numerals show rather complex morpho-syntactic properties, as discussed largely in [31] and [42]. In particular, in some genders they inflect for accomodability, i.e. in the agreeing form (*cong*) they agree with the nouns they modify, while in the governing form (*rec*) they require the noun to appear in genitive. Governing forms admit all seven cases in neutral, but only two cases in masculine human. Moreover, numerals *jeden* 'one' through *cztery* 'four' behave different from all others.

Before this behaviour finds an elegant model at the language level, it needs to be described within inflection graphs for compounds containing numerals. Recall, for instance, example (57) mentioning the popular synonym of the Monument to Brotherhood in Arms:

(65) $[Pomnik_{nom}$ $Czterech_{gen}$ $Śpiących_{gen}]_{nom}$
          'Monument of Four   Sleeping'

It belongs to the class of type *Noun [Num Noun]$_{gen}$* whose first noun *Pomnik* 'monument' holds the head position and is the only one to inflect. This fact is represented by the uppermost path in the graph in Fig. 14. The same compound admits an elliptical variant which may appear either in the agreeing or in the governing form:

(66) $[Czterej_{nom.congr}$ $Śpiący_{nom}]_{nom}$
          '[The] Four   Sleeping'

          $[Czterech_{nom.rec}$ $Śpiących_{gen}]_{nom}$
          '[The] Four   Sleeping'

In the former case the numeral and the noun agree in case, which is represented by the middle path in Fig. 14, where the unification variable c is common for the third and the fifth[7] component (*Czterech* and *Śpiących*). In the latter case the numeral imposes the genitive gender on the noun, that is why in the lowermost path in Fig. 14 the fifth component gender is assigned the constant *gen*.

5.2   Embedded Compounds

The *Multiflex* development project, at its present stage, is dedicated to the optimal handling of embedded compounds, whose quantitative importance is particularly significant within urban and institutional proper names. Note that in previous examples, compounds are assigned to flat representations, i.e. their internal syntactic structures are neglected. In compounds containing numerous components this approach may lead to considerable complexity in inflectional

---

[7] Recall that the components are numbered with respect to their position in the compound lemma, which is not necessarily the same as in the inflected form.

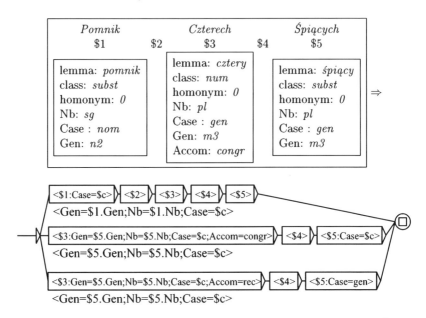

Fig. 14. Lemma annotation and inflection graph for *Pomnik Czterech Śpiących*

graphs and to some degree of redundancy within the pattern description. Consider examples (67) and (68) allowing for a considerable number of elliptical variants (each of them inflected), represented by graphs in Fig. 15 and Fig. 16, respectively.

(67)   *ulica Marii Skłodowskiej-Curie* 'Maria Skłodowska-Curie Street';
       ulica Marii Skłodowskiej; *ulica Marii Curie*; *ulica Skłodowskiej-Curie*;
       ulica Skłodowskiej; *Marii Skłodowskiej-Curie*; *Marii Skłodowskiej*, etc.

(68)   *XXIII Liceum Ogólnokształcące im. Marii Skłodowskiej-Curie*
       'Maria Skłodowska-Curie 23rd High School'
       XXIII Liceum Ogólnokształcące im. Marii Skłodowskiej
       XXIII Liceum Ogólnokształcące im. Marii Curie
       XXIII Liceum Ogólnokształcące im. Skłodowskiej-Curie
       XXIII Liceum Ogólnokształcące im. Skłodowskiej
       XXIII Liceum Ogólnokształcące
       XXIII Liceum im. Marii Skłodowskiej-Curie
       Liceum Ogólnokształcące im. Marii Skłodowskiej-Curie
       Liceum im. Marii Skłodowskiej-Curie
       Liceum Marii Skłodowskiej-Curie, etc.

The high number of these variants results in particular from the fact that the underlying person name *Maria Skłodowska-Curie* admits five variants on its own. In graphs in Fig. 15 and Fig. 16 these variants have to be described independently (by combinations of the five rightmost boxes in each graph). Thus, the morphosyntactic behaviour of *Maria Skłodowska-Curie* is represented redundantly each time a compound contains this person name.

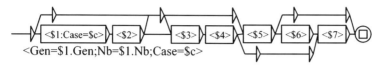

Fig. 15. A flat description of *ulica Marii Skłodowskiej-Curie*

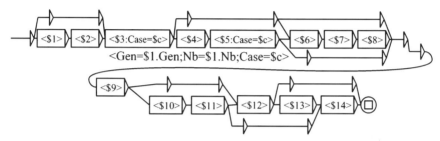

Fig. 16. A flat description of *Liceum Ogólnolsztalcące im. Marii Skłodowskiej-Curie*

Rather than describing such compounds as flat structures, it is more elegant to have a unique representation of each person and institution name, such as in Fig. 17 and Fig. 18. Then, compounds (67) and (68) are described as syntactically simpler structures of types *Noun Noun_{gen}* and *Noun im. Noun_{gen}*, in which the first and the last components may be compounds on their own, as shown in Fig. 19 and Fig. 20. Thus, the same graph can be applied to compounds with a different number of components. For instance, the graph in Fig. 19 describes not only example (67), but (50) as well, while the graph in Fig. 20 applies both to example (68) and to (29).

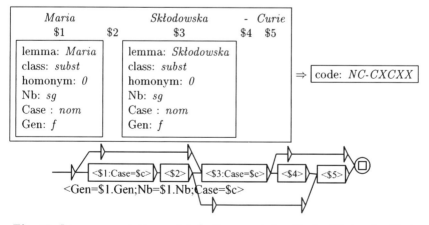

Fig. 17. Lemma annotation and inflection graph for *Maria Skłodowska-Curie*

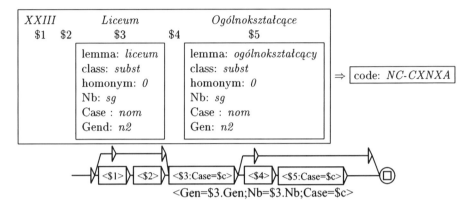

Fig. 18. Lemma annotation and inflection graph for *XXIII Liceum Ogólnokształcące*

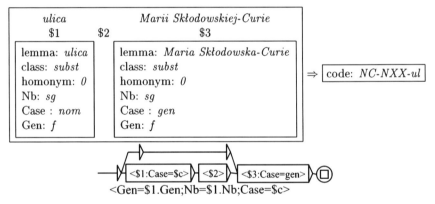

Fig. 19. Embedded lemma annotation and inflection graph for *ulica Marii Skłodow-skiej-Curie*

5.3 Extreme Elliptical Variants

The wide-spread tendency of urban toponyms to appear in elliptical variants has been discussed in the preceding sections. When the headword of a compound is omitted, as in example (66), the remaining complement usually shifts to the head position, possibly changing its gender and number, as in:

(69)   *Pomnik$_{nom.sg}$ [Czterech Śpiących]$_{gen.pl}$ [znajduje się]$_{sg}$ na Placu Wileńskim.*
'The monument of the Four Sleeping is located in Vilnius Square.'
[Czterej Śpiący]$_{nom.pl}$ [znajdują się]$_{pl}$ na Placu Wileńskim.
'The Four Sleeping are located in Vilnius Square.'

As shown in the inflection graph in Fig. 19 this 'shifting rule' is not respected in street names of type *ulica Noun$_{gen}$*. The headword *ulica* 'street' is rarely used, particularly in spoken language, however, the remaining complement keeps its original genitive form, as in the following example:

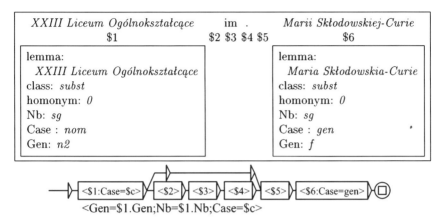

Fig. 20. Embedded lemma annotation and inflection graph for *XXIII Liceum Ogól-nokształcące im. Marii Skłodowskiej-Curie*

(70) *Ulica_{nom.fem} [Kazimierza Pułaskiego]_{gen.masc} jest zatłoczona_{fem}.*
  'Kazimierz Pułaski Street is crowded.'
  Kazimierza Pułaskiego jest zatłoczona_{fem}.
  Pułaskiego jest zatłoczona_{fem}.

(71) *Pułaskiego, Pułaski:subst:m1:sg:gen*
  Pułaskiego, ulica Kazimierza Pułaskiego:subst:f:sg:nom

This produces the surprising effect of a double morphological interpretation in the last two sentences. The sequences *Kazimierza Pułaskiego* and *Pułaskiego* as such can only be analysed as genitive singular masculine forms. However, as subjects in a sentence, they are necessarily in nominative singular feminine. In other words, depending on which lemma is assigned to the form *Pułaskiego*, it may have different morphological features, as shown in example 71.

6 Remaining Problems

We believe that the morphosyntactic framework presented above allows, in its present state, to initiate a large-scale description of Polish proper names, in particular urban toponyms. Nevertheless, we are aware of the fact that some remaining problems still await solutions.

6.1 Conflating Derivational and Semantic Variants

For applications such as information retrieval, it is crucial to be able to conflate all kinds of variants that a proper name may admit. In other words it is necessary to recognize different textual representations as variants of the same proper name. As it was shown above, conflating inflectional and syntactic (in particular elliptical) variants is possible at present. However, we still have no satisfying method of handling derivational and semantic variants.

For instance, the two following names are equivalent:

(72)   *Park Łazienki, Park Łazienkowski* 'Łazienki Park'

but we cannot describe them as variants because in the model depicted in Fig. 1 we lack morphological features allowing to say that the adjective *Łazienkowski* is a derivative of the noun *Łazienki*.

Many semantic variants, as those discussed in section 3.2, are even more difficult to recognize since they frequently contain components with no or few morpho-syntactic relations to the original compound lemma. Recall, for instance, the following synonymic pairs:

(73)   *Rondo Zgrupowania AK „Radosław" = Rondo Babka*

(74)   *aleja Jana Pawła II = aleja Juliana Marchlewskiego*

Only additional external links may solve such problems of synonymy.

6.2   Distinguishing Variants

Inflectional and syntactic variants of an inflected form frequently share the same morphological tag. Therefore, they are hard to distinguish within embedded constructions. Recall example (67) for which we suggest first to describe the person name separately as in Fig. 17, and then to represent the street name as a structure of type *Noun Noun$_{gen}$* in which the third constituent may be an embedded compound, as in Fig. 19. The latter graph allows the third constituent to appear in any variant of its genitive form, here: *Marii Skłodowskiej-Curie, Marii Skłodowskiej, Marii Curie, Skłodowskiej*, or *Skłodowskiej-Curie*. In particular, applying one of the two last variants, where the first name is omitted, gives *ulica Skłodowskiej-Curie* and *ulica Skłodowskiej*. Similarly, all the forms in example (50) can be described in the same way.

Applying the same graph to similar examples fails in some cases. If the surname is morphologically a masculine noun it remains uninflected in female person names, as in example (75). If such a name appears in a street name, the first name should not be omitted (see example (76)).

(75)   *Emilia Plater, Emilii Plater, Emilię Plater*, etc.

(76)   *ulica Emilii Plater, Emilii Plater, *ulica Plater, *Plater* 'Emilia Plater street'

This fact cannot, however, be accounted for if the generic graph in Fig. 19 is used. Since the elliptic form *Plater* has an identical morphological description as the full form *Emilii Plater* (see example (77)), no morphological feature can be used in the graph to restrict the use of the elliptic variant.

(77)   *Emilii Plater, Emilia Plater:subst:f:sg:gen*
       Plater, Emilia Plater:subst:f:sg:gen

6.3 Morpho-Graphical Problems

In many examples shown above, compound proper names contain capitalised common words, such as *Most* 'Bridge', *Aleje* 'Avenue', *Plac* 'Square', *Czterech* 'Four', etc. Currently, these words are formally described in *Morfeusz* in the lower-case forms only. Thus, when a compound like (61) is annotated as in Fig. 5, the single-word lemmas appear in lower case. While exploring the inflection graphs *Multiflex* requires of *Morfeusz* to generate particular inflected forms of these lemmas. The resulting forms are in the lower case, which produces *aleje jerozolimskie, alei jerozolimskich*, etc. Additional *ad hoc* filters are needed in order to provide the desired capitalised effect shown in Fig. 9.

In order to solve this problem more elegantly we are considering two alternative solutions. First, the lower or the upper case spelling of a word may be seen as a morphological value. Thus, an extra inflectional category *LetterCase* can be introduced to the model in Fig. 1 with possible values *low, upp, allupp* (lower, upper, all upper), etc. One problem here would be to handle examples where the intial and some following letters (but not all) are capitalized as in *PeKaO*. Second, we can admit that the letter case is a problem on a different level and is not to be handled within morphosyntactic description. If we decide so, we still need to handle possible dependencies between morphological and case variants.

A similar problem concerns abbreviations, acronyms and initialisms addressed in examples (26) through (38). It is necessary to express the fact that *al.* is a possible variant of *aleja* or *aleje*, and that *ZUS* stands for *Zakład Ubezbieczeń Społecznych* 'Social Security Office'. Then again, creating a short form can be seen as a morphological category and incorporated into the model in Fig. 1. Alternatively, it can be handled at a separate level as a semi-derivational process.

7 Perspectives

The *Multiflex-Morfeusz* platform presented in this paper allows for the treatment of inflectional morphology and variation not only in proper names, but also in various kinds of Polish multi-word units, such as compounds, complex terms, contiguous collocations, etc. At present, the tool is generation-oriented: for a given compound it allows one to generate some or all of its inflected forms and variants. Naturally enough, we also consider its integration into a fully-fledged morphological analyser in which compounding would be largely taken into account.

The particular application of our platform to the description of Polish proper names was motivated by the outcome of the LUNA project (see section 2). In the long run, we aim at a large linguistic description of Warsaw toponyms, that could serve as a part of an intelligent information system containing a model of Warsaw topography (streets, places, institutions, monuments, rivers, etc.) and transport (bus- and tram-stops, underground stations, etc.). It could be applied to a dialogue system in which a voice server answers natural language questions, e.g. how to get by public transport from one place in the city to another.

Each lexical approach such as ours is labour intensive. Proper names need to be listed and described explicitly. Maintaining such human-controlled lists becomes harder with their growing size. For better efficiency and reliability, automated ergonomic procedures are needed, thanks to which manual work can be minimized. We are presently working on a graphical encoding aid module supporting the process of annotation of compound lemmas. It will specifically include an automatic morphological look-up of simple constituents, and a simulation of the exploration of inflectional graphs.

Another interesting perspective is integration of Polish names into the multilingual ontology of proper names *Prolexbase* [18], meant in particular for machine translation. If links between the Prolex conceptual proper names and the Polish lexical items are introduced, the Polish module will be able to benefit from the language-independent relations already present in the database. Such relations (synonymy, meronymy, etc.) are useful e.g. in uni- or multilanguage information extraction and text alignment.

8   Conclusions

We have presented ongoing work on the formal description of Polish compound proper names. We have performed an analysis of their linguistic properties. We have shown how two separate morphological tools, *Morfeusz* and *Multiflex*, were brought into a collaborative framework allowing for a two-layer model of inflection and variation of compounds. Thus, graph-based rules in the latter tool describe combinations of particular inflected forms of single units controlled by the former tool. The resulting framework allows one to address not only the inflectional morphology, but also other linguistic phenomena such as ellipsis, accomodability of numerals, and embedded compounding. The inflected paradigms are represented compactly thanks to unification. Some lacking facilities include proper treatment of derivational and semantic variants, distinguishing variants which share the same morphological tag, as well as modelling letter cases and abbreviations. At its present stage the platform allows the undertaking of a project of describing urban toponyms for a model of transportation system in Warsaw, in view of dialogue systems and other natural language applications.

Acknowledgements. The work presented has been supported by the French and Polish Ministries of Research and Higher Education within the Hubert Curien Partnership POLONIUM programme.

References

1. Agafonov, C., Grass, T., Maurel, D., Rossi-Gensane, N., Savary, A.: La traduction multilingue des noms propres dans PROLEX. Meta 51(4), 622–636 (2006)
2. Alegria, I., Ansa, O., Artola, X., Ezeiza, N., Gojenola, K., Urizar, R.: Representation and Treatment of Multiword Expressions in Basque. In: Second ACL Workshop on Multiword Expressions, July 2004, pp. 48–55 (2004)

3. Bień, J.S.: Koncepcja słownikowej informacji morfologicznej i jej komputerowej weryfikacji. Rozprawy Uniwersytetu Warszawskiego t. 383. Wydawnictwa Uniwersytetu Warszawskiego, Warszawa (1991)

4. Bień, J.S., Saloni, Z.: Pojęcie wyrazu morfologicznego i jego zastosowanie do opisu fleksji polskiej (wersja wstępna). Prace Filologiczne XXXI, 31–45 (1982)

5. Chena, H.H., Huang, S.J., Ding, Y.W., Tsai, S.C.: Proper Name Translation in Cross-Language Information Retrieval. In: Proceedings of COLING-ACL-1998, pp. 232–236 (1998)

6. Cieślikowa, A. (ed.): Mały słownik odmiany nazw własnych. Rytm, Warszawa (2008)

7. Courtois, B., Silberztein, M. (eds.): Les dictionnaires électroniques du français. Larousse, Langue française, vol. 87 (1990)

8. Friburger, N., Maurel, D.: Textual similarity based on proper names. In: Proceedings of the Workshop on Mathematical Formal Information Retrieval (MFIR'2002), SIGIR'2002, Tampere, Finland, pp. 155–167 (2002)

9. Grishman, R., Sundheim, B.: Message Understanding Conference - 6: A Brief History. In: Proceedings of COLING-96, pp. 466–471 (1996)

10. Grzenia, J.: Słownik nazw własnych. Wydawnictwo naukowe PWN (2003)

11. Hajnicz, E., Kupść, A.: Przegląd analizatorów morfologicznych dla języka polskiego. Wydawnictwo IPI PAN, Warszawa (2001)

12. Handke, K.: Słownik nazewnictwa Warszawy. Slawistyczny Ośrodek Wydawniczy, Warszawa (1998)

13. Hatzivassiloglou, V., Klavans, J.L., Eskin, E.: Detecting Text Similarity over Short Passages: Exploring Linguistic Feature Combinations via Machine Learning. In: Proceedings of the 1999 Joint SIGDAT Conference on Empirical Methods in Natural Language Processing and Very Large Corpora, pp. 203–212 (1999)

14. Jacquemin, C.: Spotting and Discovering Terms through Natural Language Processing. MIT Press, Cambridge (2001)

15. Karttunen, L., Kaplan, R.M., Zaenen, A.: Two-Level Morphology with Composition. In: Proceedings of COLING-92, Nantes, pp. 141–148 (1992)

16. Krstev, C.: Processing of Serbian: Automata, Texts and Electronic Dictionaries. Faculty of Philology, University of Belgrade (2009)

17. Krstev, C., Vitas, D., Savary, A.: Prerequisites for a Comprehensive Dictionary of Serbian Compounds. In: Salakoski, T., Ginter, F., Pyysalo, S., Pahikkala, T. (eds.) FinTAL 2006. LNCS (LNAI), vol. 4139, pp. 552–563. Springer, Heidelberg (2006)

18. Maurel, D.: Prolexbase. A multilungual relational lexical database of proper names. In: Proceedings of LREC'08, Marrakech, Marocco (2008)

19. Maurel, D., Vitas, D., Krstev, C., Koeva, S.: Prolex: a lexical model for translation of proper names. Application to French, Serbian and Bulgarian. BULAG 32, 55–72 (2007)

20. Mykowiecka, A., Marciniak, M., Rabiega-Wiśniewska, J.: Proper Names in Polish Dialogs. In: Proceedings of the IIS 2008 Workshop on Spoken Language Understanding and Dialogue Systems, Zakopane, Poland, Springer, Heidelberg (2008)

21. Oflazer, K., Çetonoğlu, Ö., Say, B.: Integrating Morphology with Multi-word Expression Processing in Turkish. In: Second ACL Workshop on Multiword Expressions, July 2004, pp. 64–71 (2004)

22. Pašca, M., Lin, D., Bigham, J., Lifchits, A., Jain, A.: Names and Similarities on the Web: Fact Extraction in the Fast Lane. In: Proceedings of COLING-ACL-2006, Sydney, Australia, pp. 809–816 (2006)

140   A. Savary, J. Rabiega-Wiśniewska, and M. Woliński

23. Paumier, S.: Manuel d'utilisation du logiciel Unitex (2002),
    http://www-igm.univ-mlv.fr/unitex/manuelunitex.ps
24. Piskorski, J., Sydow, M.: Usability of String Distance Metrics for Name Matching
    Tasks in Polish. In: Human Language Technologies as a Challenge for Computer
    Science and Linguistics. Proceedings of 3rd Language & Technology Conference,
    Poland, Poznań, October 5-7 (2007)
25. Piskorski, J., Sydow, M., Kupść, A.: Lemmatization of Polish Person Names. In:
    ACL 2007. Proceedings of the Workshop on Balto-Slavonic Natural Language Pro-
    cessing 2007, Special Theme: Information Extraction and Enabling Technologies
    (2007)
26. Przepiórkowski, A., Woliński, M.: The Unbearable Lightness of Tagging: A
    Case Study in Morphosyntactic Tagging of Polish. In: Proceedings of the 4th
    International Workshop on Linguistically Interpreted Corpora, EACL (2003)
27. Rabiega-Wiśniewska, J.: Polish proper names in the corpus od spoken dialogues.
    the syntactic experiment. In: Beiträge der Europäischen Slavistischen Linguistik
    (Polyslav), vol. 12 (to appear)
28. Rzetelska-Feleszko, E. (ed.): Polskie nazwy własne. Instytut Języka Polskiego
    Polskiej Akademii Nauk, Kraków (2005)
29. Saloni, Z.: Klasyfikacja gramatyczna leksemów polskich. Język Polski LIV, z. 1,
    3-13, z. 2, 93-101 (1974)
30. Saloni, Z.: Kategoria rodzaju we współczesnym języku polskim. In: Kategorie
    gramatyczne grup imiennych we współczesnym języku polskim, pp. 41-75. Osso-
    lineum, Wrocław (1976)
31. Saloni, Z.: Kategorie gramatyczne liczebników we współczesnym języku polskim.
    Studia gramatyczne 1, 145-173 (1977)
32. Saloni, Z.: Czasownik polski. Odmiana, słownik. Wiedza Powszechna, Warszawa
    (2007)
33. Saloni, Z., Gruszczyński, W., Woliński, M., Wołosz, R.: Słownik gramatyczny
    języka polskiego. Wiedza Powszechna, Warszawa (2007)
34. Saloni, Z., Świdziński, M.: Składnia współczesnego języka polskiego. PWN,
    Warszawa (1998)
35. Sang, E.F.T.K., de Meulder, F.: Introduction to the CoNLL-2003 Shared Task:
    Language-Independent Named Entity Recognition. In: Proceedings of CoNLL-
    2003, Edmonton, Canada, pp. 142-147 (2003)
36. Savary, A.: A formalism for the computational morphology of multi-word units.
    Archives of Control Sciences 15(3), 437-449 (2005)
37. Savary, A.: MULTIFLEX. User's Manual and Technical Documentation. Version
    1.0. Technical Report 285, LI-François Rabelais University of Tours, France (2005)
38. Savary, A.: Computational Inflection of Multi-Word Units. A contrastive study
    of lexical approaches. Linguistic Issues in Language Technology 1(2), 1-53 (2008)
39. Savary, A., Jacquemin, C.: Reducing Information Variation in Text. In: Renals,
    S., Grefenstette, G. (eds.) Text- and Speech-Triggered Information Access. LNCS
    (LNAI), vol. 2705, pp. 145-181. Springer, Heidelberg (2003)
40. Savary, A., Krstev, C., Vitas, D.: Inflectional Non Compositionality and Variation
    of Compounds in French, Polish and Serbian, and Their Automatic Processing.
    BULAG 32, 73-93 (2007)
41. Silberztein, M.: NooJ's dictionaries. In: Proceedings of LTC'05, Poznań, Wydaw-
    nictwo Poznańskie, pp. 291-295 (2005)

42. Świdziński, M., Derwojedowa, M., Rudolf, M.: A Computational Account of Multi-Word Numeral Phrases in Polish. In: Kosta, P., Błaszczak, J., Frasek, J., Geist, L., Żygis, M. (eds.) Investigations into Formal Slavic Linguistics, vol. 10, Part I, pp. 405–415. Peter Lang, Frankfurt (2003)
43. Thompson, P., Dozier, C.: Name Searching and Information Retrieval. In: Proceedings of the Second Conference on Empirical Methods in Natural Language Processing, Providence, Rhode Island (1997)
44. Tokarski, J.: Schematyczny indeks a tergo polskich form wyrazowych (ed. Zygmunt Saloni), 2nd edn. Wydawnictwo Naukowe PWN, Warszawa (2002)
45. Tran, M., Maurel, D.: Prolexbase: Un dictionnaire relationnel multilingue de noms propres. Traitement Automatiques des Langues 47(3), 115–139 (2006)
46. Woliński, M.: System znaczników morfosyntaktycznych w korpusie IPI PAN. Polonica XXII–XXIII, 39–55 (2003)
47. Woliński, M.: Morfeusz — a Practical Tool for the Morphological Analysis of Polish. In: Kłopotek, M., Wierzchoń, S., Trojanowski, K. (eds.) Intelligent Information Processing and Web Mining, IIS:IIPWM'06 Proceedings, pp. 503–512. Springer, Heidelberg (2006)

A New Formal Definition of Polish Nominal Phrases

Marek Świdziński[1] and Marcin Woliński[2]

[1] Institute of Polish, Warsaw University,
m.r.swidzinski@uw.edu.pl,
[2] Institute of Computer Science, Polish Academy of Sciences,
wolinski@ipipan.waw.pl

Abstract. In the paper, a new formal definition of Polish nominal phrases is presented. Based upon a certain formal grammar of Polish (FGP) that applies a formalism of metamorphosis grammar, it is the first step towards redesigning the entire grammar. It makes use of the results of experiments with implementation of the grammar. After a report on empirical data a large set of parameters that formalize various grammatical features is introduced. Some of those parameters are really new, others are to be reinterpreted and improved. A number of rules are written down to illustrate the way empirical expressions are accounted for. The paper ends in formulating some postulates that the new version of FGP is expected to fulfil.

Key words: noun phrase/nominal phrase, Polish, formal grammar, metamorphosis grammar

1 Introduction

The aim of the paper is to discuss a redefinition of a formal grammar of Polish (FGP), given in [7]. The account presented here makes use of the results of experiments with the implementation of the grammar, which was designed as a pure theoretical study. We have decided to take one syntactic unit as a testbed of our efforts to redesign the whole FGP. Our choice has been NP or, as we call it, nominal phrase (**nomp**).

The metamorphosis grammar formalism [4] applied in FGP makes the grammar a convenient basis for a real parser, which does not hold true, e.g., for GB-based descriptions. FGP has been implemented by M. Woliński who designed an automatic syntactic analyser Świgra (cf. [13,14,12]). Some adjustments to FGP had to be made during implementation; others follow from our theoretical considerations.

FGP has an ambition to define most syntactic structures of Polish. The grammar is fairly extensive and contains almost 500 rules. Parsing experiments have revealed a number of types of structures that have either been overlooked (sometimes deliberately), or described inaccurately. One can say that FGP has been tested both for its explicitness and exhaustiveness.

M. Marciniak and A. Mykowiecka (Eds.): Bolc Festschrift, LNCS 5070, pp. 143–162, 2009.

What follows is that the Polish language plays an important role here. Its grammatical (mainly syntactic) setup, individual idiosyncrasies included, provides stimuli for various formal solutions. Therefore, our paper is, to some extent, Polish-oriented, rather than theoretically linguistic or computational.

The new version of FGP discussed in this paper will be abbreviated as FGPn further on.

* * *

The fact we dedicate this paper to Professor Leonard Bolc is by no means accidental. On the contrary, it seems to have a symbolic meaning. It was exactly 30 years ago that Allain Colmerauer's article introducing a new apparatus of metamorphosis grammar appeared in the volume edited exactly by Professor Bolc (Natural Language Communication with Computers. Lecture Notes in Computer Science, vol. LXIII, cf. [4]).

2   FGP and Its Philosophy

FGP is a context-free phrase structure grammar. Actually, the grammar deals with nothing but surface syntax; neither semantics, nor Chomsky-style principles and parameters are taken into consideration. Language expressions, understood as mere sequences of words, are represented by terms with parameters that formalize various grammatical features of those units. Except for function words, there are two types of syntactic units: sentences and phrases. Rules of the grammar define particular units as sequences of other units and establish correspondences between their grammatical features via unification.

It should be emphasized that the grammar is "restrictive" in some sense. Not only does it try to catch all well-formed language expressions, but also takes care of the overt blocking of some structures or some grammatical features no matter if they appear in real text corpora.

Trees defined by the grammar differ from those the generativists work with. FGP operates with multi-level hierarchies: the trees it provides are as a rule long-branched. FGP does not contain a lexicon, the only terminals introduced by rules actually being function words: complementizers, conjunctions, pronouns, and some prepositions.

Syntactic connections are accounted for mainly by parameter matching. An extensive set of parameters is introduced, most of them being general, i.e., valid for the majority of syntactic units. The values a given unit is assigned, be it from the top ("syntactic" features) or from the bottom ("lexical" features), infect the whole syntactic tree, reaching most of its constituents.

In the grammar rules, parameter values can be fixed or non-fixed. Fixed values are either explicitly given by the rule, or are the result of a calculation. Non-fixed values are used to express the fact that a syntactic unit agrees for a given parameter with another syntactic unit. If both syntactic units are constituents of one construction we have to do with what one can call *horizontal unification* (at one level); if one unit corresponds to the construction and the other is its constituent it is the case of *vertical unification* (between levels).

Vertical unification provides tools for bottom-up and top-down transmission of grammatical features. It is important that rules defining different syntactic units (sentences or phrases) follow, in a sense, one format. Many features are typical of a majority of those units and appear in a majority of rules. In other words, various language expressions turn out to be structurally similar, which should be, and really *is*, modeled by FGP.

The account given in FGP is capable of providing a variety of interpretations for a given expression, accounting this way for syntactic ambiguity phenomena so typical of natural languages.

3   Świgra—An Implementation of FGP

Since the formalism of FGP is a variation of metamorphosis grammar it would seem natural to attempt an implementation in Prolog. However, due to extensions of the formalism specific to FGP, it has turned out impossible to use the interpreter of grammars built into most Prolog implementations. Instead, we have decided to build a parser from scratch. (For details of the implementation see [13].)

The analyser works in a manner similar to that of a chart parser. It builds a shared parse forest [3], which lets the parser avoid unnecessary re-computation and limits the analysis time to polynomial. Compared to a typical chart the shared forest contains only inactive edges. The active edges are hidden in the recursion stack and reclaimed when no longer used, which results in smaller memory requirements of the algorithm. A gain in speed results from the fact that rules are not interpreted at the runtime by the parser but they are compiled to Prolog clauses. Hence, the parser proved to be more efficient for FGP than a naïve chart parser.

The bottom-up parsing strategy is hardwired into the parser. For highly inflecting languages like Polish it seems to be the best choice since the inflectional features of particular words come into play early in the process when going bottom-up, thus blocking many impossible paths in the search space.

The result of a parse is a shared parse forest which is polynomial in size even when the number of trees assigned to a sentence is exponential. Since sets of parse trees are sometimes quite large even for sentences that are neither long, nor particularly complicated it is an interesting challenge to navigate in the set and process it as a whole without generating all possible trees (cf. [15,2]).

The parser is available to the public under GNU General Public License at the address: http://nlp.ipipan.waw.pl/~wolinski/swigra

To build an analyser for FGP a number of changes and adjustments in the grammar have been indispensable.

The first challenge has been caused by the fact that the grammar does not contain a lexicon. For implementation purposes FGP has been coupled with the morphological analyser Morfeusz SIaT [16]. (The program is available for download at http://nlp.ipipan.waw.pl/~wolinski/morfeusz and can be used for non-commercial research purposes.) About fifty simple rules were added to the grammar to interface it with the morphological component.

Another problem to cope with, more difficult indeed, has been that of void recursion in the grammar rules. FGP contains five cycles containing rules which allow to rewrite a single nonterminal on the left hand side to a single nonterminal on the right hand side. For example, FGP introduces a multi-level hierarchy of nominal units, which resembles somewhat, though unintentionally, GB treatment of NP, derived from Abney's DP-hypothesis [1]. Each unit in this chain can be realized by the next one. Preterminals have, *inter alia*, a recursive realization by a topmost unit, i.e., **nomp**. This means that for each nominal phrase (and, in fact, four other syntactic units) accepted by the grammar it is possible to build alternative trees with arbitrarily long branches by going several times through the cycle. In order to avoid it, the recursion has been blocked by a technical trick (counting the length of branches being created). FGPn proposes a different solution; see section 4.

Unfortunately, the rules defining elementary (i.e., simple) sentences in FGP allow to permute their constituents too freely. If an analysis were to be performed exactly the way given in FGP, one would get plenty of interpretations even for simple sentences which we would like to consider unambiguous. In Świgra, only the skeleton of the original rules was kept; the method to give account of the free word order of Polish was designed anew.

A technical problem has arisen connected with the fact that for bottom-up analysis rules with empty right hand side are rather difficult to implement. Fortunately, most of those rules in the grammar are no longer necessary after reformulating the rules for the elementary sentence; few other instances are simple to eliminate.

FGP contains five context-sensitive rules which describe empty realizations of a comma in the context of adjacent punctuation marks. In the implementation those rules have been replaced with a special mechanism of multiplying commas in the text. However, those rules force the grammar to accept sentences with doubled commas. The comma issue is general: a number of parts of the grammar dealing therewith will be radically rebuilt in FGPn, as described in section 5.3.

The parameter called 'dependency' in the original version of FGP (somewhat misleading, indeed) is the most difficult to interpret. It makes the structure of an expression depend on the context in which a given unit is placed. The parameter is used to confront elements of the unit (such as interrogative particles or relative pronouns) with that context.

The usage of this parameter hardly seems consistent. In some parts of the grammar it marks a phrase as being acceptable for some context just because it does not contain any element that disagrees with such context. In other places it signals that a phrase has to be placed in a specific context due to its constituents. This leads to the over-generation of trees for some sentences.

In FGPn, the parameter will be replaced with parameters which consistently give account of the inner structure of the phrase and not of the context. So as long as the phrase being built does not contain any 'special' element its context-boundedness remains 'neutral' (see section 5.3).

Another implementation problem is posed by conditions in the rules. They are taken for static constraints on the values of parameters. Sometimes the program has been unable to compute values that are needed in conditions at the time the condition is to be evaluated (in some of those cases it does not even depend on parsing strategy). Such conditions have been reformulated or moved to a place where all needed values are known.

It should be admitted that the resulting parser has a rather low corpus coverage of about 30% of sentences. This is partly due to morphological description, and partly to a number of shortcomings of FGP. On the other hand, FGP quite often over-generates. A source of unwanted structures is, for example, a rule that allows realization of a nominal phrase by an adjectival phrase. This type of a realization will be strongly restricted further on (cf. section 6).

4 Taxonomy of Nominal Phrases

Nominal phrases are syntactic units equivalent to a word form of a noun. Unlike practised by the generative school since Steven Abney, our view on NP is purely distributional. For us, nominal phrases are headed by the subordinating constituent they are reducible to—it is a noun in most cases. We will not deal here with coordinate nominal phrases. The structure of coordinate constructions is specific and requires non-standard means to account for. Actually, nominal phrases of that type are exceptionally idiosyncratic among coordinate constructions. There are exocentric phrases in which gender, number, and person agreements must be calculated in a way hardly shared by other types of phrases.

Let the example below illustrate a nominal phrase which is maximally complicated:

(1) jego pięcioro nowych studentów z   tej grupy, w których imieniu
      his five   new   students from this group on whose   behalf
      przyszła najmłodsza,
      came   the youngest
      'his five new students from this group on whose behalf the youngest (girl)
      has come'

Nominal phrases appear in a variety of syntactic contexts. A nominal phrase can first of all be a complement. We take subjects for a special type of complement. It is syntactically implied (required) by (a) verbs, either within an elementary (simple) sentence, or within a verb phrase—or by (b) prepositions, within a prepositional phrase. There are also (c) adjuncts realized by a nominal phrase. Nominal phrases happen to be (d) constituents of other nominal phrases, such as a genitive or appositive nominal phrase (the latter will not be discussed here), or a nominal component of the nominal phrase headed by a numeral phrase. At last, nominal phrases are sometimes, rather rarely, (e) constituents of adjectival phrases.

Consider the examples:

(2)   **Jan** kupił **łódkę**. (ad (a))
      'Jan bought a boat.'
(3)   Chcę kupić **łódkę**. (ad (a))
      'I want to buy a boat.'
(4)   To było w **kościele**. (ad (b))
      'This happened in the church.'
(5)   **Godzinami** dziecko płakało. (ad (c))
      'The kid used to cry for hours$_{instr,pl}$.'
(6)   Widziałem lekarza **matki**. (ad (d))
      'I saw the mother's doc.'
(7)   Widziałem lekarza **dentystę**. (ad (d))
      'I saw a dentist (lit. physician dentist).'
(8)   Widziałem pięciu **lekarzy**. (ad (d))
      'I saw five docs.'
(9)   Pokój był pełen **ludzi**. (ad (e))
      'The room was full of people.'

Nominal phrases get the case value from the outside. The value is motivated either lexically (imposed on such a phrase by a verb, preposition, or subordinating adjective; cf. examples (2) and (3), (4), (9), respectively), or syntactically (the fixed genitive value; cf. (6), (8)), or the agreement value shared with another constituent; cf. (7)). Nominal phrases in adjunct positions are treated here, somewhat carelessly, as not getting the case value from the top.

Nominal phrases may have various dependents. The nominal head can introduce (a) a clause (or, as we call it, a sentential phrase), (b) a nominal phrase, (c) a prepositional-nominal phrase, or (d) an adjectival phrase:

(10) chłopiec, **którego lubię,** (ad (a))
      '(the) boy I like'
(11) **pielęgniarki** kolegom (ad (b))
      '(this) nurse's colleagues$_{dat}$'
(12) dwoma **facetami** (ad (b))
      '(with) two guys$_{instr}$'
(13) muzyk **instrumentalista** (ad (b))
      '(the) musician instrumentalist'
(14) dom **za rogiem** (ad (c))
      '(the) house around the corner'
(15) **małych** problemów (ad (d))
      'small problems$_{gen,pl}$.'

Terminals represent a variety of classes, including nouns, pronouns of various types (also adjectival), certain adjectival word forms, numeral phrases (realised by a single numeral or a sequence of numerals), etc. Besides case, gender, and number, the network of syntactic connections within a nominal phrase involves a galore of syntactic features. As far as we know, the whole package has not been taken into account in descriptions of Polish nominal phrases up to now.

The contents and the structure of nominal phrases (including word order) depend mainly on the head of such phrases. Nominal phrases inherit many features after their head.

Another important classification of nominal phrases (and, in fact, of many other syntactic units, from sentences to phrases) gives us four types of syntactic units differing in word order: standard (or neutral), interrogative, relative, and indirect-question type. Some syntactic features are connected with the initial or final position within a nominal phrase.

There are two main structural types on nominal phrases. To the former belong phrases that contain a numeral phrase which is regarded as the head of the whole nominal phrase, usually accompanied by a standard nominal phrase (cf. examples (8) and (12)); the latter includes standard noun-headed nominal constructions (all remaining examples).

5   Reformulation of FGP

5.1   Hierarchy

The new version of FGP applies the metamorphosis grammar apparatus in its DCG variant (cf. [6]). Our new definition of **nomp** operates practically with one syntactic unit: **nomp**, which means that the hierarchy is maximally flat (one-level). We get rid then of the original FGP hierarchy (cf. [7]):

Nominal phrase
Nominal construction with genitive
Nominal construction with prepositional phrase
Nominal construction with attribute
Nominal construction with incorporation
Nominal construction

where the lowest unit has a recursive realization by the topmost one.

Typical (subordinated) nominal phrases are sequences containing a **nomp** plus something else (= head plus dependent). Simple realizations of **nomp** provided by FGPn are the following: nouns, pronouns of various kinds, or numerals. Nominal phrases can also be realized by adjectival phrases, verbal phrases, and sentential phrases. Of those three, the first will be discussed later on, unlike the remaining two which are not important here. The last two "phrasal" realizations are the source of some technical problems. First, the verbal realization is a gerundial one. The verbal phrase has some parameters that nominal phrases do not share; up to now no decision has been made as to how to include gerunds in FGPn. Second, there are "headless" relative clauses distributionally equivalent to a word form of a noun (cf. *Ktokolwiek przyjdzie, wygra.* 'Whoever will come will (surely) win.'); their distribution is not clear to us up to now.

5.2   The Nominal Phrase and Its Dependents

The nominal phrase as a syntactic unit has the following format:

nomp(*Case, Gen/Num, Pers, Neg, I, Cb, Cl, Imod, Sub, Comma* **)**

Parameters of **nomp** will be presented in 5.3 below. The set of dependents of **nomp** is as follows:

1. sentential phrase **sentp** (i.e., clause):
   sentp(*Type, Corr, Asp, Tense, Neg, I, Sub, Comma* **)**
2. nominal phrase **nomp** accompanying the numeral head (**nump**):
   nump(*Case, Gen/Num, Pers, Neg, I, Cb, Cl, Imod, Sub, Comma* **)**
3. nominal phrase **nomp** in the genitive or agreed for case with the head;
4. prepositional-nominal phrase (**prepnomp**):
   prepnomp(*Prep, Case, Neg, I, Cb, Cl, Imod, Sub, Comma* **)**
5. adjectival phrase (**adjp**):
   adjp(*Case, Gen/Num, Deg, Neg, I, Cb, Cl, Imod, Sub, Comma* **)**

where *Type*—type of a clause, *Corr*—correlate, *Asp*—aspect, and *Tense*—tense; *Prep*—preposition; *Deg*—degree.

Practically, all rules that reveal that type of structuring are recursive, which does not result in vicious circles.

5.3   Parameters

We enrich the set of parameters for **nomp**, getting the number of 11 features. The description of agreement phenomena, although subtle and detailed in the former version of FGP, is significantly developed here. Morphological features notwithstanding, we introduce such parameters as negation, initiality, class, context-boundedness, subordination, comma blockage etc., all carefully controlled through the whole grammar. Some of them are new (New), some other have either been renamed, or reinterpreted (Re).

The complete list is as follows:

case	*Case*	
gender	*Gen*	
number	*Num*	
person	*Pers*	
negation	*Neg*	(Re)
initiality	*I*	(Re)
context-boundedness	*Cb*	(Re)
class	*Cl*	(Re)
internal modifier	*Imod*	(Re)
subordination	*Sub*	(New)
comma blockage	*Comma*	(New)

The first four do not require any comment, the remaining will be discussed in the following subsections.

Negation. The parameter of negation *Neg* plays an important role throughout the whole grammar. Grammatical impact of negativity concerns sentences (elementary and coordinate) and phrases, as well. Out of such phenomena as genitive of negation, negative concord, negation in verb clusters, or negativity in coordinate sentences with the conjunction *ani* '(neither...,) nor' (cf. [9]), there is a syntactic restriction on negation within non-verbal phrases. Nominal phrases containing a negative pronoun as their constituent (not necessarily immediate) bear the no value of *Neg*. The value is to be calculated by means of a condition called *phrasal negativity calculation*, which is first introduced in FGPn:

phrneg_ calc(*Neg*, *Neg1*, *Neg2*)

where *Neg* is the value of the whole nominal phrase, *Neg1* and *Neg2* feature the former and the latter constituent of that phrase, respectively. If either *Neg1* = no, or *Neg2* = no then *Neg* is so; otherwise *Neg* can get any value. Cf. examples below:

(16) **Nikogo** nie lubię. / * Nikogo lubię.
     'I don't like anybody.'
(17) **Nikogo** z nich nie lubię. / * Nikogo z nich lubię.
     'I don't like anyone of them.'
(18) Książki przyjaciół **żadnego spośród nich** nie lubię. / * Książki przyjaciół żadnego spośród nich lubię.
     'I don't like books (written by) friends of anyone of them.'

The condition supplements the original description of negation given in FGP, where there are two other conditions that help calculate negation (see also [11,8]).

Initiality. The parameter of initiality *I* is multifunctional. Its role is to take care of transmission of various syntactic features onto the initial component in a variety of constructions; hence its new name (it was called *incorporation* in the previous version of FGP). Features of that type drop down within that component to reach its initial component, and so forth.

First, the parameter stores the name of a peculiar two-place conjunction that cannot appear between conjuncts it joins; rather, it is embedded in the initial component of the second conjunct. There are four conjunctions of that type: *bowiem* 'because', *natomiast* 'on the other hand', *więc* 'so', and *zaś* 'whilst'; the first one can hardly be regarded as coordinate. A special condition *incorporation calculation* establishes the distribution of incorporational conjunctions within nominal phrases: inc_ calc(*I*, *I1*, *I2*).

It says that for *I* equal to bowiem, natomiast, więc, or zaś (those "lexical" values come from the outside), the respective value either drops on the initial component as *I1* and, consequently, *I2* = ni (ni 'not incorporational'), or it reaches the final one as *I2* and, consequently, *I1* = ni. Cf. examples below:

(19) Przyjdę, **kolega bowiem matki** mnie zaprosił.
      'I'll come because my mother's colleague has invited me.'
(20) Przyjdę, **kolega matki bowiem** mnie zaprosił.
(21) * Przyjdę, **bowiem kolega matki** mnie zaprosił.

Sentences like (21), though possible, are hardly tolerable in formal Polish. The formulation of this condition is also new in FGPn.

Second, the parameter serves as a device to transmit the number-gender characteristics of the head of the nominal phrase onto the initial component of a relative clause; usually, in a number of steps. This assignment is accounted for by special rules in definitions of all types of sentences and, practically, all types of phrases. Cf. examples below and [10]:

(22) chłopca, który / * która śpi,
      '(the) boy$_{gen/acc,mhum,sg}$ who$_{nom,mhum / nom,fem,sg}$ is sleeping'
(23) chłopca, którego / * których rodzice żyją,
      '(the) boy$_{gen/acc,mhum,sg}$ whose$_{gen,mhum,sg / gen,pl}$ parents are still alive'
(24) chłopca, o którym / * której książkę napisano,
      '(the) boy$_{gen/acc,mhum,sg}$ about whom$_{loc,mhum,sg / loc,fem,sg}$ a book has been written'

Third, a special fixed value c_less of this parameter is provided by FGPn. It also comes from the outside as given by the rule defining utterance, i.e., the highest syntactic unit in our hierarchy. Its duty is to prevent various syntactic units from beginning with a comma at the beginning of the utterance. This particular property will be elaborated in detail in other publications.

Context-Boundedness. The parameter of *context-boundedness* (*Cb*) replaces the one that was called, somewhat awkward, *dependency* in FGP. Of the two functions of the parameter therein, only one remains in FGPn. Actually, there are four fixed values of this parameter: neutral (**neut**), interrogative (**q**), relative (**rel**), and indirect-question (**indq**), which gives a four-class typology of a majority of syntactic units.

The parameter is responsible for the distribution of lexically interrogative or relative components within various syntactic units.

First, syntactic units bearing the value **q** must contain at least one interrogative component (not necessarily as an immediate constituent and as the initial one), the remaining component(s) being either **q**, or **neut**:

(25) **Z tej półki którą książkę** kupimy?
      'Of the books (stored) on this shelf, which one will we buy?'
(26) **Po ile książek którego z nich** kupimy?
      'How many books will we buy, (written) by which of them, respectively?'
(27) **Kupimy książkę którego z nich**?
      'We'll buy a book by which of them?'

It is worth adding that the three sentences given above sound perhaps strange but are perfectly correct. They illustrate the fact that a direct question in Polish can contain many interrogative pronouns and their location is really looser than, say, in English. The above examples exhibit a marked topic-comment structure.

Second, the value rel percolates down onto the initial component, all the remaining being neut (cf. examples (22)–(24)). Cf. [10].

Third, the value indq reaches the initial component, the remaining being either q (not indq), or neut:

(28)   Wiem, **którą jego książkę** kupimy.
        'I know which of books by him we will buy'
(29)   * Wiem, **jego którą książkę** kupimy.
(30)   Wiem, **co którego z nich** kupimy.
        'I know what we will buy from whom'

Finally, neut syntactic units have got only neut components.

Class. The parameter of *class* (*Cl*) is introduced in FGPn to account for restrictions on the structure and contents of nominal phrases imposed by the head of those phrases. Standard nominal phrases headed by a noun-class component do not exhibit any restrictions of that sort. Pronominal nominal phrases of the co 'what' and kto 'who' values differ in blocking some types of relative clauses. The former allows but for co-type relative clauses, the latter—kto-type ones:

(31)   coś, **czego** nie znam,
        'something what I don't know'
(32)   * coś, kogo / którego nie znam,
        'something whom / which I don't know'
(33)   komuś, **kogo** zaprosiliśmy
        'somebody$_{dat}$ whom / * what / * which we have invited'
(34)   * komuś, co / którego zaprosiliśmy,
        'somebody$_{dat}$ what / which we have invited'

Furthermore, if nominative or accusative co-type nominal phrases contain an adjectival phrase, it does not agree with the head of such a phrase for case; rather, the phrase is in the genitive:

(35)   To jest **coś nowego** / * **nowe**.
        'This is something$_{nom,neu,sg}$ new$_{gen,neu,sg}$'
(36)   To jest **ktoś** / **pracownik nowy**.
        'This is one$_{nom,mhum,sg}$ / an employee$_{nom,mhum,sg}$ just hired$_{nom,mhum,sg}$'

Special values agr and nonagr characterize nominal phrases with a numeral head. There are two subclasses in the set of numeral word forms. To the former belong numerals agreeing with the nominal dependent for case, gender, number, and person; the latter consists of genitive-governing numerals some of which syntactically behave as if they were 3rd-person singular neuter noun word forms. Those subclasses also allow us to classify numeral phrases and nominal phrases with a numeral head onto two subsets—an agreement and non-agreement one:

(37)   Przyszły **dwie dziewczyny** (agr)
       'Two$_{nom,fem}$ girls$_{nom,fem,pl}$ have come$_{fem,pl}$'
(38)   Znam **pięć kobiet** (nonagr)
       'I know five$_{acc,fem}$ women$_{gen,fem,pl}$'
(39)   **Pięć kobiet** przyszło (nonagr)
       'Five$_{nom,fem}$ women$_{gen,fem,pl}$ have come$_{neu,sg}$'

Finally, nominal phrases headed by the numeral do not allow their nominal dependent to be "numeral" (cf. [5]):

(40)   * Przyszły **dwie trzy dziewczyny** (agr)
       'Two three girls have come'
(41)   * Znam **pięć dwudziestu kobiet** (nonagr)
       'I know five twenty women'

Internal Modifier. The next parameter (*Imod*) formalizes a mere detail of Polish syntax. Aside from nominal phrases, it is a syntactic feature of adjectival and adverbial phrases. Two fixed values **tk** and **tyle** are lexical, typical of pronouns *tak* 'to such an extent (that)' or *taki* 'such... (that)' and *tyle* 'so many/much... (that)', respectively. The value **tyle** is a source of a special value of the type of a clause—the **ile** one (*ile* 'how many'). The *Imod* parameter works similar to the machinery of phrasal negation mentioned above. In order for a given phrase to get the value **tk** or **tyle** it is sufficient that at least one constituent of a given nominal phrase, not necessarily immediate, bears that value. A special condition *internal modifier calculation* allows one to compute the value *Tk*:

   imod_ calc(*Tk*, *Tk1*, *Tk2*)

If either *Tk1* or *Tk2* is equal to **tk** or **tyle** then *Tk* is so; otherwise *Tk* = std. Cf. examples:

(42)   Znam **takiego / tak miłego faceta, że trudno uwierzyć**,
       'I know such a guy / so nice a guy that it is hard to believe'
(43)   * Znam faceta, że trudno uwierzyć,
       'I know a guy that it is hard to believe'
(44)   Nie mamy **tylu pieniędzy, ile chcielibyśmy**,
       'We haven't got so much money (how much money) we would like to have'
(45)   * Nie mamy pieniędzy, ile chcielibyśmy,
       'We haven't got money how much we would like to have'

Subordination. The parameter *Sub* is of technical importance. It has nothing in common with the dependency parameter in FGP which has been partly replaced by context-boundedness. Its two fixed values **head** and **dep** are explicitly assigned to constituents of a given syntactic unit, the former value standing for the head, and the latter for the dependent. The *Sub* value is not subject to vertical unification. We want to be able to automatically get dependency trees for any phrase structure tree our parser generates. It seems attractive to design such a device for those who prefer dependency structures over ours.

Comma Blockage. The last parameter *Comma* is designed also for technical purposes. FGP takes punctuation for a purely syntactic phenomenon which means that punctuation mistakes make expressions grammatically incorrect. Particular attention is given to comma, which appears as a constituent of a number of syntactic units. Moreover, there is a coordinate comma-conjunction in Polish, as well as in many other languages. It may happen that a syntactic unit ending in the comma proper (i.e., punctuational) immediately precedes another comma (either "punctuational", or "coordinating") within another unit. FGP in its original version introduces five special rules to account for an empty realization of a comma in the neighbourhood of another comma or another punctuation mark (period, exclamation mark, question mark, etc.). Unfortunately, the rules are context-dependent.

FGPn, which is context-free, proposes the parameter *Comma* to solve the problem. The idea is that a set of contexts should be listed in which no comma is allowed. There are syntactic units that have a comma or other punctuation mark as their final immediate constituent: utterances (always ending in a punctuation mark), coordinate sentences and phrases of all types, adjunct phrases (often comma-separated), nominal phrases, adjectival phrases, adverbial phrases, and sentential phrases. Rules defining those syntactic units take care of the constituent that immediately precedes the comma, be it a proper or "coordinating" one, or perhaps other punctuation mark. They fix the value of *Comma* of that constituent as c_less, while the other constituent of that syntactic unit gets the value standard (std).

Another classification of realizations of practically all syntactic units (sentences and phrases) results therefrom. Those of them that end in the comma get the c_less value, others being stamped with the value std. One can say they are impervious to comma. The *Comma* value a given syntactic unit gets from the outside is transmitted onto its final constituent and so forth. The parameter is somewhat analogous to that of initiality, both controlling boundary positions of the structure of syntactic units, the final and initial one, respectively.

6   Rules of the Grammar

Rules of our grammar, fairly numerous, have a term corresponding to the defined syntactic unit on their left hand side and a sequence of terms that correspond to immediate constituents of that unit on their right hand side. The sequence is 1 to 5 elements long; unlike in FGP, no empty element is possible as a realization of the defined construction. The right side is often supplemented with conditions that restrict values of various parameters.

As shown above, we use a number of notational conventions applied in the Edinburgh variant of DCG grammars. Names of terms, followed by a sequence of parameters in brackets, are given in lower case, same as fixed values of parameters. Unfixed values are abbreviated by a sequence of letters beginning with an upper case one. Conditions are given in curly brackets.

The set of rules accounting for the structure of nominal phrases comprises about 150 items. We group them generally according to dependents: nominal phrases with a sentential, adnumeral nominal (the head of the whole **nomp** is a numeral phrase), adnominal nominal, prepositional, and adjectival phrase. Within such subsets our rules are classified by context-boundedness for interrogative, relative, indirect-question, and neutral. For each of those subclasses we give separate rules that cover all possible permutations. Various realizations of the preterminal are listed, including terminals understood morphologically and certain phrases—adjectival or verbal.

Instead of presenting an arbitrary and accidental subset of rules of FGPn, we will take example (1) to show only rules that account for its structure. Our aim is to reveal all mechanisms formalizing the network of agreements nominal phrases are involved in. Actually, the sequence of rules we will put down is by no means the only one: the examined **nomp** could freely be described by a different sequence of rules of our grammar.

Let us recall the example:

(46)   jego pięcioro nowych studentów z tej grupy, w których imieniu przyszła najmłodsza, (= 1)

We will assume the phrase is embedded in the following sentence:

(47)   [Zaprosiłem] jego pięcioro nowych studentów z tej grupy, w których imieniu przyszła najmłodsza, [na kolację.]
'[I invited] his five new students from this group on whose behalf the youngest (girl) had come [for dinner]'

where it is an accusative realization of the syntactic unit called complement.

The structure of (46) is revealed by the following rule, one of the four rules defining nominal phrases with a constituent clause:

nomp(*Case, Gen/Num, Pers, Neg, I, Cb, Cl, Tk, Sub, Comma*) $\longrightarrow$ (no1)
    nomp(*Case, Gen/Num, Pers, Neg, I, Cb, Cl, Tk*, head, c_less),
    comma,
    sentp(ktory, no_corr, *Asp, Tense, Mood, Neg2, Gen/Num*, dep, *Comma*),
    { not_equal(*Cl*, co) }.

The rule gives this example the following interpretation:

(48)   [jego pięcioro nowych studentów z tej grupy] [,] [w których imieniu przyszła najmłodsza,]

In (47), the phrase (48) receives accusative as case value by virtue of one of the rules defining complements, i.e., from the outside. It is a syntactic motivation of the value. On the other hand, the value can be treated as originating from the inside. The accusative value is one of the three possible morphological readings of the word form *pięcioro*, which means it originates from the lexicon. If the nominal phrase just discussed contained, say, the adjectival word form *dobre* 'good$_{acc,pl}$' the accusative value would also be regarded as following from, or prompted by, the adjectival phrase that agrees for the case with the numeral head; it is an instance of the internal syntactic motivation. Actually, automatic syntactic analysis consists in checking if "external" and "internal" interpretations of a given syntactic unit match or not.

Case, gender, number, person, negation, initiality, context-boundedness, class, and internal modifier values are shared by the initial constituent nominal phrase. Of those features, the first four come from the lexicon, while negation and context-boundedness are syntactically motivated (the sentence (47) is affirmative and neutral, i.e., neither interrogative, nor relative, nor indq). The class value is not equal to co (in fact, the head is a real noun), which allows sentential phrases of the *który* type but blocks other. The internal modifier value comes from the inside and depends on the structure and contents of the whole phrase, which does not contain any tk or tyle terminal.

The rule, moreover, assigns the head value of the subordination parameter of the former constituent, the latter being dep. As the whole phrase immediately precedes the punctuational comma, the value of its parameter *Comma* is not equal to c_less: the value is shared by the final component, i.e., by a sentential phrase.

The former constituent of (46) (example (48)) is a nominal phrase that works as the head of the whole:

(49)   jego pięcioro nowych studentów z tej grupy

Its structure is revealed by the rule given below, one of four defining nominal phrases with genitive:

nomp($Case$, Gen/Num, $Pers$, Neg, I, neut, Cl, Tk, Sub, $Comma$**)** $\longrightarrow$ (no47)
   nomp(gen, $Gen1/Num1$, 3, $Neg1$, I, neut, $Cl1$, $Tk1$, dep, $Comma1$**),**
   nomp($Case$, Gen/Num, $Pers$, $Neg2$, $I2$, neut, Cl, $Tk2$, head, $Comma$**),**
   { not_equal($Cl1$, kto),
   inc_calc(I, $I1$, $I2$),
   phrneg_calc(Neg, $Neg1$, $Neg2$),
   not_equal(Cl, pers),
   imod_calc(Tk, $Tk1$, $Tk2$) **}.**

which treats (49) as a sequence of two nominal phrases, the former being a genitival dependent, and the latter—the nominal head:

(50)   [jego] [pięcioro nowych studentów z tej grupy]

We can see a vertical agreement for case, number, person, context-boundedness, and class here. The values of negation, initiality, and internal modifier are calculated by respective conditions: neither of the two constituents is negative, so the whole nominal phrase does not get the value no (actually, the value yes is inherited from the sentence (47)); both constituents bear the value ni (not incorporational) of initiality; none of them has the value tk or tyle of the internal modifier parameter. The value of the *Comma* parameter, fixed as c_less by rule (no1), reaches the final constituent by virtue of (no47). Two conditions not_equal reject constituent nominal phrases with pronominal heads of certain types—kto and pers (personal).

The final component of (47):

(51)   pięcioro nowych studentów z tej grupy

is structured by the following rule:

nomp(*Case, Gen/Num, Pers, Neg, I*, neut, nagr, *Tk, Sub, Comma***)** ⟶   (no20)
    nump(*Case, Gen/Num, Pers, Neg1, I1*, neut, nagr, *Tk1*, head, *Comma1***)**,
    nomp(gen, *Gen/Num, Pers2, Neg2, I2*, neut, *Cl2, Tk2*, dep, *Comma***)**,
    { imod_calc(*I, I1, I2*),
    phrneg_calc(*Neg, Neg1, Neg2*),
    not_equal(*Case*, nom),
    not_equal(*Cl2*, agr.nagr),
    imod_calc(*Tk, Tk1, Tk2*) **}**.

Again, we have got a two-component construction with a numeral phrase at the beginning and a nominal phrase at the end:

(52)   [pięcioro] [nowych studentów z tej grupy]

Vertical agreements look similar to those in (50), same as other calculations. The head of this **nomp** is a numeral phrase. Since the numeral head represents the nagr subset of numeral word forms the whole phrase gets the non-agreement value of *Cl*. The final constituent is a standard nominal phrase, not a numeral-nominal one, as guaranteed by rejection of **agr** or **nagr** value of the class parameter. The value of the *Comma* parameter touches the final component.

The final component of (52):

(53)   nowych studentów z tej grupy

gets its structure by means of this rule:

nomp(*Case, Gen/Num, Pers, Neg, I*, neut, *Cl, Tk, Sub, Comma***)** ⟶   (no69)
    adjp(*Case, Gen/Num, Deg, Neg1, I1*, neut, *Cl1, Tk1*, dep, *Comma1***)**,
    nomp(*Case, Gen/Num, Pers, Neg2, I2*, neut, *Cl, Tk2*, head, *Comma***)**,
    { inc_calc(*I, I1, I2*),
    phrneg_calc(*Neg, Neg1, Neg2*),
    imod_calc(*Tk, Tk1, Tk2*) **}**.

which looks like below:

(54)   [nowych] [studentów z tej grupy]

As is easy to see, case, gender, and number are unified both vertically and horizontally. The value of *Comma* drops on the final nominal phrase.

The final phrase in (54):

(55)   studentów z tej grupy

is defined by this rule:

nomp(_Case, Gen/Num, Pers, Neg, I,_ neut, _Cl, Tk, Sub, Comma_**)** ⟶   (no56)
    nomp(_Case, Gen/Num, Pers, Neg1, I1,_ neut, _Cl, Tk1,_ head, _Comma1_**)**,
    prepnomp(_Prep, Case2, Neg2, I2,_ neut, _Cl2, Tk2,_ dep, _Comma_**)**,
    { inc_ calc(_I, I1, I2_),
    phrneg_ calc(_Neg, Neg1, Neg2_),
    imod_ calc(_Tk, Tk1, Tk2_) **}**.

which gives us the following structure:

(56)   [studentów] [z tej grupy]

The agreement machinery works the same way as sketched above. There is a
special parameter _Prep_ that stores the information about preposition. _Case1_
stands for the case implied by a given preposition. _Comma_ lands on the final
constituent, as well.

The prepositional phrase given below:

(57)   z tej grupy

is accounted for by this rule, one of two defining prepositional-nominal phrases:

prepnomp(_Prep, Case, Neg, I,_ neut, _Cl, Tk, Sub, Comma_**)** ⟶   (pr1)
    prep(_Prep, Case, Neg1, I1,_ neut, _Cl1, Tk1,_ head, _Comma1_**)**,
    nomp(_Case, Gen/Num, Pers, Neg, I,_ neut, _Cl, Tk,_ dep, _Comma_**)**,
    { inc_ calc(_I, I1, I2_),
    phrneg_ calc(_Neg, Neg1, Neg2_),
    imod_ calc(_Tk, Tk1, Tk2_) **}**.

The structure is the following:

(58)   [z] [tej grupy]

The final component of (58) is covered by rule (no69).
    Let us analyse the final constituent of (48):

(59)   w których imieniu przyszła najmłodsza

It is a sentential phrase (we will not give here respective rules which would
reveal the structure of (59)). Rule (no1) fixes the value **ktory** 'who' of the type
of sentential phrase. Moreover, it transmits the gender and number values of the
whole **nomp** onto **sentp** as a value of initiality. Within **sentp** or, equivalently,
within a sentence by which the clause is realized the value percolates down to
reach the initial constituent. It is an adjunct phrase expressed by a prepositional-
nominal phrase:

(60)   [w] [których imieniu] (cf. 49)

The gender and number values land on the nominal constituent of **prepnomp** and on its initial **nomp** in the genitive *których* 'whose$_{gen,pl}$'.

The nominal complement in (60) is a **nomp** which has got an adjectival realization accounted for by the following rule:

nomp(Case, Gen/Num, 3, Neg, I, Cb, Cl, Tk, Sub, Comma**)** ⟶   (no86)
    adjp(Case, Gen/Num, Deg, Neg, I, Cb, Cl, Tk, head, Comma**),**
    { equal(Deg, comp.sup) **}**.

Unlike in FGP where such realizations were quite unrestricted (which resulted in getting many unwanted interpretations), FGPn lists but a number of specific structures, including comparative and superlative adjectival phrases, ordinal numerals, and certain adjectival pronouns.

Finally, the value of *Comma* of the sentential phrase is equal to std. This means that the punctuational comma is a final component of **sentp**. Were our nominal phrase (46) followed, e.g., by a period, its constituent **sentp** would have another realization—a c_less one.

What remains are rules that define preterminals. Let us give some examples:

nomp(Case, Gen/Num, Pers, tak, ni, neut, person, Tk, Sub, Comma**)** ⟶
                                                        (no114)
    pers_pron(nominal, Lemma, Case, Gen/Num, Pers, person**)**.

nomp(Case, Gen/Num, 3, tak, ni, rel, Cl, Tk, Sub, Comma**)** ⟶   (no120)
    rel_pron(nominal, Lemma, Case, Gen/Num, 3, Cl**),**
    { equal(Cl, co.kto.ktory) **}**.

nomp(Case, Gen/Num, 3, Neg, ni, neut, noun, Tk, Sub, Comma**)** ⟶   (no119)
    noun(Case, Gen/Num**)**.

for *jego*, *których*, and nouns, respectively.

7   Conclusion

The new account of **nomp** shown above significantly differs from that given in the original version of FGP. The hierarchy is designed as maximally flat, which, fortunately, does not result in any empirical limitations. Thanks to that type of redesigning, vicious circles no longer appear in the process of analysis. Our new set of parameters works better. The account of negation (parameter *Neg*) makes the structure and grammatical setup of nominal phrases depend on the presence of a negative terminal anywhere within such phases ("phrasal" negation); originally, FGP only transmitted the no value of the elementary (simple) sentence or of the verbal phrase onto their constituents ("sentence" negation). The parameter of context-boundedness now plays only one role, classifying nominal phrases according to whether a given phrase contains an interrogative, relative, or indq-type constituent. The class parameter (*Cl*) does more than that in FGP. It allows us to reveal the network of syntactic correspondences nominal phrases with numerals are involved in. The issue was left untouched in the previous version of FGP. Finally, an advanced mechanism of two parameters (comma blockage and initiality) is introduced to account for punctuational properties of nominal phrases; in fact, almost all phrases in FGPn.

Several of the adjustments we have made in the definition of **nomp** should be generalized to other syntactic units, both phrases and sentences: the *Comma* parameter is a good example thereof.

The reformulation of FGP which will follow what has been proposed for **nomp** seems to change its original philosophy. Intended as a theoretical account of the wealth of Polish syntax, FGPn does switch to practice. Flattening of the hierarchy makes automatic analysis more effective. Parsing results get simpler, clearer, and easier to read. All adjustments to the grammar have been made for theoretical purposes: to enrich the formal description of Polish nominal phrases. However, we hope that our work may also significantly increase corpus coverage of our grammar.

References

1. Abney, S.: The English noun phrase in its sentential aspect. Ph.D. thesis, MIT, Cambrigde (1987)
2. Bień, J. S.: Wizualizacja wyników analizy syntaktycznej. Poradnik Językowy 9, 24–29 (2006). http://www.mimuw.edu.pl/~jsbien/publ/PJ06/JSB-PJ06-9.pdf
3. Billot, S., Lang, B.: The Structure of Shared Forests in Ambiguous Parsing. In Meeting of the Association for Computational Linguistics (1989), 143–151
4. Colmerauer, A.: Metamorphosis grammar. In Natural Language Communication with Computers, L. Bolc, editor, Springer-Verlag, Lecture Notes in Computer Science 63, 133–189 (1978)
5. Derwojedowa, M., Rudolf, M., Świdziński, M.: A computational account of multi-word numeral phrases in Polish. In Linguistik International Band 10. Investigations into Formal Slavic Linguistics. Contributions of the Fourth European Conference on Formal Description of Slavic Languages—FDSL IV (2003), 405–415
6. Pereira, F., Warren, D. H. D.: Definite clause grammars for language analysis—a survey of the formalism and a comparison with augmented transition networks. Artificial Intelligence 13, 231–278 (1980)
7. Świdziński, M.: Gramatyka formalna języka polskiego. Rozprawy Uniwersytetu Warszawskiego. Wydawnictwa Uniwersytetu Warszawskiego, Warszawa (1992)
8. Świdziński, M.: Negativity Transmission in Polish Constructions with Participles and Gerunds. Journal of Slavic Linguistics 8, 263–293 (2000)
9. Świdziński, M.: Transmisja oddolna i odgórna negacji w zdaniu polskim: konstrukcje ze spójnikiem negatywnym. In Nie bez znaczenia... Prace ofiarowane Profesorowi Zygmuntowi Saloniemu z okazji jubileuszu 15000 dni pracy naukowej, Białystok, 275–285 (2001)
10. Świdziński, M.: A DCG account of Polish relative constructions. In Proceedings of the 3rd Language and Technology Conference, October 5-7, 2007, Z. Vetulani, editor. Poznań (2007), 478–482

11. Świdziński, M., Przepiórkowski, A.: Polish Verbal Negation Revisited: A Metamorphosis vs. HPSG Account. Technical Report 829, Instytut Podstaw Informatyki PAN, Warszawa (1997)
12. Świgra — komputerowa implementacja gramatyki formalnej Marka Świdzińskiego (2005). URL http://nlp.ipipan.waw.pl/~wolinski/swigra/
13. Woliński, M.: Komputerowa weryfikacja gramatyki Świdzińskiego. Ph.D. thesis, Instytut Podstaw Informatyki PAN, Warszawa (December 2004)
14. Woliński, M.: An efficient implementation of a large grammar of Polish. Archives of Control Sciences 15(LI), 3, 251–258 (2005)
15. Woliński, M.: Jak się nie zgubić w lesie, czyli o wynikach analizy składniowej według gramatyki Świdzińskiego. Poradnik Językowy 9, 102–114 (2006)
16. Woliński, M.: Morfeusz—a Practical Tool for the Morphological Analysis of Polish. In Intelligent Information Processing and Web Mining, IIS:IIPWM'06 Proceedings, M. Kłopotek, S. Wierzchoń, K. Trojanowski, editors. Springer (2006), 503–512

Morphosyntactic Constraints in the Acquisition of Linguistic Knowledge for Polish

Maciej Piasecki[1] and Adam Radziszewski[1]

Institute of Applied Informatics, Wrocław University of Technology, Poland
{maciej.piasecki,adam.radziszewski}@pwr.wroc.pl

Abstract. Many approaches to the construction of language tools and acquisition of linguistic knowledge from corpora assume the application of some robust shallow parser. Construction of such a parser is difficult in the case of inflective languages with relaxed word order like Polish. The goal of the work presented here is to analyse the extent of knowledge that can be expressed in the form of morphosyntactic constraints referring to morphological properties of word forms, and its applications in the automatic extraction of syntactic and semantic knowledge. Basic properties of an extended version of the language of morphosyntactic constraints called JOSKIPI are briefly presented. The application of morphosyntactic constraints as background knowledge for extraction of disambiguation rules for Polish is discussed. A new approach to extraction of lexical semantic relations is presented: it relies on the constraints in identifying lexico-morphosyntactic dependencies among word forms in the text. Finally, a combination of the constraints and statistical analysis in the acquisition of multiword expressions is outlined.

Key words: morphosyntactic constraints, morphosyntactic tagging, measures of semantic relatedness, decision trees, extraction of multiword expressions, annotated corpus, Polish

1 Introduction

In construction of many language tools and in methods of automatic extraction from corpora of language resources such as thesauruses it is typically assumed that a robust shallow parser exists. Moreover, the methods are often dependent on the syntactic structures previously identified in the text. The idea of using a shallow parser originates from English, which is a positional language. However, the implementation of a parser for inflectional languages with relaxed word order poses several difficulties. Indeed, there is no robust, publicly available shallow parser for Polish. On the other hand, *word forms* (henceforth WFs) in inflectional languages encode rich morphosyntactic information constraining the possible syntactic structures of language expressions. The general aim of our research is to utilise this type of information in the extraction of linguistic knowledge by means of Machine Learning and Statistical Learning methods without resorting to parsing. The goal of the work presented here was to analyse the extent of knowledge that can be expressed in the form of morphosyntactic

M. Marciniak and A. Mykowiecka (Eds.): Bolc Festschrift, LNCS 5070, pp. 163–190, 2009.

constraints referring to WF properties and its applications in the automatic extraction of syntactic and semantic knowledge.

First, we survey types of syntactic dependencies that can be identified without the application of a parser and we briefly present the basic properties of a language of morphosyntactic constraints called JOSKIPI.

Next, we analyse the application of morphosyntactic constraints as background knowledge for the extraction of disambiguation rules for Polish. The constraints are used to capture some necessary morphosyntactic dependencies and regularities supporting disambiguation decisions, but the constraints are not rules of disambiguation themselves. The constraint set transforms the example space by aggregating some combinations of values under new symbolic names and improves the performance of the learning algorithm (namely C4.5) as measured by tagger accuracy.

Given a morphosyntactically disambiguated corpus, the constraints can be employed to identify instances of selected syntactic dependencies. The dependencies associating occurrences of a given word with other words in the corpus are then used for the extraction of lexical semantic relations. Interestingly, the accuracy achieved by the whole process is reasonably high in spite of the medium accuracy of many morphosyntactic constraints themselves (their accuracy estimated on the basis of selective manual evaluation is presented in Table 3—Section 3). The positive outcome of relation extraction shows that the errors introduced by constraints during the analysis can be partially compensated by statistical regularities of the data collected from the whole corpus. Similar effects are observed when applying morphosyntactic constraints to collocation extraction and automatic extraction of syntactic structures of multiword expressions.

The compared examples of applications show that morphosyntactic constraints can be successfully applied to the extraction of different types of language knowledge from corpora. Existing approaches are based on the combination of Machine Learning and Statistical Learning methods, as well as the analysis of the corpora done with the help of more advanced language tools. Constraints, which are constracted with relative ease, can be considered as a possible replacement for the tools whose development requires more effort.

2   Language of Morphosyntactic Constraints

A complete description of syntactic dependencies, which could be extremely useful in many applications even if limited to intra-sentential information, is difficult to be obtain robustly for unrestricted text due to the limitations of deep parsers in coverage and processing efficiency. For many inflectional languages there are no shallow parsers at all; for others, the existing ones exhibit much higher error rate than shallow parsers constructed for English. However, WFs in inflectional languages like Polish deliver substantial morphosyntactic information constraining the syntactic structures in which they can possibly participate.

This information concerns different possible types of *agreement* and *government* associations based on grammatical categories. We will use a joint name of *compatibility relation* to refer to both types of associations. The reason for treating these types as one is that on the level of morphological description both are manifested in the same way: the values of particular grammatical categories are unified. For example, in the case of Polish there are three grammatical categories that participate in the majority of the compatibility relations, namely:

- *case*: governed by nouns, verbs, verbal participles, gerunds, prepositions and numerals, also agreed inside nominal phrases,
- *gender*: governed by nouns and agreed inside nominal phrases and usually in subject–verb constructions,
- *number*: governed by numerals, agreed inside nominal phrases and usually in subject–verb constructions.

The fourth one: *person* is limited to combinations of a nominal subject and a verb, but only pronoun subjects can express values different than the third person.

In the case of different WF pairs subsequent useful types of *complex compatibility* on several grammatical categories can be identified, e.g.:

1. *adjective–noun* compatibility on: case (agreement), gender, and number (both governed by the noun), inside a nominal phrase or in the attributive uses of *być* (*to be*),
2. *adjective–adjective* compatibility on: case, gender, and number, where the source of the compatibility is the relation of both adjectives to the head of a nominal phrase,
3. *numeral–noun* or *numeral–adjective*: case, gender, number (in a nominal phrase); however in the case of many numerals, case is assigned by the numeral,
4. *noun (subject)–verb*: person, number and gender in the case of some verb forms, together with case set to *nominative* in noun.

Only the government relation inside pairs of the general scheme: *preposition–a constituent of a nominal phrase*, and *verb–a constituent of nominal argument* are based exclusively on the category of case, whose value is imposed by the preposition and the verb (together with its polarity), respectively.

Several types of *complex compatibility* take place over proper sequences of words of a given language, e.g. adjective–noun combined with adjective–adjective across a nominal phrase. A complex compatibility is decomposable to a number of simple compatibility instances on the level of word pairs.

A particular compatibility type can be captured with the help of a formally expressed constraint applied to word pairs or subsequent words in sequences. During morphosyntactic tagging, such constraints can be a source of knowledge used in the decision process, e.g. if the given word pair: $\langle w_1, w_2 \rangle$ fulfils the constraint *adjective–noun*, then the disambiguation of w_1 as an adjective WF is supported. The number of possible constraints for word pairs is limited in every inflectional language, but open in the case of possible word sequences. However, we argue that the manual definition of constraints for word sequences is still a valuable solution in the case of Machine Learning methods.

Moreover, assuming that the input text has already been disambiguated morphosyntactically, constraints can also be applied to signal possible syntactic relations.

The primary purpose of JOSKIPI was to provide means for writing and extracting highly expressive disambiguation rules. Formally expressed *morphosyntactic constraints* and *derived morphosyntactic values* are used in order to describe relations between surface WFs and underlying syntactic structural relations.

Thus, a *morphosyntactic constraint* is a function: $S \rightarrow \{0, 1\}$, where S is a sequence of words, each being an instance of one or more WFs annotated by *grammatical classes* and the values of grammatical categories. Following the IPIC tagset [30], grammatical classes express a more fine grained division of WFs than Parts of Speech, e.g. verb WFs are divided into 12 grammatical classes.

Given a sequence of words satisfying some specified morphosyntactic constraint, one can retrieve values of the grammatical categories characterising them, e.g. the value of case on which the whole sequences: preposition–nominal phrase constituents, are compatible. We will call such values *derived morphosyntactic values*. Derived values partially characterise the properties of selected higher level constituents of the syntactic structure, but still we do not assume that any part of this structure is overtly or unambiguously identified in text.

The idea of the formalism originates from Constraint Grammar [16]. We are therefore focused on *sentences* and *positions*; a sentence is processed one word at a time, allowing for access to the central position (the word being processed) as well as its local context. JOSKIPI operators allow for (cf. [25,4]):

Expressing constraints (test operators). Simple tests are similar to that of Constraint Grammar; they allow for instance to check whether the preceding word has a reading belonging to the class of nouns and whether its case values include accusative. Complex tests can for instance verify if a compatibility holds according to the given morphological features (typically number, gender and case) extending through words between two given positions. Tests can be combined by means of boolean operators.

Retrieving values (simple operators, filters). It is useful for some processing tasks to gain access to values of selected attributes concerning a given position (for instance possible grammatical classes of the word at the given position). To cope with the inherent ambiguity of natural language, all values are treated as sets. The values returned can be compared with another set in order to express a constraint. There is also one complex operator here, the recently introduced *compatibility filter*: it checks against compatibility between two given positions and, if it holds, returns only the values of the selected attributes that satisfy the compatibility requirements. This helps reduce ambiguity and has been proved useful in morphosyntactic disambiguation.

Conditioning by constraints. Conditional operators are used to retrieve a value of a particular expression under a given condition (a constraint). If the constraint is not satisfied, an empty set is returned.

Constraint satisfaction search. Similarly to Constraint Grammar, JOSKIPI provides variables over positions. Such variables can be set by means of search operators—if a word satisfying a given constraint is found, the variable is set to its position. It can be then referred to, directly or by specifying a position relative to that variable. The search is driven by specifying the initial and final position, being a number, a variable or the sentence beginning/ending.

We will now present some selected JOSKIPI operators and their usefulness by providing an example sentence[1]. More details can be found in [25,4].

Sztuka utraciła swoją moc pobudzającą.
The art has lost its stimulating power.

Fig. 1 presents the words of the sentence along with their tags resulting from morphological analysis. Ambiguous words have more than one tag attached and represent multiple WFs. The tags are given using their original mnemonics from the IPI PAN Corpus [30] (henceforth IPIC) tagset. Each tag consists of a *grammatical class* symbol followed by values of *grammatical categories* applicable to this grammatical class. For instance, subst:sg:nom:f is the noun (subst) in singular number (sg), nominative case (nom) and feminine gender (f). The numbers given in the first row are relative position indices numbered relatively to the central word which is being disambiguated at the given moment; all words preceding the central one get the subsequent negative offset, the ones following receive positive offsets. For the sake of this example the word *moc* has been chosen for the central position (hence 0).

−3	−2
Sztuka	*utraciła*
art	lost
subst:sg:nom:f	praet:sg:f:perf

−1	0	1	2
swoją	*moc*	*pobudzającą*	.
its	power	stimulating	
adj:sg:acc:f:pos	subst:sg:acc:f	pact:sg:acc:f:imperf:aff	interp
adj:sg:inst:f:pos	subst:sg:nom:f	pact:sg:inst:f:imperf:aff	

Fig. 1. Example sentence with morphosyntactic tags.

---

[1]   It is actually the first sentence of the manually disambiguated part of the IPI PAN Corpus [30].

We will begin with simple operators returning values of a given attribute at a given position. The `flex`[2] operator returns the set of grammatical classes at a given position. Similarly, `nmb`, `gnd` and `cas` return respectively the values of: number, gender and case. For instance, the expression:

```
flex[0]
```

will evaluate to {subst} as the word *moc* is unambiguous with respect to its grammatical class. The expression:

```
cas[-1]
```

will evaluate to {acc, inst}, corresponding to possible cases of the word *swoją*—accusative and instrumental. For each grammatical category from the tagset there exists a corresponding JOSKIPI operator, returning values for the specified position.

Using these simple operators, we can construct complex constraints, e.g.:

```
and(
    equal(flex[0], subst),
    inter(cas[-1], cas[0])  )
```

This expression will evaluate to true if the central word is unambiguously a noun and the set of possible values of previous word's case has a non-empty intersection with the cases of the central word (in our example it is true as both words have common accusative reading).

A constraint satisfaction search can be used in order to find the boundaries of a phrase, on which another constraint will be subsequently applied. Although it is a heuristic approach with no strong syntactic support, interesting effects can be derived with quite simple expressions. Consider the following example.

```
and(
    rlook(0, end, $RBound,
          in(flex[$RBound], {adj,pact,ppas}) ),
    llook($-1RBound, begin, $LBound,
          in(flex[$LBound], {adj,pact,ppas}) ),
    agr($LBound, $RBound, {nmb,gnd,cas}, 3)      )
```

This constraint will be satisfied if:

1. there exists a word to the right with an adjective or adjectival participle reading (within sentence boundaries),
2. there exists a word to the left of the previously found word with similar readings,

---

[2]   The name comes from the notion of *flexeme*, which is the basis for the division of WFs into grammatical classes in the IPIC tagset [30].

3. there holds a strong compatibility according to the values of number, gender and case for all the words in the range; that is, there exists a path of tags throughout all of the words in the range that have the same values for all given attributes.

Given our example sentence, this constraint would be satisfied with the variable $LBound instantiated to position -1 and $RBound to position 1. It is because the words *swoją moc pobudzającą* constitute indeed a nominal phrase and the compatibility constrains possible readings to accusative singular feminine.

Sometimes this type of compatibility seems to pose too strong a condition for a phrase. There may appear an intervening word such as an adverb or a particle, which has no value for any of the attributes in question. One may also want to consider grammatical classes with, say, values of number provided leaving case and gender unspecified (various verb types). For such situations *weak compatibility* (wagr) is suitable. It is satisfied with a path of tags that either have the same values of given attributes or have some of the values unspecified, provided that the values of the first and last word in the range are fully specified (this additional constraint was introduced after experiencing problems caused by possible compatibility with full stops or commas). There is also a *point-to-point compatibility* operator (agrpp), which operates on two given positions, regardless of what words are in-between.

Although such a low-level means of expressing syntactic knowledge is prone to errors, it appears to capture regularities and subregularities well. Even if a constraint sometimes mispredicts a type of construct, it is very likely that it still acts in a consistent manner. It renders such heuristic constraints useful for Machine Learning tasks: having the JOSKIPI expressions evaluated, we pass their values to the classifier, which makes the final decision.

3 Extraction of Disambiguation Rules

In order to apply machine learning techniques to the extraction of disambiguation rules, a homogeneous representation of training examples was required. The baseline approach assumed selecting several attributes (namely part of speech, case, number and gender; consult [26] for details) and providing their values for each token in a fixed-length window. Selecting an appropriate window length is not a trivial decision. A typical length of 1–3 tokens both to the left and to the right limits the number of attributes to a reasonable number; on the other hand the resolution of some ambiguities is impossible without considering long distance dependencies.

Instead of making a difficult compromise between the small window length and the expressiveness of the rules, we applied a form of feature selection. A set of hand-crafted *patterns*, expressed using JOSKIPI operators, was provided for most common types of ambiguity. The patterns transform the set of features (values defined for each token in the window limited to the bounds of the whole analysed sentence) into a fixed-length vector of high-order features, being the output of JOSKIPI operators.

This transformation is conceptually similar to feature selection methods for machine learning (a survey may be found in [1]), with the difference that all the transformations result from manually written patterns instead of statistically acquired combinations of features. It follows the idea of *specialized features* introduced by Ratnaparkhi for "the words that are especially problematic for the model" [33]. The specialized features have a form of a conjunction of simple tests on the values of POS tags and words in the local context. Ratnaparkhi provided such specialized features for 29 "difficult words", but no significant improvement was observed.

It is worth stressing that a constraint satisfaction search allows for handling of unbounded dependencies, which results in an expressive power beyond that of any conjunction of simple tests on local context. Finding the gender of the closest possible anaphoric antecedent for a personal pronoun is an example of its use.

3.1   Patterns for Ambiguity Classes

Our experiments are aimed at improving the accuracy of the TaKIPI tagger [26]. It performs the task of morphosyntactic disambiguation in the following manner:

1. Input text is subjected to morphological analysis, resulting in a set of tags assigned to each word.
2. Those sets are reduced to tags appropriate for the context of the appearance of the word. Most often the tagger leaves one tag per word (1.03 tags on average).

An important assumption of the tagger architecture is the division of data into *layers* and *ambiguity classes*. Layers correspond to three subsequent stages of disambiguation; some of the tags are eliminated at each. The first layer deals with grammatical class ambiguity, leaving only tags belonging to one class per word. The second one disambiguates number and gender, the last one is responsible for case. Ambiguity classes are subsets of the attributes of a given layer; an example ambiguity class of the layer one is {subst, fin} (represents words that are ambiguous between noun and finite verb reading); {nom, acc, voc} is an example ambiguity class at layer three (case). The stages of processing an example sentence are presented in Fig. 2. A more detailed technical description of the TaKIPI architecture can be found, e.g., in [26].

Ambiguity classes are important here, because for each of them a separate classifier is constructed. What is more, patterns (describing high-level features for the classifier) are prepared for each ambiguity class separately. The training data is therefore divided into parts for each of the ambiguity classes; each training example consists of the values of each of the high order attributes as well as the decision—disambiguating the ambiguity class. Below is an example of the ambiguity class definition from the second layer (number and gender).

```
: sg pl m3 # <
   !nmbstd,
   !CoordNumber >
```

Input text

Wszystkie	kopie	zrobiono	z	rana
all	*copies*	*were made*	*in*	*the morning*

Tags after application of hand-crafted disambiguation rules

```
adj:sg:nom:n:pos  subst:sg:loc:m3  imps:perf  prep:gen  subst:sg:nom:f
adj:sg:acc:n:pos  subst:sg:voc:m3             prep:inst subst:sg:gen:n
adj:pl:nom:m2:pos subst:pl:nom:f                        subst:pl:nom:n
adj:pl:acc:m2:pos subst:pl:acc:f                        subst:pl:acc:n
adj:pl:nom:m3:pos subst:pl:voc:f                        subst:pl:voc:n
adj:pl:acc:m3:pos fin:sg:ter:imperf
adj:pl:nom:f:pos
adj:pl:acc:f:pos
adj:pl:nom:n:pos
adj:pl:acc:n:pos
```

Grammatical class disambiguation (Layer 1)

```
adj:sg:nom:n:pos  subst:sg:loc:m3  imps:perf  prep:gen  subst:sg:gen:n
adj:sg:acc:n:pos  subst:sg:voc:m3
adj:pl:nom:m2:pos subst:pl:nom:f
adj:pl:acc:m2:pos subst:pl:acc:f
adj:pl:nom:m3:pos subst:pl:voc:f
adj:pl:acc:m3:pos fin:sg:ter:imperf
adj:pl:nom:f:pos
adj:pl:acc:f:pos
adj:pl:nom:n:pos
adj:pl:acc:n:pos
```

Number and gender disambiguation (Layer 2)

```
adj:sg:nom:n:pos  subst:sg:loc:m3  imps:perf  prep:gen  subst:sg:gen:n
adj:sg:acc:n:pos  subst:sg:voc:m3
adj:pl:nom:m2:pos subst:pl:nom:f
adj:pl:acc:m2:pos subst:pl:acc:f
adj:pl:nom:m3:pos subst:pl:voc:f
adj:pl:acc:m3:pos
adj:pl:nom:f:pos
adj:pl:acc:f:pos
adj:pl:nom:n:pos
adj:pl:acc:n:pos
```

Case disambiguation (Layer 3)

```
adj:pl:nom:f:pos  subst:pl:nom:f  imps:perf  prep:gen  subst:sg:gen:n
adj:pl:acc:f:pos  subst:pl:acc:f
                  subst:pl:voc:f
```

Fig. 2. The stages of disambiguation.

```
!nmbstd (
  flex[-3],flex[-2],flex[-1],flex[0],flex[1],flex[2],
  cas[-3], cas[-2], cas[-1], cas[0], cas[1], cas[2],
  gnd[-3], gnd[-2], gnd[-1],          gnd[1], gnd[2],
  nmb[-3], nmb[-2], nmb[-1],          nmb[1], nmb[2]    )
```

Fig. 3. Definition of the *standard vector* consisting of the basic attributes.

This class represents those words that are ambiguous with respect to number (singular or plural) and of unambiguously masculine inanimate gender (m3). The pattern definition employs two macros, whose definitions are presented in Figures 3 and 4.

The first of the macros defines the standard vector—the basic features from the local context (positions from -3 to 2). !CoordNumber is an example of advanced use of JOSKIPI operators. It employs the *compatibility filter* operator in order to retrieve the value of number, which is common for both coordinated phrases. The operator is evaluated under the complex condition (the code following the second question mark in Fig. 4):

- a conjunction has been found, either to the left or to the right of the position being disambiguated,
- the range between the central word and the word next to the conjunction contains no prepositions (as they may introduce change of case).

If the condition is not satisfied, an empty set will be returned.

The *standard vector* macro defines 22 attributes, !CoordNumber defines one; thus each training example for this ambiguity class consists of 23 values (each of which is a binary-encoded set[3] followed by the class to select (the decision).

During the tagger's work, at first a set of hand-crafted disambiguation rules are applied (currently 35), partially reducing the input ambiguity. Then the

---

[3]   In the implementation of JOSKIPI a compact data representation is used, which allows for handling various domains of attributes in a homogeneous way. Each possible value of any grammatical category is assigned a unique bit; a set of such values is represented as a 64-bit word. Boolean values are simply represented as numbers; string literals are stored in a dictionary, which makes it possible to use the same 64-bit word to store the string index. The only case when an array of such records is necessary is when a set of string literals is to be used. For this reason every value is represented as a vector of 64-bit words, which most often contains exactly one element. This is also the reason why one can include operators of any domain in the pattern. After evaluating the pattern, the classifier is provided with an integer representation of the first element of the vector. This simplification makes it impossible to handle sets of string literals correctly as the direct input for classifiers; it is yet based on the assumption that it would hardly make any sense to feed the classifier with the infinite domain of strings.

```
!CoordNumber ( // evaluate agrflt under condition
    ? agrflt($BeforeConj,$AfterConj,cas,nmb,gnd,3,nmb) ? and(
      or(
            and( // look left for a conjunction
                llook(-2,-5,$BeforeConj,
                    inter(flex[$+1BeforeConj],conj)
                ), // set AfterConj to position 0
                rlook(0,0,$AfterConj,equal(nom,nom))
            ),
            and( // look right for a conjunction
                rlook(2,5,$AfterConj,
                    inter(flex[$-1AfterConj],conj)
                ), // set BeforeConj to position 0
                llook(0,0,$BeforeConj,equal(nom,nom))
            )
        ), // constrain the range not to contain prepositions
        only($BeforeConj,$AfterConj,$What,not(
            in(prep,flex[$What]) ))
))
```

Fig. 4. Definition of the *number of coordinated phrase* macro.

tagger enters stage one; a sentence is processed from left to right, an ambiguity class is selected for the word being disambiguated. The values of each of the attributes defined for the ambiguity class are calculated (by evaluating JOSKIPI expressions) and the classifier decides on which class to select.

3.2   Evaluation

To measure the impact of feature transformation, we prepared two versions of the tagger: one with a complete set of hand-crafted patterns and one only with the standard vector for each ambiguity class (consisting of the values of grammatical class, number, gender and case for each word in the local context $\langle -3, +2\rangle$). To compare these results with the benefits of hand-written rules, we decided to introduce one more distinction. Hence the four versions of the tagger:

T+R+P – full version, including rules and patterns.
T+R – tagger with hand-written rules, no patterns.
T+P – tagger with patterns, no hand-written rules.
T – plain tagger: only machine learning with default attributes.

The results[4] are presented in Table 1. The column *All words* provides the general accuracy of the tagger, while *Ambiguous* reports the accuracy calculated on

---

[4]   The figures differ from those published in [26,25] for of the following reasons: (i) the rules and patterns have been slightly improved (besides introducing new

Table 1. Comparison of the accuracy of different versions of the tagger.

Tagger	All words	Ambiguous
T+R+P	93.2608%	85.9305%
T+R	93.0085%	85.4038%
T+P	93.0820%	85.5573%
T	92.8999%	85.1771%

ambiguous words only (roughly half of the words in the benchmark corpus). The improvement is definitely not overwhelming, yet visible. It seems that the improvement resulting from patterns outweighs the benefit of hand-written rules. However, we cannot draw a firm conclusion from this observation as both rules and patterns have been designed assuming their cooperation. This hypothesis is supported by the numbers: the benefit of rules and patterns working together over rules only or patterns only is higher than either the improvement of introducing rules only or patterns only.

In Section 2 we discussed various compatibility-related operators. Two of them, namely the weak compatibility (wagr) and the compatibility filter (agrflt) have been introduced in the version of JOSKIPI presented here—they were absent in [25,4]. We have swapped strong compatibility for the weak compatibility operator in some patterns. Additionally, the compatibility filter has been used to develop three new macros. As a result 20 out of the 147 ambiguity class definitions now contain the weak compatibility operator and 51 employ the new macros based on the compatibility filter.

To verify the usefulness of these language extensions, we compared the accuracy of the tagger equipped with the new patterns (and no hand-written rules) with the accuracy of the tagger without rules and using old patterns — that is without the new macros and using strong compatibility only. The results are presented in Table 2. Again, the improvement is not impressive. It is however encouraging that the new patterns appeared to introduce a positive tendency and were applied in classifiers built by the C4.5 learning algorithm. The low improvement ratio was caused by the limited application of the new operators. In spite of the raw number of ambiguity classes affected, the new operators were added as an additional source of information competing with the basic set of operators established during the long-term development of TaKIPI.

3.3   Proposed Procedure for Developing Disambiguation Rules

Employing the described model of patterns and rules has interesting consequences for the whole process of developing disambiguation rules. First of all, it allows for an elegant combination of manually provided knowledge with machine

---

operators), (ii) some corrections in the annotation of the benchmark corpus have been made and (iii) experiments have been carried out without the optimisation of pruning confidence level of decision trees [12], which is a very time-consuming process. The evaluation procedure is described in [26].

Table 2. Comparison of the accuracy of the tagger with different pattern sets.

Patterns	All words	Ambiguous
Without `wagr` or `agrflt`	93.0732%	85.5389%
With `wagr` and `agrflt`	93.0820%	85.5573%

learning mechanisms. Further, one can extend the usual procedures for writing a disambiguation grammar (cf. Chapter 14 of [13] for a comprehensive description of such a process) in the following manner:

1. Apply existing disambiguation rules to the benchmark corpus.
2. Analyse the remaining ambiguities and formulate new rules.
3. Evaluate the new rules on the benchmark corpus.
4. Analyse the mispredictions and try to correct the rules.
5. If the rules are accurate enough, add them to the rule set and start over.
6. If the rules mispredict too often to be reliable, yet still reduce a considerable amount of ambiguity, consider using their conditions as patterns for Machine Learning. It is likely that their inclusion will improve the classifier's accuracy.

The described procedure is based on vague notion of what is reliable enough and, moreover, leaves the form of the patterns unspecified (the rule conditions can be used directly as patterns or as a condition under which some value is returned, as in the compatibility filter example above). Nevertheless, it seems a good starting point for writing patterns.

4   Extraction of Lexical Semantic Relations

In the previous section we discussed the application of morphosyntactic constraints in the extraction of disambiguation rules. This section deals with using the same formalism in order to extract semantic relations.

As the purely manual construction of large semantic lexicons requires large linguistic workload, one can recently observe a tendency to introduce automatic support. The applied methods and tools range from intelligent browsers, which extract *lexical units*[5] of related meaning from a corpus, to tools suggesting partial substructures of lexical units linked by *lexical semantic relations* (synonymy, hypernymy, meronymy, etc.). In further considerations we will focus only on the browsing methods as a representative example of a morphosyntactic constraint-based approach to lexical semantics.

The goal is to automatically extract a *Measure of Semantic Relatedness* (henceforth MSR) exclusively on the basis of a large text corpus. MSR is a

---

[5]   A lexical unit is understood here as a one word or multiword lexeme represented by its *lemma* (called also base form or entry form), i.e. a WF of the arbitrarily chosen values of grammatical categories, e.g. nominative case and singular number in the case of nouns. A lexical unit groups all WFs of the same meaning.

function: $L \times L \to R$, where L is a set of lemmas representing lexical units[6], and R is the set of real numbers. MSR should produce the higher values the closer semantically the given two lemmas are. A typical process of MSR extraction is based on the Distributional Hypotheses of Harris [14], according to which the lexical meaning is determined by contexts in which the given lemma (or WF) occurs in text. The types of contexts considered in literature differ in granularity, e.g. whole documents vs. the nearest surrounding words, and richness of their description, e.g. lemmas observed in text vs. pairs: a lemma and the type of syntactic dependency. Here, we limit ourselves to discussing the case where context is identified with a *lexico-syntactic relation*, e.g. the noun *ptak* (*bird*) can occur in the context which can be described as: 'being the subject of the verb *śpiewać* (*sing*)', henceforth written: subject_of_śpiewać (*sing*). Instances of lexico-syntactic relations are typically recognised in text by the application of a shallow parser. Later, we will show that morphosyntactic constraints can be used instead.

The MSR extraction process is performed typically in four general steps presented below.

1. The corpus is preprocessed by a shallow parser (morphosyntactic tagger in our case).
2. A *coincidence matrix* **M** is constructed, such that rows correspond to lemmas, columns to contexts and a cell stores the number of occurrences of the given lemma in the given context.
3. The matrix is transformed in order to decrease the influence of statistical noise in data.
4. Semantic relatedness of lemmas on the basis of similarity of the corresponding rows is calculated.

Many methods of MSR extraction assume employing a shallow parser, e.g. in [20,11] the MiniPar parser [19] is applied to identify instances of different types of syntactic relations. But, a robust shallow parser is not available for Polish. The dgp parser [23] has limited coverage and produces all candidate parses without any ordering. However, we assumed that in order to grasp the meaning of a lemma w (as representing some lexical units) on the basis of its occurrences one needs only to identify those lexico-syntactic relations that contribute to the basic information characterising the meaning w. For example, in the case of a noun w: its modification by adjectives and nouns gives hints about possible features that can be attributed to objects denoted by lexical units corresponding to w; the position of w in the argument structures of subsequent verbs reveals the role played by w-objects in the eventualities represented by those verbs. It seems possible to describe occurrences of these basic contexts by relatively

---

[6]   In some approaches L is simply a set of WFs; in the case of inflectional languages such an approach would be unfeasible as the number of different WFs can be enormous. The lexical units themselves cannot be distinguished in the MSR results, when it is constructed on the basis of a corpus which has not been annotated by word senses.

simple morphosyntactic constraints, especially when one accepts lower accuracy of recognition than the one expressed by a robust parser.

The goal of our experiment was to extract an MSR for Polish nouns on the basis of the morphosyntactically tagged corpus (using TaKIPI). Four types of basic contexts were identified and described with constraints:

A — modification by a *specific adjective* or a *specific adjectival participle*
Nc — co-ordination with a *a specific noun,*
Vsb — co-occurrence of *a specific verb* for which a given noun can be its subject,
Nmg — modification by *a specific noun* in the genitive case.

The A constraint is presented in a simplified form in Fig. 5. It is assumed that the constraint is run with the noun being described located in the centre of the context, i.e. at the position 0. In the first step, we are looking to the left (the llook operator) for a *lexical element of the context,* in this case we look for an adjective or adjectival participle, specified by its lemma equal to some specific "*lemma name*". Only lexical elements that express compatibility according to number, gender and case are taken into account. In the second step we must check (to some extent) whether the found lexical element is really associated with the noun in the centre or, instead, with some other word in between. In this part of the constraint we check some selected aspects of the syntactic structure, but still the level of detail, and precision, is significantly lower than the one of a parser. The manual evaluation of the precision of the constraints is presented in Table 3 and discussed below.

The core functionality of the Nmg constraint presented on Fig. 6 does not depend on compatibility as there is no compatibility means between the head noun and a noun in genitive case modifying the head. In Nmg we look for the specified nominal lemma represented by the WF in the genitive case, which can possibly be a part of the same nominal phrase. The search is performed only in the right context according to the unmarked order of the Polish nominal phrase constituents. Obviously the modifying genitive WF can also precede the head, but such cases are intentionally not recognised by the constraints in order to increase the precision at the cost of decreased completeness. Only position is important, as the genitive nominal WF presents no compatibility with the head, and there is no syntactic means to differentiate between the head and the noun modifier when both are in the genitive case. The modifier can be separated by adjectives, adjectival participles and numerals which are compatible with the head and other nominals in genitive, as well. Unfortunately, the problem of the semantic analysis of the nominal phrase constituents in order to distinguish between the real noun modifiers and the head is very difficult. So, any potential genitive modifier is recognised as the proper one; the constraint precision given in Table 3 was calculated bearing this assumption in mind.

Most of the observed errors resulted from tagger errors; other reasons were: incorrect separation of subsequent nominal phrases in genitive (e.g. title plus the full name of a person), incorrect recognition of adjectival participle arguments as genitive modifiers of the nominal phrase head. However, any attempts to make the constraints more precise could reduce its coverage, while our intention was

```
or(
  and( llook(-1,-5,$A,and(
                         in(flex[$A],adj,pact,ppas),
                         equal(base[$A],"lemma name"),
                         agrpp(0,$A,nmb,gnd,cas,3)  )),
      or(
          only adjectives, numerals adjectival participles, adverbs,
                 adverbial participles and punctuation
                 between the position -1 and $A
          and(
              there is not other verb than ''być'' (to be)
                         between -1 and $A
              only words without case or in case different from genitive
                         between -1 and $A
              there is no preposition before a possible noun
                         between -1 and $A ) ) ),
  and(a similar symmetrical condition for the right context)
)
```

Fig. 5. Example of a constraint: an *specific adjective* or an *specific adjectival participle* as a modifier of the given nominal lemma (A).

to use the constraint for collecting information describing different contexts in which given lemmas can be used. So, we assumed that the lower precision of the constraint can be compensated, to some extent, on the level of MSR extraction by the constraint's increased coverage.

The constructed MSR was limited to the description of 13 285 Polish nominal lemmas (one-word or two-words), corresponding to the at least as large a number of lexical units. A larger number of lemmas described would cause problems with the efficiency of matrix processing. Moreover, we used the complete set of nominal lemmas planned to be the basis for the first version of a Polish wordnet called *plWordNet* (*Słowosieć* in Polish, henceforth plWN) [6]. They consist of:

- nominal lemmas described in the *plWN core* constructed manually,
- and lemmas acquired from:
    - the small Polish-English dictionary [29],
    - two word lemmas from the general dictionary of Polish [32],
    - and the lemmas occurring more than 1 000 times in in the IPI PAN Corpus (IPIC) [30] — the largest corpus of Polish publicly available for research applications.

Besides IPIC, we also used two other corpora:

- the corpus of the electronic edition of *Rzeczpospolita* (a Polish newspaper), about 113 million tokens (atomic segments), full editions from the years: January 1993 to March 2002,

```
and(
   rlook(1,end,$B,and(  in(flex[$B],subst,ger,depr),
                                equal(base[$B],''base form''),
                        equal(cas[$B],gen) )),
      only(1,$-1B,$Ad, or(
                          the $Ad variable points to adverb
                              or adverbial participle
                          $Ad points to a noun in genitive
                          $Ad points to an adjective, an adjectival
                              participle or a numeral which is compatible
                              with respect to number, gender
                              and case with the noun in centre
                      ))//only
   )
```

Fig. 6. Example of a constraint: a *specific noun in genitive* as a modifier of the given noun (Nmg).

- and a corpus of large electronic text documents written in Polish collected from the Internet — only texts in which majority of tokens were Polish proper WFs and which did not occur in any of the other corpora were kept, 214 million tokens in total.

In the coincidence matrix, the rows corresponded to the selected nominal lemmas and columns to the four lexico-syntactic constraints instantiated by lexical units acquired directly from the used corpora. Only lemmas occurring more than 5 times in the joint corpus were kept for processing.

The accuracy of the constraints was evaluated manually according to the sets of lemmas used as lexical elements of the constraints. The results are presented in Table 3. For each constraint its number of matches in IPIC was established and next a sample of matches was randomly drawn. Each match was stored and evaluated in the form of the whole sentence with the instances of the two lemmas: described and lexical elements, marked. The sample sizes were chosen according to the method described in [15], in such a way that the results of the sample evaluation can be ascribed to the whole set with 95% confidence level.

As one could expect, the highest accuracy was achieved for the A constraint, which is based strongly on compatibility. The majority of errors are due to tagger mispredictions. In some cases an adjective located between two nouns of the

Table 3. Accuracy of the lexico-morphosyntactic constraints.

	Constraints			
	A	Nc	Nmg	Vsb
Precision [%]	97.39	67.78	92.36	80.36

same values of the analysed grammatical categories was mistakenly associated with the wrong one. The satisfactory result of Nmg was, to a large extent, artificially increased by the aforementioned loose definition of the genitive nominal modifier assumed in Nmg and its evaluation, e.g. we did not distinguish genitive arguments of a gerund which modifies the head from the proper genitive modifiers of the head. However, the relatively good result of subject identification, i.e. Vsb, achieved with the help of quite a simple mechanism of the constraint, is worth noticing.

The constructed coincidence matrix was next transformed by the *Rank Weight Function* (RWF). RWF has been previously presented in detail and discussed, e.g. [27,28], here we briefly summarise only the main steps. A constraint instantiated by a lemma will be henceforth called a *feature*.

1. *Global filtering*: entropy of features (columns) is calculated and all features exceeding the maximal entropy threshold or occurring less than 5 times are eliminated; also the MSR value is calculated only for nominal lemmas occurring at least 5 times in the joint corpus.
2. *Weighting*: some function f_w measuring the relevance of a feature for the given lemma is applied and the returned values are stored in cells; next cells with values below the threshold τ are set to 0.
3. *Transformation to ranks*: cells in a row (corresponding to a lemma) are sorted in descending order, and next their values are set to the respective *ranks*; two schemes of rank values are used: *absolute* — the highest rank for every lemma is equal to the parameter r, and *relative* — the number of non-zero cells is taken as the highest rank.

After the application of RWF(f_w) pairs of row vectors from the transformed matrix are compared by the cosine measure in order to compute the MSR values for the respective lemmas.

Several different functions were tested as f_w and the best results were obtained with *Lin's weighting function* [20] and *z-score* measure.

The transformation to ranks reduces the dependency of MSR on the exact frequencies collected from a corpus and the weight values expressing the statistical relevance of the lemma–feature associations. The values depend on the frequencies of lemmas and features which are strongly influenced by the corpus, which is never ideal and always biased. Low frequency co-occurrences, which can include some possible errors produced by constraints are filtered out by thresholds. However, the large statistical mass of the joint corpus gives us the possibility to omit them without any loss in the accuracy of the final MSR. Possible constraint errors among co-occurrences of higher frequency are masked to some extent by the RWF transformation according to which the exact value of the weighted feature is not as important as the order of the feature values. The above assumptions are supported by the observed behaviour of RWF which increases the accuracy of an MSR in comparison to the application of the weighting function itself, e.g. [28,3].

Evaluation of MSR accuracy is always a serious problem. Several different strategies can be applied, e.g. [40,28]. For the experiments discussed here we applied a version of the *WordNet Based Synonymy Test* (WBST) introduced in [10], adapted to Polish and further developed in [28]. WBST is a kind of evaluation by application and an MSR is used to distinguish between a synonym and three non-synonyms of the given question lexical unit q, e.g. (the proper answer is marked by the bold case)

Q: *mebel (piece of furniture)*,
A: *enzym (enzyme)*, *paliwo (fuel)*, *płyta (plate, board, panel, record)*,
     ława *(bench)*.

A WBST consists of several thousand such questions generated automatically on the basis of a wordnet: q and the proper answer are selected randomly from the same *synset*[7], wrong answers are drawn randomly from the synsets including neither q nor the correct answer. For our experiments we used plWN. Because plWN includes a substantial number of singleton synsets, we used a modified version of WBST called HWBST (see [28]), in which a correct answer can be also a hypernym of q if q belongs to a singleton synset.

WBST-like evaluation is indirect, by application, but the task is very similar to the identification of semantically related lexical units to the given one in the corpora, which is intended to be a tool supporting linguists in development of plWN. Thus, it seems plausible that better support can be expected from MSRs which achieve higher results in HWBST.

On the basis of the version of plWN from the August 2008, two HWBSTs were generated. The first including 10 428 question–answers pairs for all processed lemmas, and the second one, including 1 898 QA pairs for lemmas occurring more than 1000 times in the joint corpus (henceforth called *frequent lemmas*). In order to compare the influence of subsequent constraints we constructed several MSRs: on the basis of one of the four constraints separately, and all four jointly. The results of the evaluation for all lemmas and frequent lemmas are presented in Table 4. The exact values are lower than presented previously, e.g. in [28], but here the newer, improved version of plWN was used to produce HWBSTs.

We can notice in Table 4 that the application of RWF improves the accuracy in comparison to Lin's measure alone. The difference is statistically significant on the basis of McNemard's test, e.g. [7]. The difference RWF(z-score) vs Lin's

Table 4. Accuracy [%] of different MSRs based on RWF together with different constraints and two weighting functions in comparison to Lin's measure alone.

	A		Nc		Nmg		Vsb		all	
	≥ 1000	all	≥ 1000	all	≥ 1000	all	≥ 1000	all	≥ 1000	all
Lin	79.28	77.55	65.76	61.79	75.88	72.74	63.27	59.83	80.87	79.25
RWF(Lin)	81.02	80.24	68.44	64.58	78.51	75.64	66.06	62.84	**82.29**	**81.78**
RWF(z-score)	80.49	79.84	68.92	65.01	77.31	74.99	64.72	61.37	81.25	81.13

---

[7]   A synset is a set of nearly synonymous lexical units and is the basic building element of a wordnet.

measure is not significant, but the overall tendency is in favour of RWF(z-score). The combination of all constraints slightly improves the accuracy, but in all cases an MSR based on all constraints is significantly better than an MSR based on any constraint alone, e.g. the smallest differences are between MSRs based on *all* vs A constraints, but even those differences are significant. However, the most important thing is that the results for all and frequent lemmas come closer to each other after we introduce more constraints. Thus, the additional constraints improve the description of the less frequent lemmas.

A high accuracy MSR is typically built on the basis of parsed texts in which several types of lexico-syntactic relations are identified. There is no such possibility in the case of Polish, and the proposed mechanism of morphosyntactic constraints is selective, i.e. oriented only on a few specified relations, limited in its expressive power in comparison to a parser and provides average accuracy (with the exception of A whose high accuracy is due to the rich compatibility required). Nevertheless, the extracted MSR provides very good accuracy, allowing for practical applications. Cumulative errors of TaKIPI and the constraints are seldom observed in the MSR results. The statistical mass of collected data successfully masks the errors. For the task of MSR extraction a simple tool of morphosyntactic constraints appeared to be a robust solution. Moreover, the development of one constraint takes about two working days including tests. Development of a shallow parser takes much longer time.

5   Extraction of Collocations

According to the first approximation, the structure of natural language sentences is determined by the lexicon of lexemes and morphological, syntactic and semantic rules. All language expressions matching them should be correct and met in a large corpus. However, some sequences of WFs seem to be bound closer than others: their constituents co-occur more often and changes (if any) in such structures are very restricted. There is no general name for this broad class of 'non-atomic language units', yet subsets of the class (varying in the scope of their semantic properties) are called: *collocations, fixed expressions, terms* or *proper names*. Here, we call them jointly *multiword expressions*[8] (MWEs). MWEs introduce a kind of fixed points into the space of possible language expressions and thus are very important for Natural Language Processing, as they strongly limit the number of probable language constructions. MWEs form a significant part of the lexicon as the majority of them are lexical units. Unfortunately, dictionaries list only a limited number of idioms (and almost no collocations), partly because MWE lists are very large and many MWEs are domain-dependent. A possible solution can be automatic recognition of MWEs.

There are plenty of methods for the recognition of MWEs, inspired by the seminal paper [35]. Most of them are based on the statistical measures of likelihood of the co-occurrence of two WFs in texts. This general scheme works

---

[8]   In literature, the term *collocation* is also used in a broad sense.

Table 5. Examples of lists (G — *good* and Acc — *accidental*) of the 20 most similar lemmas to the given one according to the MSR based on RWF(z-score).

złudzenie (*illusion*) (G)		klub (*club*) (G)	
iluzja (*illusion*)	0.1979	drużyna (*team*)	0.2306
mrzonka (*fiction*)	0.1650	zespół (*team*)	0.2161
nadzieja (*hope*)	0.1367	liga (*league*)	0.2135
omam (*hallucination*)	0.1298	szkoła (*school*)	0.2078
urojenie (*delusion*)	0.1235	ośrodek (*centre*)	0.1969
halucynacja (*hallucination*)	0.1231	sekcja (*section, division*)	0.1947
wyobrażenie (*imagination*)	0.1209	klub poselski (*parliamentary club*)	0.1801
fantazjowanie (*fantasizing*)	0.1180	działacz (*activist*)	0.1662
ułuda (*delusion*)	0.1171	federacja (*federation*)	0.1659
mit (*myth*)	0.1119	koło (*circle, wheel*)	0.1614
fantazja (*fantasy*)	0.1066	środowisko (*environment*)	0.1600
fikcja (*fiction*)	0.1051	placówka (*outpost*)	0.1597
lęk (*anxiety, fear*)	0.1003	uczelnia (*university, college*)	0.1583
utopia (*utopia*)	0.0983	liceum (*secondary school*)	0.1578
wymysł (*fabrication*)	0.0955	stowarzyszenie (*association*)	0.1568
kłamstwo (*lie*)	0.0938	towarzystwo (*assoc., company*)	0.1561
wiara (*faith, belief*)	0.0922	turniej (*tournament*)	0.1560
przywidzenie (*visual hallucination*)	0.0918	grupa (*group*)	0.1549
słuch absolutny (*absolute pitch*)	0.0904	impreza (*party, event*)	0.1538
sen (*dream, sleep*)	0.0884	szpital (*hospital*)	0.1512
książka (*book*) (G)		**port lotniczy (*airport*) (Acc)**	
powieść (*novel*)	0.311693	lotnisko (*airport*)	0.2754
publikacja (*paper*)	0.245929	dworzec (*railway station*)	0.1743
czasopismo (*journal*)	0.228347	port (*port*)	0.1515
dzieło (*work*)	0.224221	konserwatorium (*conservatoire*)	0.1373
broszura (*brochure*)	0.207965	terminal (*terminal*)	0.1237
tekst (*text*)	0.202943	uniwersytet (*university*)	0.1175
utwór (*piece of art*)	0.196941	stacja (*station*)	0.1111
film (*film*)	0.192953	uczelnia (*university, college*)	0.1100
podręcznik (*handbook*)	0.188993	rozgłośnia (*broadcasting station*)	0.1081
artykuł (*article*)	0.188355	szpital (*hospital*)	0.1045
album (*album*)	0.187306	giełda (*market*)	0.1034
wydawnictwo (*publisher, p. item*)	0.184067	bazar (*bazaar*)	0.1027
tomik (*volume*)	0.183207	ogród zoologiczny (*zoo*)	0.0998
literatura (*literature*)	0.181711	socjeta (*society*)	0.0994
esej (*essay*)	0.181679	gazeta (*newspaper*)	0.0977
proza (*prose*)	0.181049	hotel (*hotel*)	0.0976
monografia (*monograph*)	0.176563	spedycja (*shipping*)	0.0955
księga (*large book*)	0.175754	dziennik (*newspaper, diary*)	0.0951
biografia (*biography*)	0.175241	filharmonia (*concert hall*)	0.0923
fotografia (*photography*)	0.175029	autostrada (*motorway*)	0.0918

fine for languages with fixed word order and limited morphology like English, but it has two significant drawbacks in the case of inflective languages like Polish. Firstly, a Polish MWE is represented in text by several different combinations of constituent WFs depending on the value of grammatical categories of the whole. Secondly, the relaxed word order of Polish causes the number of observed WF sequences to multiply further. For example, the following two sequences: *tłusty druk* (lit. 'fat print or type', *bold face font*$_{case=nom, number=sg}$) and *tłustym drukiem* (*bold face font*$_{case=inst, number=sg}$), represent the same MWE. Thus, the statistical significance of the co-occurrences of MWE constituents must be measured across different possible combinations.

However, most methods of MWE recognition are based on the identification of such sequences of WFs that are more frequent than can be expected on the basis of the probabilistic distributions of their constituents. Many statistical and heuristic measures based on statistics have been proposed, e.g. see extensive lists of measures surveyed in [8,5,24]. Statistical identification of significantly frequent sequences is often accompanied by additional pre- and post-processing, especially in the case of languages with rich inflection. During preprocessing, the text is first filtered against *stop lists* of meaningless, too general or unknown WFs. In the following step, the text is analysed morphosyntactically in order to annotate it with PoS and values of morphosyntactic categories, e.g. case, gender, number, tense. Moreover, lemmas (or morphological base forms) can also be assigned to WFs in text. In the case of Serbian (cf. [22]), the preprocessing stage was extended with syntactic filters (implemented as regular expressions) identifying potential terminology.

There is a limited number of works on Slavic languages ([22,34,37]) and only one for Polish: Buczyński's *Kolokacje* system (cf. [5]) is based exclusively on the statistical recognition of significantly frequent two-word sequences of WFs.

One of the possible solutions is to identify different possible forms of an MWE during the collection of frequencies from corpora. In general, such a process would require the syntactic analysis of word sequences in the text, in order to identify those that form proper expressions. As the computational cost of this analysis would be high in the case of a very large corpus, we proposed a three-step method based on the combination of morphosyntactic constraints and statistical processing, cf. [2]. Here, only the general scheme and the most characteristic features of the approach will be briefly presented to complete the picture of the applications of morphosyntactic constraints.

The MWE recognition process is divided into three phases:

1. *lemmatisation of WFs* — all WFs are reduced to lemmas,
2. *statistical recognition* — frequent sequences of lemmas are identified and marked as *potential MWEs*,
3. *statistical syntactic filtering* — data about frequencies of potential collocations satisfying specified syntactic constraints is collected from the corpus, tested for statistical significance and a list of structurally annotated collocations is generated.

The goal of the first step is to estimate frequencies of MWEs by a kind of generalisation from particular WFs — all are reduced to lemmas, e.g. the pair of lemmas: *tłusty* (*fat*) and *druk* (*type*) represents all occurrences of this pair of lexemes. The base form disambiguation of ambiguous WFs was performed with the help of TaKIPI. The accuracy of lemmatisation was 99.31% as measured on the manually disambiguated part of IPIC.

In the second step, a list of potential MWEs is produced according to the selected statistical measure. We tested several measures implemented in the ready-to-use system *Kolokacje* [5], achieving the best results (according to the selective manual evaluation) for the *Frequency Biased Symmetric Conditional Probability* (FSCP) [5]. All tests were performed on IPIC and only two word potential MWEs were extracted according to the limitations of the tool.

The reduction of MWE occurrences to lemmas enables collecting and counting of different occurrences of given MWE, but also introduces some substantial statistical noise, e.g. the sequence:

... *tłusty*$_{case=nom, number=sg}$ *drukiem*$_{case=instr, number=sg}$ ...

will be reduced to *tłusty druk* and wrongly counted as an occurrence of this MWE, while both WFs are not compatible with respect to case, so it cannot be an MWE instance. For this type of MWE, we can write a morphosyntactic constraint expressed in JOSKIPI defining the necessary syntactic relations between the constituents (the position 0 is the first word in a pair):

```
and(
    in(flex[0],adj,adja,adjp),
    in(flex[1],subst,depr),
    agrpp(0,1,nmb,gnd,cas,3) )
```

Such a constraint is called a *constructional constraint* (CC) and must be satisfied by any instance of the given MWE class. After the manual inspection of potential collocations we identified several syntactic collocation types, namely: `Adj-Noun`, `Noun-Adj`, `Noun-Noun`, `Verb-Noun`, `Noun-Verb`, `Adj-Verb`, and `Verb-Adj`. Each type is characterised by one or more patterns of necessary syntactic relations[9]. For each pattern a corresponding CC was defined. During the step 3 we check if the number of the WF pairs satisfying the appropriate CC, written $CC(\langle b_i, b_j \rangle)$, is significantly large in comparison to some accidental value. For that purpose we used the standard *z-score test*. During that process, CCs selected on the basis of grammatical classes of WFs in corpus are run only against occurrences of the potential MWEs. In that way we reduced the problem of syntactic filtering to satisfaction of simple morphosyntactic constraints and limited the amount of processing necessary.

Moreover, not all MWEs occur in all their possible forms, e.g. all WFs of *Wysoki Sejm* (an expression used by the parliament members during speech while referring to the audience) occur only in singular number, i.e.

---

[9] For example, there are three possible patterns for `Noun-Noun`: one for names and two genitive patterns.

Wysoki Sejmie. In order to identify such properties we introduced an additional set of constraints for each class of potential collocations, called *specifying constraints* (SCs). Each SC is defined indirectly by a template of all possible significant syntactic regularities of WF pairs corresponding to potential MWEs. The template is written in the form of a sequence of JOSKIPI operators — $o_1, \ldots, o_k$. The regularities are statistically significant patterns of values of the operators, i.e. $o_1 = v_{1,i}, \ldots, o_k = v_{k,j}$. All possible instances of SC (of any subsequence of operators) for the given CC and potential collocation are generated and tested. The instances satisfying the *z-score* test (with 99.5% confidence) are stored as significant regularities, e.g. for *wysoki sejm* SC = and(cas[1],nmb[1]) and the SC instance equal(nmb[1],sg) was automatically found as statistically significant.

The precision of MWE extraction was evaluated manually by two linguists, and ranges from 24.46% to 75.63%. However, the inter-judge agreement which was measured by Fleiss's kappa [9] was either very low or low (depending on the granularity of evaluation classes). Thus, the criteria of MWE identification are vague and the achieved precision expresses at least a positive tendency. Manual inspection showed that the number of proper MWEs among those extracted is high enough to use the list of extracted MWEs as a supporting tool for a lexicographer.

The mechanism of CCs was also successfully applied in reverse for the automatic extraction of the syntactic structure of new lexical units being added to the dictionary. New multiword lexical units, which were acquired from a dictionary, e.g. [29], in a form of their lemmas without any annotation, required a description of their syntactic structure to be recognised properly during corpus processing. However, manual description of several thousands of new lexical units could be laborious. Instead, we applied the following procedure; for each new multiword lexical unit w:

1. identify sets of grammatical classes corresponding to subsequent lemmas of w;
2. for each possible sequence of grammatical classes identify the appropriate CCs;
3. statistically check each CC against the corpus iterating across potential occurrences of w;
4. choose the CC of the highest statistical significance as the description of w.

The procedure was then extended with heuristic rules for the recognition of lemma syntactic head and how fixed is the linear structure of the given lemma. The latter is used in optimising the efficiency of lemma recognition during corpus processing. Besides errors caused by mistakes of morphological analysis [39], no errors were observed in the acquired descriptions. As CCs are morphosyntactic constrains expressed in JOSKIPI defining the necessary properties, they can be used directly for multiword lemma recognition.

6 Conclusions

In inflective languages with relaxed word order, the linear sequences of language expression is a secondary means of encoding syntactic structures in comparison to morphological properties of WFs. This trivial observation is supported by the role played by the recognition of morphosyntactic compatibility (i.e. agreement and government) in applications of JOSKIPI. Even in the case of ambiguous annotation, compatibility operators deliver the most reliable knowledge. In the case of lexical semantic acquisition the worst results are achieved with the help of the Nmg operator based almost exclusively on word sequences.

Most of the morphosyntactic operators discussed earlier work on word level — more complex syntactic structures are neither recognised nor further used. Nevertheless, the recognition of morphosyntactic binary relations among WFs in a sentence, appeared to be a rich source of information when combined with large scale statistical processing, e.g. a prominent example of the good results achieved in lexical semantics acquisition. Morphosyntactic constraints are also a sufficient tool in the acquisition and recognition of MWEs. Constraints can be written and tested in a much shorter time than even a shallow parser.

Morphosyntactic constrains can be used to express fundamental linguistic knowledge, a kind of necessary conditions. They are easy enough to be quickly encoded by a linguist, and in combination with Machine Learning techniques they can be used to create complex models, e.g. TaKIPI. The constraints define stable point in the learning process, while automatically acquired rules can approach an approximated description of elusive mutual dependencies in the language structures as perceived from the limited lexico-syntactic-only perspective, e.g. TaKIPI.

The version of JOSKIPI presented in Sec. 2 differs from the earlier versions mainly in the extended set of the compatibility operators. We plan to extend control structures of JOSKIPI, e.g. with explicit variable assignment and functions with arguments. Further development should also introduce a possibility to describe syntactic structures and manipulate them from within the language, either by using labels denoting the syntactic function of a word (an idea from Constraint Grammar [16]) or by grouping words together and labelling such groups (cf. *syntactic words* and *syntactic groups* used in Spade [31]). However, the core of JOSKIPI language (i.e. the present state) is to be maintained and preserved.

The increasing expressive power of JOSKIPI brings about a need to reconsider performance issues. Possibilities of partial translation to some version of transducers will be examined.

Acknowledgement. Work financed by the Polish Ministry of Education and Science, project No. 3 T11C 018 29.

References

1. Blum, A., Langley, P.: Selection of relevant features and examples in machine learning. Artificial Intelligence 97(1-2), 245–271 (1997)
2. Broda, B., Derwojedowa, M., Piasecki, M.: Recognition of structured collocations in an inflective language. In: Proceedings of the International Multiconference on Computer Science and Information Technology — 2nd International Symposium Advances in Artificial Intelligence and Applications (AAIA'07), pp. 237–246 (2007)
3. Broda, B., Derwojedowa, M., Piasecki, M., Szpakowicz, S.: Corpus-based semantic relatedness for the construction of Polish WordNet. In: (ELRA), E.L.R.A., (ed.), Proceedings of the Sixth International Language Resources and Evaluation (LREC'08), Marrakech, Morocco (May 2008)
4. Broda, B., Piasecki, M., Radziszewski, A.: Towards a set of general purpose morphosyntactic tools for Polish. In: Kłopotek, M.A., Przepiórkowski, A., Wierzchoń, S.T., Trojanowski, K. (eds.) Intelligent Information Systems XVI. Proceedings of the International IIS'08 Conference held in Zakopane, Poland, 2006, June 2006. Advances in Soft Computing, pp. 441–450. Academic Publishing House EXIT, Warsaw (2006)
5. Buczyński, A.: Pozyskiwanie z internetu tekstów do badań lingwistycznych. Master's thesis, Wydział Matematyki, Informatyki i Mechaniki, Uniwersytet Warszawski (2004)
6. Derwojedowa, M., Piasecki, M., Szpakowicz, S., Zawisławska, M., Broda, B.: Words, Concepts and Relations in the Construction of Polish WordNet. In: Tanács, A., Csendes, D., Vincze, V., Fellbaum, C., Vossen, P. (eds.) Proc. Global WordNet Conference, Seged, Hungary University of Szeged, January 22-25 2008, pp. 162–177 (2008)
7. Dietterich, T.G.: Approximate statistical tests for comparing supervised classification learning algorithms. Neural Computation 10(7), 1895–1924 (1997)
8. Evert, S.: The Statistics of Word Cooccurrences: Word Pairs and Collocations. PhD thesis, University of Stuttgart (2004)
9. Fleiss, J.L.: Measuring nominal scale agreement among many raters. Psychological Bulletin 76(5), 378–382 (1971)
10. Freitag, D., Blume, M., Byrnes, J., Chow, E., Kapadia, S., Rohwer, R., Wang, Z.: New Experiments in Distributional Representations of Synonymy. In: Proc. Ninth Conference on Computational Natural Language Learning (CoNLL-2005), Ann Arbor, Michigan, Association for Computational Linguistics, June 2005, pp. 25–32 (2005)
11. Geffet, M., Dagan, I.: Vector quality and distributional similarity. In: Proceedings of the 20th international conference on Computational Linguistics, COLING2004, pp. 247–254 (2004)
12. Godlewski, G., Piasecki, M.: Optimisation of Polish tagger parameters. In: Kłopotek, M.A., Tchórzewski, J. (eds.) Proceedings of Artificial Intelligence Studies, vol. 3, pp. 157–164. Publishing House of University of Podlasie (2006)
13. van Halteren, H. (ed.): Syntactic Wordclass Tagging. Kluwer Academic Publishers, Dordrecht (1999)
14. Harris, Z.S.: Mathematical Structures of Language. Interscience Publishers, New York (1968)
15. Israel, G.D.: Determining sample size. Tech. Rep. PEOD6, University of Florida (1992)

16. Karlsson, F., Voutilainen, A., Heikkilä, J., Anttila, A. (eds.): Constraint Grammar: A Language-Independent System for Parsing Unrestricted Text. Mouton de Gruyter, Berlin and New York (1994)
17. Kłopotek, M.A., Przepiórkowski, A., Wierzchoń, S.T., Trojanowski, K. (eds.): Intelligent Information Systems XVI. Proceedings of the International IIS'08 Conference held in Zakopane, Poland, June 2006. Advances in Soft Computing. Academic Publishing House EXIT, Warsaw (2006)
18. Kłopotek, M.A., Wierzchoń, S.T., Trojanowski, K. (eds.): Intelligent Information Processing and Web Mining — Proceedings of the International IIS: IIPWM'06 Conference held in Wisła, Poland, June 2006. Advances in Soft Computing. Springer, Berlin (2006)
19. Lin, D.: Principle-based parsing without overgeneration. In: Annual Meeting of the ACL. Proceedings of the 31st annual meeting on Association for Computational Linguistics, pp. 112–120 (1993)
20. Lin, D.: Automatic retrieval and clustering of similar words. In: International Conference On Computational Linguistics (COLING'98). Proceedings of the 17th International Conference on Computational Linguistics, vol. 2, pp. 768–774. ACL (1998)
21. Matoušek, V., Mautner, P. (eds.): TSD 2007. LNCS (LNAI), vol. 4629. Springer, Heidelberg (2007)
22. Nenadić, G., Spasić, I., Ananiadou, S.: Morpho-syntactic clues for terminological processing in Serbian. In: Proceedings of Workshop on Morphological Processing of Slavic Languages, EACL 2003, Budapest, Hungary, pp. 79–86 (2003)
23. Obrębski, T.: An all-path parsing algorithm for constraint-based dependency grammars of cf-power. In: Matoušek, V., Mautner, P. (eds.) TSD 2007. LNCS (LNAI), vol. 4629, pp. 139–146. Springer, Heidelberg (2007)
24. Pecina, P.: An extensive empirical study of collocation extraction methods. In: Proceedings of the ACL Student Research Workshop, Ann Arbor, Michigan, June 2005, pp. 13–18. Association for Computational Linguistics (2005)
25. Piasecki, M.: Hand-written and automatically extracted rules for Polish tagger. In: Sojka, P., Kopeček, I., Pala, K. (eds.) TSD 2006. LNCS (LNAI), vol. 4188, pp. 205–212. Springer, Heidelberg (2006)
26. Piasecki, M., Godlewski, G.: Effective architecture of the Polish tagger. In: Sojka, P., Kopeček, I., Pala, K. (eds.) TSD 2006. LNCS (LNAI), vol. 4188, pp. 213–220. Springer, Heidelberg (2006)
27. Piasecki, M., Szpakowicz, S., Broda, B.: Automatic selection of heterogeneous syntactic features in semantic similarity of polish nouns. In: Matoušek, V., Mautner, P. (eds.) TSD 2007. LNCS (LNAI), vol. 4629, pp. 99–106. Springer, Heidelberg (2007)
28. Piasecki, M., Szpakowicz, S., Broda, B.: Extended similarity test for the evaluation of semantic similarity functions. In: Vetulani, Z. (ed.) Human Language Technologies as a Challenge for Computer Science and Linguistics, 3rd Language & Technology Conference, Poznań, Poland, October 5–7 2007, pp. 104–108. Wydawnictwo Poznańskie Sp. z o.o (2007)
29. Piotrowski, T., Saloni, Z.: Kieszonkowy słownik angielsko-polski i polsko-angielski. Wyd. Wilga, Warszawa (1999)
30. Przepiórkowski, A.: The IPI PAN Corpus: Preliminary version. Institute of Computer Science PAS (2004)

31. Przepiórkowski, A., Buczyński, A.: ♠: Shallow parsing and disambiguation engine. In: Vetulani, Z. (ed.) Human Language Technologies as a Challenge for Computer Science and Linguistics, 3rd Language & Technology Conference, Poznań, Poland, October 5–7, 2007, pp. 340–344. Wydawnictwo Poznańskie Sp. z o.o (2007)

32. P.W.N.: Słownik języka polskiego, May 2007. Published on the web page (2007), http://sjp.pwn.pl/

33. Ratnaparkhi, A.: Maximum Entropy Models for Natural Language Ambiguity Resolution. PhD thesis, University of Pennsylvania, Philadelphia, PA, USA (1998)

34. Sharoff, S.: What is at stake: a case study of Russian expressions starting with a preposition. In: Tanaka, T., Villavicencio, A., Bond, F., Korhonen, A. (eds.) Second ACL Workshop on Multiword Expressions: Integrating Processing, Barcelona, Spain, July 2004, pp. 17–23. ACL (2004)

35. Smadja, F.: Retrieving collocations from text: Xtract. Computational Linguistics 19(1), 143–177 (1993)

36. Sojka, P., Kopeček, I., Pala, K. (eds.): TSD 2006. LNCS (LNAI), vol. 4188. Springer, Heidelberg (2006)

37. Spasić, I.: A Machine Learning Approach to Term Classification. PhD thesis, Information Systems Research Centre School of Computing, Science and Engineering University of Salford, Salford, UK (May 2004)

38. Vetulani, Z. (ed.): Human Language Technologies as a Challenge for Computer Science and Linguistics, 3rd Language & Technology Conference, Poznań, Poland, October 5–7, 2007. Wydawnictwo Poznańskie Sp. z o.o (2007)

39. Woliński, M.: Morfeusz — a practical tool for the morphological analysis of Polish. In: Kłopotek, M.A., Wierzchoń, S.T., Trojanowski, K. (eds.) Intelligent Information Processing and Web Mining — Proceedings of the International IIS: IIPWM'06 Conference held in Wisła, Poland, June 2006. Advances in Soft Computing, pp. 511–520. Springer, Heidelberg (2006)

40. Zesch, T., Gurevych, I.: Automatically creating datasets for measures of semantic relatedness. In: Proceedings of the Workshop on Linguistic Distances, Sydney, Australia, July 2006, pp. 16–24. Association for Computational Linguistics (2006)

Towards the Automatic Acquisition of a Valence Dictionary for Polish

Adam Przepiórkowski[1,2]

[1] Institute of Computer Science, Polish Academy of Sciences,
ul. Ordona 21, 01-237 Warszawa, Poland,
adamp@ipipan.waw.pl,
http://nlp.ipipan.waw.pl/~adamp/
[2] Institute of Informatics, University of Warsaw,
ul. Banacha 2, 02-097 Warszawa, Poland

Abstract. This article presents the evaluation of a valence dictionary for Polish produced with the help of shallow parsing techniques and compares those results to earlier results involving deep parsing. We show that the valence dictionary obtained with the use of shallow parsing attains higher quality when it is measured on the basis of a corpus of valence frames, while the dictionary produced with the help of deep parsing seems superior when the results are compared to existing valence dictionaries.

Key words: valence acquisition, arguments of verbs, evaluation of valence dictionaries, partial parsing, IPI PAN Corpus of Polish

1 Aim and Scope

The valence of a given lexeme is, in general terms, its combinatorial potential, i.e., its ability to combine with other constituents of the utterance. In practice the term *valence* (or *valency*) usually refers to verbal lexemes, and it denotes the number and morphosyntactic makeup of the arguments of the verb. Hence, a valence dictionary will contain the information that the verb CHRAPAĆ, 'snore', combines only with the nominative subject, while the verb CHOWAĆ 'hide' also takes an accusative complement (*chować coś*, 'to hide something') and, optionally, a prepositional group with the preposition PRZED governing the instrumental case (*chować coś przed kimś*, 'to hide something from somebody'). Such dictionaries have various theoretical linguistic, psycholinguistic and educational uses, and they are a valuable resource in deep parsing and generation.

While there exist Polish dictionaries containing valence information,[3] including [15], [27], [2], and [14], the automatic acquisition of such information from corpora has many advantages when compared to the process of manual dictionary compilation. First, the automatic method is much faster and cheaper.

---

[3] A short comparison of a few of them in the context of Natural Language Processing may be found in [17], with some desiderata concerning such dictionaries put forward in [19].

M. Marciniak and A. Mykowiecka (Eds.): Bolc Festschrift, LNCS 5070, pp. 191–210, 2009.

Second, as it is based on naturally occurring texts, it is immune to prescriptive influences and to conflicting intuitions of a team of lexicographers. Hence, this method may be considered more objective. Third, the automatic procedure does not only extract valence frames, but also assigns them relative frequencies. For example, we may learn how often the verb DZIWIĆ 'make (one) wonder, surprise' combines with a nominal subject (*Dziwiło mnie takie postępowanie*, 'Such behaviour made me wonder'), and how often it co-occurs with a sentential subject (*Dziwiło mnie, że tak postąpił*, 'That he behaved like that made me wonder'). Such quantitative information is indispensable in probabilistic parsers (cf., e.g., [4], as well as [1]), which assign probabilities to particular parses, and it is also relevant in psycholinguistic research [13]. Fourth, as has been repeatedly noted (e.g., [28], [24], [12,11] and [9]), valence changes not only with time, but also with genre and topic. Once developed, automatic valence acquisition algorithms may be applied to various sets of texts in order to quickly and cheaply construct diachronic or thematic valence dictionaries. Fifth, automatic valence extraction may be used not only for the development of a new valence dictionary, but also for the verification and extension of an existing manually created dictionary (cf., e.g., [25]). [26] shows, for an automatically acquired valence dictionary of German, that the quality of such dictionaries may rival the quality of traditional dictionaries.

This paper focuses on the most basic type of valence information, which is found in all valence dictionaries, i.e., on morphosyntactic information regarding the grammatical class (part of speech) of the argument (e.g., CHOWAĆ, 'hide', combines with a nominal complement and not with a verbal complement), its grammatical case (e.g., CHOWAĆ combines with the accusative, not with the instrumental), etc. Some valence dictionaries, e.g., [15], also contain certain semantic information. Although the acquisition of such semantic valence information is beyond the scope of the current article, some work towards this end is currently being carried out within another ICS PAS project.[4]

2 Algorithm

As in virtually all previous work on valence acquisition (cf. [22], § 10.2, for an overview), the experiments described below proceed in two stages. First, at the linguistic stage, syntactic groups are identified which may be arguments of verbs. The result of this stage is a set of observations, where each observation consists of a verb and its observed potential arguments within a given sentence. Obviously, such observations will be noisy, with errors due to the inadequacies of morphosyntactic and syntactic processing. Second, the set of linguistic observations is subjected to statistical inference rules whose task is to decide which observations are reliable. Only thus filtered observations are considered valid valence frames.

---

[4] A Ministry of Science and Higher Education research grant (number NN516016533) "Automatyczne wykrywanie zależności semantycznych w strukturze argumentowej czasowników w dużych korpusach tekstów anotowanych syntaktycznie".

The main general steps of the algorithm, described in more detail below, are:

1. pre-process the empirical material, i.e., an appropriate subcorpus of the IPI PAN Corpus of Polish;
2. shallow process all sentences within that subcorpus and select fully parsed sentences for further statistical processing;
3. apply statistical filtering techniques, namely, the techniques proposed in [5].

2.1   Empirical Material

The main empirical material for the work reported here is the 2nd edition of the IPI PAN Corpus (http://korpus.pl/; [18]) containing about 255 million segments (over 200 million traditionally understood orthographic words, i.e., words delimited by white spaces and punctuation). Only sentences of less than 16 segments were selected from this corpus for further processing.[5] This restriction was imposed in order to reduce the time needed to process the massive amount of data, and similar restrictions are imposed in earlier work on the extraction of Polish valence, reported in [5]. Only candidates for true sentences, i.e., those containing verbal segments, were included in the processing chain.

As a result of this selection procedure, an IPI PAN subcorpus containing 25 647 017 segments (2 724 353 sentences) was created.

2.2   Shallow Parsing

The corpus obtained as described in the previous section was shallow parsed with the Spejd (http://nlp.ipipan.waw.pl/Spejd/; [23]) grammar presented in ch. 8 of [22], i.e., maximal nominal, prepositional, adjectival, verbal and other groups were automatically identified. Because of the partial nature of the grammar and the parser, not all sentences were fully parsed; after syntactic processing, some sentences contained sequences of lexical segments not assigned to any syntactic constituents. Only fully parsed sentences underwent the statistical processing. There were 1 137 014 (41.74%) such sentences and they contained 8 516 676 segments. One such sentence, *Kto się wstrzymał od głosu?* ('Who abstained?', literally: 'Who self abstained from voice?'), is presented below:

```
<chunk type="s">
<group id="a106ac9" rule="(1) NG between verbs/groups/aby/etc."
       synh="a106abf" semh="a106abf" type="NG">
<tok id="a106abf">
<orth>Kto</orth>
<lex disamb="1">
 <base>kto</base><ctag>subst:sg:nom:m1</ctag>
 </lex>
```

---

[5] 16 is the average length of a sentence in the 30-million "varied" (roughly balanced) subcorpus of the IPI PAN Corpus, 2.sample.30 (cf. http://korpus.pl/index.php?page=download).

```
</tok>
</group>
<group id="a106ac7" rule="sie" synh="a106ac0" semh="a106ac0"
      type="sie">
<tok id="a106ac0">
<orth>się</orth>
<lex disamb="1"><base>się</base><ctag>qub</ctag></lex>
</tok>
</group>
<syntok id="a106ac6" rule="czasownik niezanegowany 2a">
<orth>wstrzymał</orth>
<lex disamb="1">
 <base>wstrzymać</base><ctag>praet:sg:m1:perf:aff</ctag>
</lex>
<lex>
 <base>wstrzymać</base><ctag>praet:sg:m2:perf:aff</ctag>
</lex>
<lex>
 <base>wstrzymać</base><ctag>praet:sg:m3:perf:aff</ctag>
</lex>
<tok id="a106ac1">
<orth>wstrzymał</orth>
<lex disamb="1">
 <base>wstrzymać</base><ctag>praet:sg:m1:perf</ctag>
</lex>
<lex><base>wstrzymać</base><ctag>praet:sg:m2:perf</ctag></lex>
<lex><base>wstrzymać</base><ctag>praet:sg:m3:perf</ctag></lex>
</tok>
</syntok>
<group id="a106ac8"
      rule="(1) Dobre PrepNG na koncu zdania lub nawiasu"
      synh="a106ac2" semh="a106ac3" type="PrepNG">
<tok id="a106ac2">
<orth>od</orth>
<lex disamb="1"><base>od</base><ctag>prep:gen:nwok</ctag></lex>
</tok>
<tok id="a106ac3">
<orth>głosu</orth>
<lex disamb="1">
 <base>głos</base><ctag>subst:sg:gen:m3</ctag>
</lex>
</tok>
</group>
<ns/>
<tok id="a106ac5">
```

```
<orth>?</orth>
<lex disamb="1"><base>?</base><ctag>interp</ctag></lex>
</tok>
</chunk>
```

The syntactic representation exemplified above was subsequently translated to the format accepted by the statistical module, as proposed in [20] (and slightly modified in [5]). In the case of the above sentence, the result of this conversion is as follows:

```
% 'Kto się wstrzymał do głosu ?'
wstrzymać :np:nom: :prepnp:do:gen: :sie:
```

In the general case, the translation from the output format of Spejd to the input format of the statistical stage consists of the following steps:

1. each immediate constituent of a sentence, i.e., each XML child of the `<chunk type="s">` element, is assigned to one of the following three classes: grupa (i.e., a syntactic group), czasownik (i.e., a verb), inny token (neither, i.e., a token which is not a verb and does not belong to a recognised syntactic group); in particular, each token containing a verbal interpretation is assigned to the verbal class czasownik;
2. if, as a result, the number of elements in the verbal czasownik class is different than 1, the processing of the current sentence is aborted, and the algorithm moves to the next sentence;
3. since the only sentences that entered this stage of processing consisted of groups, verbs and punctuation marks, the class inny token must — after the previous steps — contain only punctuation marks and, as such, it is ignored in further processing;
4. the orthographic makeup of the sentence is retrieved for the purpose of a comment in the resulting file (starting with a %; cf. the example above);
5. the base form of the single verb in the sentence is retrieved; if this segment has a number of verbal interpretations with different base forms, the first of them is arbitrarily assumed to be the correct one;
6. each syntactic group belonging to the grupa class is translated into a list of morphosyntactic interpretations of the syntactic head of the group, e.g., `:np:nom:`, `:np:acc:`, `:prepnp:do:gen:`, `:infp:imperf:`; as a head may contain a number of morphosyntactic interpretations, the result is a list rather than a single such representation;
7. the Cartesian product of the lists (treated as sets) of representations of all elements of the grupa class is taken as the set of potential observations adduced by the currently processed sentence;
8. the potential observations are sorted and printed out.

Despite the fact that the shallow Spejd grammar used in the experiments reported in this paper contains some morphosyntactic disambiguation rules, not all segments are fully disambiguated, so one sentence may be the basis of a number of different potential observations, calculated in steps 6–7 of the above algorithm. For example, 4 potential observations were obtained for the sentence *Składam te podziękowania na ręce szefowej komisji pani senator Genowefy Grabowskiej* ('I thank the head of the commission, senator Genowefa Grabowska', literally: 'I-put these thanks onto hands boss.gen commission.gen Mrs. senator Genowefa Grabowska'):[6]

```
%  'Składam te podziękowania na ręce szefowej komisji
%  pani senator Genowefy Grabowskiej .'
składać :np:acc:  :prepnp:na:acc:
składać :np:acc:  :prepnp:na:loc:
składać :np:nom:  :prepnp:na:acc:
składać :np:nom:  :prepnp:na:loc:
```

The group *te podziękowania* 'these thanks' is not fully disambiguated and it retains both the accusative and the nominative case interpretations, and similarly the prepositional group, *na ręce...* 'onto hands', is not disambiguated as to the accusative or locative case value, which results in 4 potential observations. Note that, unlike in the case of the deep parser Świgra (http://nlp.ipipan.waw.pl/~wolinski/swigra/; [30,31]) utilised in [5], the observations may only differ in the values of morphosyntactic categories, not in the number or extent of syntactic groups; following the general shallow parsing principles, Spejd outputs a unique parse of the sentence.

A more crucial difference between the current experiments and the approach proposed by Dębowski consists in our refraining from any further linguistic processing: all linguistic knowledge is contained in the grammar, and the resulting observations correspond directly to the groups found by the parser. This should be contrasted with the algorithm described in [5], where the results of the grammar are subject to some further linguistic processing, including the following steps:

- a nominal group is added to an observation in case an elided subject (so-called *pro*-drop) is detected;
- the nominal genitive group, if any, is removed from an observation in case of a negated sentence, as this genitive group may actually be a Genitive of Negation (cf. [16]) realisation of an otherwise accusative argument of the verb;
- nominal phrases suspected of having the grammatical function of (temporal) adjuncts are removed from observations.

It is not clear to what extent such further transformations influenced the final results of [5], but they probably played a role in producing results more comparable to valence frames found in existing valence dictionaries. Such *a posteriori* modifications of observations must also lead to less accurate data concerning the actual text frequencies of particular realisations of valence frames.

---

[6] The orthographic form of the sentence, given here as a comment, was broken for typographical reasons.

2.3   Statistical Processing

The pre-processing step of the statistical stage is the selection of a single observation in case of sentences with multiple potential observations. A simple EM-type (Expectation Maximisation) algorithm described in [5] is used to this end. In the case of the example sentence given above, the observation actually selected for *Składam te podziękowania na ręce szefowej komisji pani senator Genowefy Grabowskiej* correctly assumes the accusative case of the nominal group, but incorrectly identifies the case within the prepositional group as locative:

```
% 'Składam te podziękowania na ręce szefowej komisji
% pani senator Genowefy Grabowskiej .'
składać :np:acc: :prepnp:na:loc:
```

Observations collected and further selected this way are the first version of the resulting valence dictionary, the so-called proto-dictionary [7]. For example, the lexical entry for the verb WYPŁYWAĆ, 'flow out, emerge, follow' in the proto-dictionary created within the current shallow parsing experiments is given below:[7]

```
'wypływać' => {
  'np(nom),z+np(gen)' => 9,
  'adv,np(nom),z+np(gen)' => 2,
  '' => 1,
  'adj(nom),adv,dla+np(gen),np(acc)' => 1,
  'adj(nom),np(acc),z+np(gen)' => 1,
  'adj(nom),z+np(gen)' => 1,
  'adv' => 1,
  'dla+np(gen),np(nom),z+np(gen)' => 1,
  'do+np(gen),np(nom),z+np(gen)' => 1,
  'np(acc),o+np(loc),z+np(gen)' => 1,
  'np(acc),od+np(gen),z+np(gen)' => 1,
  'np(dat),np(nom)' => 1,
  'np(nom),np(voc),z+np(gen)' => 1,
  'np(nom),o+np(loc)' => 1
}
```

According to this entry, forms of the verb WYPŁYWAĆ were observed 9 times with a nominative nominal group and a prepositional group headed by the genitive-taking preposition z, twice with an additional adverbial group, once without any accompanying groups, etc.

The two main steps of the statistical stage make use of an approximate representation of valence proposed in [6], where a valence frame is described as a *set* of possible arguments of the verb (the set of all arguments in all possible frames of that verb) and an additional table specifying whether any two possible

---

[7] The format of such lexical entries is actually the representation of hash tables in the Perl programming language.

arguments always co-occur, never co-occur, unidirectionally imply one another, or seem independent of each other.

First, all possible arguments of a given verb are collected. A subset of that argument set is identified as those arguments which occur in all possible frames of that verb. An argument type a is considered a possible argument of a verb v in case the inequality in (1) holds; $c(v)$ denotes here the number of occurrences of the verb v, $c(v, a)$ — the number of observed co-occurrences of v with the argument a. The argument is additionally considered a necessary argument of v in case (2) holds.

(1) $c(v, a) \geq p_a c(v) + 1$
(2) $c(v) - c(v, a) < p_{\neg a} c(v) + 1$

The parameters p_a and $p_{\neg a}$ occurring in the inequalities above are trained — separately for each argument type — on the basis of the manually created dictionary [27], as well as on the basis of lexical entries of around 200 verbs in [15] and [2]. The exact parameter estimation procedure for p_a and $p_{\neg a}$ is described in [5].

This first step ends with projecting information of estimated possible and necessary arguments into actually observed frames: in each observed frame only those arguments are retained which are "possible" in the sense above and, moreover, in case the observation does not contain the "necessary" argument, it is artificially added to the frame.[8] As a result of this step, the lexical entry of WYPŁYWAĆ is reduced as follows:

```
'wypływać' => {
  'np(nom),z+np(gen)' => 15,
  'np(nom)' => 4,
  'np(acc),np(nom),z+np(gen)' => 3,
  'np(acc),np(nom)' => 1
}
```

It follows from the comparison of this lexical entry with the corresponding lexical entry in the proto-dictionary that adv, np(voc), np(dat) and various prepositional argument types were rejected as possible arguments of the verb, so the set of possible arguments is reduced to {np(nom), np(acc), z+np(gen)}. Further, the nominative nominal group was classified as a necessary argument. Hence, the four "observations" 'np(nom)' in the lexical entry above actually correspond to the original observations: 'np(dat),np(nom)', 'np(nom),o+np(loc)', 'adv' and the empty observation ''.

In the second step, full frames obtained in the first step are evaluated. Again, on the basis of existing valence dictionaries, possible relationships between arguments are estimated: do the two given arguments usually co-occur within frames of various verbs, do they have a complementary distribution, does one imply the other, or are they independent. For any two argument types, such a

---

[8] This description is based on the observation of the algorithm at work and it differs a little from the description in [5].

relationship is calculated on the basis of the whole dictionary, independent of particular verbs. For a given verb, this default relation between two arguments is assumed, unless there are strong reasons to override it. For example, the 15 "observations" of the nominative nominal group np(nom) co-occurring with the z+np(gen) prepositional group in the vicinity of WYPŁYWAĆ were not sufficient to retain that frame of that verb (note that the frame is correct here, but it is generally rather rare in the corpus), so the final lexical entry for WYPŁYWAĆ looks as follows:

```
’wypływać’ => {
  ’np(nom)’ => 4,
  ’np(acc),np(nom),z+np(gen)’ => 3,
  ’np(acc),np(nom)’ => 1
}
```

3 Results

Three dictionaries were created as a result of the algorithm described in the previous section: the proto-dictionary, which contains the actual observations (perhaps selected from alternative potential observations with the help of a simple EM algorithm), the dictionary resulting from the first step of statistical processing (the intermediate dictionary), and the final dictionary created in the second step of statistical processing. Table 1 gives the sizes of these dictionaries.

Table 1. Sizes of automatically obtained valence dictionaries

dictionary	entries	f r a m e s	
		tokens	types
proto	6,845	1,084,286	20,894
intermediate	6,845	1,084,286	517
final	4,166	863,731	141

The proto-dictionary obtained as described in the previous sections contains 6 845 entries. As a result of the second step of statistical processing, this number is reduced to 4 166 entries with 207.33 observations per entry on average. The full dictionary would consume around 430 pages, so — given the space limits — it must suffice to present some of its characteristics here.[9]

The final number of different "observed" valence frames is 141. This number is substantially reduced with respect to the number of 20 894 different realisations of frames actually observed, and also with respect to the 517 types of "observations" remaining after the first step of statistical processing. The most

---

[9] Appendix A contains a fragment of the dictionary resulting from the simplification of the statistical stage, as described in § 5.

frequent frame was the intransitive frame (only a nominative nominal group; 232 034 occurrences) and the empty frame (129 720), while the actually very frequent transitive frame (nominative and accusative nominal groups) is the 4th most frequent frame in the resulting final dictionary (84 611 occurrences). Out of the three frames with single occurrences: 'do+np(gen),np(gen),np(nom),sie' (for the verb UŻYWAĆ 'use'), 'np(dat),np(nom),o+np(loc),sie' (MARZYĆ 'dream') i 'np(nom),o+np(loc),z+np(inst)' (POROZMAWIAĆ 'talk'), the first two are erroneous, and the last one seems to be correct.

4   Evaluation

Two evaluation procedures were applied to the dictionaries obtained as described above: the dictionary-based evaluation (at the level of frame types) and the corpus evaluation (at the level of frame occurrences, or tokens).

4.1   Dictionary-Based Evaluation

Dictionary-based evaluation consists in finding the ratio of automatically extracted valence frames also present in manually constructed dictionaries (precision), and the ratio of frames in such previously built dictionaries also present in the automatic results (recall). In the current experiments, these values were estimated on the basis of a sample of 202 verbs randomly selected from [27]. For all these 202 verbs, their entries were also extracted from two other manually constructed dictionaries [15,2] and converted to the "least common denominator" format.[10] Since the dictionaries differed a little in the scope and character of the valence information, the conversion process was to some extent interpretative.

Precision (P), recall (R) and their harmonic mean, called F-measure (F), were computed for the final dictionary obtained in the current experiments, as well as for both dictionaries reflecting earlier stages of processing: for the proto-dictionary and for the intermediate dictionary. In each case automatically obtained valence frames were compared to various gold standards, that is, to each of the three manually constructed valence dictionaries, marked below as "Bań." [2], "Pol." [15] and "Świ." [27], and to two dictionaries compiled from these three manually constructed dictionaries by taking their set-theoretical sum ("SUM") and by majority voting ("MV"; i.e., a frame is present in the MV dictionary, if it is present in at least two manually constructed dictionaries). In each comparison, only the frames of those verbs were considered which were present both in the automatically obtained dictionary and in the gold standard. The results of these comparisons are contained in Tables 2–4.

The comparison of Tables 2 and 3 shows the great importance of the first step of statistical processing. Although it resulted in a significant drop of recall (from 40.40% to 29.80%, for the MV dictionary), that decrease in recall was

---

[10] These data were prepared by Witold Kieraś and Łukasz Dębowski, whose scripts were used for calculating precision and recall given below.

Table 2. Dictionary-based evaluation of the proto-dictionary

	Bań.	Pol.	Świ.	SUM	MV
P	3.97	3.04	3.05	5.11	**3.15**
R	37.33	31.03	34.83	28.60	**40.40**
F	7.17	5.54	5.62	8.68	**5.85**

Table 3. Dictionary-based evaluation of the intermediate dictionary

	Bań.	Pol.	Świ.	SUM	MV
P	50.69	41.70	39.95	57.80	**44.94**
R	24.68	22.00	23.56	16.72	**29.80**
F	33.20	28.81	29.64	25.94	**35.84**

Table 4. Dictionary-based evaluation of the final dictionary

	Bań.	Pol.	Świ.	SUM	MV
P	63.81	52.23	51.34	70.41	**57.58**
R	22.04	19.51	21.49	14.42	**27.07**
F	32.77	28.41	30.30	23.94	**36.83**

more than compensated by the dramatic increase in precision (from 3.15% to 44.94%), which resulted in the clear increase in the harmonic mean of these measures (from 5.85 to 35.84). The second step of the statistical stage wasn't so significant: although the F-measure for the MV dictionary is higher in Table 4 than in Table 3, the difference is relatively minor (36.83 to 35.84) because the increase in precision was to a large extent annulled by the decrease in recall. It is interesting to note that, for two gold standards, Bań. and Pol., the second step of statistical processing was slightly detrimental, if the quality were evaluated with the F-measure.

Let us also note at the end of this section that, although the results are far from perfect, they constitute a clear improvement over the reasonable baseline consisting in the assignment of two frames to each verb: the intransitive frame 'np(nom)' and the transitive frame 'np(acc),np(nom)'.[11] A "dictionary" constructed this way would have relatively high precision (47.41% when measured against the MV dictionary), but very low recall (15.15%). Complete results of the evaluation of such a baseline dictionary are given in Table 5.

4.2   Corpus-Based Evaluation

Also corpus-based evaluation shows that, after shallow processing at the linguistic stage, it may be beneficial to stop statistical processing after the first step,

---

[11] Experiments were also performed for other baselines, including: only the intransitive frame, only the transitive frame, the empty frame, the infinitival frame, and various combinations of these frames. In each case the dictionary-based evaluation gave worse (in terms of F-measure) results than the results for the baseline given below.

Table 5. Dictionary-based evaluation of the baseline, i.e., a dictionary created artificially by assuming two frames for each verb: the transitive frame and the intransitive frame

	Bań.	Pol.	Świ.	SUM	MV
P	54.66	42.49	43.52	59.33	**47.41**
R	12.83	10.80	12.37	8.27	**15.15**
F	20.78	17.23	19.27	14.52	**22.96**

before frames are removed on the basis of the low co-occurrence of the arguments in manually constructed dictionaries.

The evaluation was performed for 12 verbs selected on the basis of their frequencies in the corpus resulting from the linguistic processing. These are 4 very frequent verbs (tens of thousands of occurrences): WSTRZYMAĆ 'stop', CHCIEĆ 'want', STWIERDZAĆ 'conclude', MIEĆ 'have', 4 verbs of medium frequency (around 4 000 occurrences): MUSIEĆ 'must', ZABRAĆ 'take (away)', PRZY-POMINAĆ 'remind, remember', and ZGŁOSIĆ 'report', and 4 relatively rare verbs (around 400 occurrences): STAWIAĆ 'put', USŁYSZEĆ 'hear', ZAUWAŻYĆ 'notice' and USTALIĆ 'establish'.

For each verb, 120 sentences containing that verb were randomly selected. These sentences were linguistically annotated[12] on the basis of brief guidelines of syntactic annotation [21] with the help of the Anotatornia annotation tool developed at ICS PAS [8].

Some of these sentences were lacking full morphosyntactic analysis, some were the result of erroneous segmentation of text into sentences. The remaining 985 sentences were annotated for maximal syntactic groups. Hence, the result of the annotation was a set of fully correct valence frame observations. Obviously, just as would be the case in fully correct shallow processing, these observations contained information of all observed dependents of the main verb: arguments and adjuncts alike. Also, these observations were not further processed linguistically in any way; in particular, no information about missing (elided) arguments was added.

The corpus prepared this way was the basis for calculating token recall, i.e., the ratio of manually annotated frames also found by the algorithm. The result was 89% for the proto-dictionary, 32% for the intermediate dictionary and 22% for the final dictionary. Table 6 presents the results in more detail, while Table 7 presents analogous results for the approach based on deep linguistic processing, described in [5]. It is interesting to note that, although the proto-dictionary based on shallow processing with Spejd contains many more valence frames observed in manually annotated texts than the proto-dictionary based on the deep parser Świgra (89% compared to 39%), this difference reduces significantly for the intermediate dictionary (32% to 27%) and reverses for the final dictionary (22% to 27%). This effect is probably to some extent caused by the greater

---

[12] By linguists: Monika Czerepowicka, Hanna Maliszewska, Marta Nazarczuk-Błońska, Marta Piasecka and Izabela Will.

Table 6. The number of observations of valence frames for the 12 verbs for which the appropriate frame is also present in the valence dictionary automatically obtained with the use of the Spejd parser and the grammar presented in ch. 8 of [22]

| verb | frames (tokens) | | | |
| | | in dictionary | | |
	in texts	proto	intermediate	final
USTALIĆ	73	54	6	11
ZABRAĆ	103	100	1	1
STAWIAĆ	78	34	8	6
CHCIEĆ	91	89	19	19
ZAUWAŻYĆ	65	48	6	11
WSTRZYMAĆ	88	88	88	88
MIEĆ	86	84	28	28
MUSIEĆ	84	83	34	34
PRZYPOMINAĆ	108	93	0	0
STWIERDZAĆ	119	119	114	0
ZGŁOSIĆ	73	70	15	15
USŁYSZEĆ	17	13	0	0
total	985	875	319	213
percent	100	88.83	32.39	21.62

Table 7. The number of observations of valence frames for the 12 verbs for which the appropriate frame is also present in the valence dictionary automatically obtained with the use of the Świgra deep parser

| verb | frames (tokens) | | | |
| | | in dictionary | | |
	in texts	proto	intermediate	final
USTALIĆ	73	23	8	10
ZABRAĆ	103	93	53	53
STAWIAĆ	78	22	8	6
CHCIEĆ	91	23	19	19
ZAUWAŻYĆ	65	18	7	11
WSTRZYMAĆ	88	88	88	88
MIEĆ	86	43	28	28
MUSIEĆ	84	37	34	34
PRZYPOMINAĆ	108	8	7	4
STWIERDZAĆ	119	0	0	0
ZGŁOSIĆ	73	23	18	10
USŁYSZEĆ	17	2	0	0
total	985	380	270	263
percent	100	38.58	27.41	26.70

dispersion of data in the current approach (many more different valence frame types are found in shallow processing), but it may also be a result of the close fit of the statistical approach proposed by Dębowski and the deep parsing with Świgra, on the basis of which that approach was developed and fine-tuned [6,5].

As shown in the next section, the rather disappointing results given above improve when the second step of statistical processing is simplified. Nevertheless, already these modest results are well above the baseline described above: in the case where every verb is assigned the transitive frame and the intransitive frame, the resulting valence dictionary would cover only 101 (10.25%) corpus observations.

5   Simplification of Statistical Processing

The previous three sections describe some experiments in valence extraction, where linguistic processing is performed with the Spejd implementation of the shallow grammar presented in ch. 8 of [22], while the statistical processing follows the ideas described in [5]. In the preceding section we noted that the second step of the statistical stage, where frames with uncommon combinations of arguments are rejected, is a mixed blessing at best: it improves the results of the dictionary-based evaluation only slightly (and in fact has a detrimental effect, if Polański's or Bańko's dictionaries are taken as gold standards), and it causes a clear drop in the quality measured via corpus-based evaluation.

On the other hand, as noted in various earlier works on valence acquisition for other languages (e.g., [3,10,11,26,5]), simpler methods of rejecting rare observations often give results comparable to more complicated statistical techniques. Hence, it would be interesting to find out whether applying such simpler methods in the current linguistic setup also gives results comparable to or better than the techniques proposed in [5].

Table 8. Dictionary-based evaluation (for the MV dictionary) of valance information acquired by rejecting observations rare in the intermediate dictionary; for comparison, the table also recalls previous results for the intermediate and final dictionary

| $c_{min}(v)$ | 10 | 10 | 13 | **d i c t i o n a r y** | |
$p_{min}(r, v)$	0	2	2	**intermediate**	**final**
P	45.49	53.08	53.01	44.94	57.58
R	32.45	30.94	31.45	29.80	27.07
F	37.88	39.09	39.48	35.84	36.83

To this end, further experiments based on shallow linguistic processing were conducted, where the second step of statistical processing was replaced with a simpler rejection of rare observations. Two parameters, or cutoff points, were used: the sheer number of occurrences of the verb, $c_{min}(v)$, and the ratio of the number of co-occurrences of a given frame with a given verb to the numer

Table 9. Dictionary-based evaluation of valance information acquired with shallow linguistic processing and with cutoff points $c_{\min}(v) = 13$ and $p_{\min}(r, v) = 2$

	Bań.	Pol.	Świ.	SUM	MV
P	58.53	49.00	46.66	66.05	**53.01**
R	25.13	23.05	25.09	16.98	**31.45**
F	35.16	31.35	32.63	27.02	**39.48**

of all occurrences of that verb, $p_{\min}(r, v)$ (expressed as percent points). The requirement that valence frames be acquired only for verbs occurring at least $c_{\min}(v) = 10$ times in the parsed corpus improved F-measure (as computed for the MV dictionary) to 37.88, and further rejection of observations less frequent than $p_{\min}(r, v) = 2$ (i.e., 2%) increased the value to 39.09. In various experiments performed, the best F-measure, 39.48, was achieved for $c_{\min}(v) = 13$ and $p_{\min}(r, v) = 2$. The results are summarised and compared to earlier results in Table 8, while more complete results for the best cutoff points are given in Table 9.

Let us note that significant improvements as measured by dictionary-based evaluation were achieved with practically no decrease in the quality measured with corpus-based evaluation (cf. Table 10). The number of corpus observations corresponding to automatically identified frames is 318, i.e., almost the same as in the intermediate dictionary (319; cf. Table 6 on p. 203), and much higher than in the final dictionary (213).

Table 10. The number of observations of valence frames for the 12 verbs for which the appropriate frame is also present in the valence dictionary automatically obtained with the use of the Spejd parser and the grammar presented in ch. 8 of [22] (simplified statistical processing)

verb	frames (tokens) in texts	in dictionary
USTALIĆ	73	6
ZABRAĆ	103	1
STAWIAĆ	78	7
CHCIEĆ	91	19
ZAUWAŻYĆ	65	6
WSTRZYMAĆ	88	88
MIEĆ	86	28
MUSIEĆ	84	34
PRZYPOMINAĆ	108	0
STWIERDZAĆ	119	114
ZGŁOSIĆ	73	15
USŁYSZEĆ	17	0
total	985	318
percent	100	32.28

Table 11. Dictionary-based evaluation of valance information acquired with deep linguistic processing [5]

	Bań.	Pol.	Świ.	SUM	MV
P	63.53	54.56	55.17	74.01	**59.88**
R	25.39	23.63	26.71	17.58	**32.59**
F	36.28	32.98	35.99	28.41	**42.21**

Table 12. Dictionary-based evaluation of valance information acquired with deep linguistic processing and with cutoff points $c_{\min}(v) = 17$ and $p_{\min}(r, v) = 2$

	Bań.	Pol.	Świ.	SUM	MV
P	59.26	49.66	49.66	69.82	**54.18**
R	27.98	25.86	29.87	20.00	**35.55**
F	38.01	34.01	37.30	31.09	**42.93**

It should also be noted that replacing the second statistical step with cutoff points in the original methodology — based on deep parsing with the Świgra parser — described in [5] also brings about certain, but less significant, improvements in the values of the F-measure. In this case the best cutoff points were $c_{\min}(v) = 17$ and, as above, $p_{\min}(r, v) = 2$. The results of dictionary-based evaluation for these cutoff values are given in Table 12, while the original results presented in [5] are cited in Table 11. When compared to the results in Table 7, the result of corpus-based evaluation is practically the same: there were 264 corpus observations corresponding to automatically acuired valence frames.

Taking into consideration both evaluation methodologies, the results based on shallow linguistic processing are comparable to Dębowski's ([5]) results based on deep processing; such a comparison is presented in Table 13. While the current experiments produce inferior results, when measured as the similarity to the majority voting dictionary, they are clearly superior when measured with reference to actually occurring frame realisations of the 12 verbs of varying frequencies.

6   Summary

The aim of this article was to present a practical application of the formalism and the grammar described in [22] to the task of automatic valence acquisition from morphosyntactically annotated corpora. The quality of the results of valence acquisition with shallow parsing and simplified statistical processing turns out to be comparable to the best results for Polish found in the literature, and much higher when measured against frames actually observed in texts. Also, the simplification of the statistical stage alone makes it possible to slightly improve

Table 13. A comparison of final results of three approaches: [5], the approach presented there with the second step of statistical processing replaced by simple cutoff points, and the approach presented in [22] and summarised in this article, also with simple cutoff points instead of the second step of statistiacal processing; P, R and F are precision, recall and their harmonic mean, as measured in dictionary-based evaluation, and C is the corpus-based token recall; the best results are in boldface

	Dębowski ([5])	Dębowski ([5]) $c_{min}(v) = 17$ $p_{min}(r, v) = 2$	Przepiórkowski ([22]) $c_{min}(v) = 13$ $p_{min}(r, v) = 2$
P	**59.88**	54.18	53.01
R	32.59	**35.55**	31.45
F	42.21	**42.93**	39.48
C	26.70	26.80	**32.28**

the results of the dictionary-based evaluation, when compared to the earlier best results described in [5].

There are many possible ways the approach presented above may be developed further and improved. The most obvious concern linguistic processing: both the morphological analyser and the shallow grammar could be extended in various ways. Also the empirical basis could be improved, not only by increasing the corpus size, but also by making better use of the current corpus: at the moment evidence provided by subordinate clauses and less than fully parsed sentences is lost in the process. The evaluation of the results obtained using different linguistic and statistical methods also suggests that the novel approach to the statistical stage proposed in [5], promising in combination with deep processing at the linguistic stage, may be less adequate when coupled with shallow linguistic processing. We hope to continue work both on the empirical basis and on linguistic and statistical methodologies of valence acquisition within subsequent projects carried out at ICS PAS.

Acknowledgements. This article contains some of the material of chapter 10 of the monograph [22], summarising the main results of a valence acquisition project carried out at the Institute of Computer Science of the Polish Academy of Sciences (ICS PAS) in 2005–2008.[13] The acknowledgements therein carry over to this paper, with additional thanks for comments to Małgorzata Marciniak.

A   An Extract from the Valence Dictionary

This appendix contains an extract from the valence dictionary automatically acquired with the use of the shallow Spejd grammar presented in ch. 8 of [22], combined with the simplified statistical processing described in § 5.

[13] A Ministry of Science and Higher Education research grant (number 3T11C00328) "Automatyczna ekstrakcja wiedzy lingwistycznej z dużego korpusu języka polskiego", http://nlp.ipipan.waw.pl/PPJP/.

```
'gadać' => {
 'np(nom)' => 58,
 'np(acc),np(nom)' => 8,
 'np(nom),PZ' => 5
}
'gasić' => {
 'np(nom)' => 12,
 'np(acc),np(nom)' => 5
}
'gasnąć' => {
 'np(nom)' => 10,
 'nad+np(inst),np(nom)' => 3
}
'generować' => {
 'np(nom)' => 12,
 'np(acc),np(nom)' => 6
}
'ginąć' => {
 'np(nom)' => 49
}
'gniewać' => {
 'np(nom),sie' => 22,
 'np(nom),sie,ZE' => 3
}
'godzić' => {
 'na+np(acc),np(nom),sie' => 42,
 'inf,np(nom),sie' => 28,
 'np(nom),w+np(acc)' => 26,
 'np(nom),sie' => 17,
 'np(acc),np(nom),w+np(acc)' => 4,
 'np(acc),np(nom)' => 4,
 'np(acc),np(nom),sie' => 4,
 'np(nom)' => 3
}
'gonić' => {
 'np(acc),np(nom)' => 17,
 'np(nom)' => 15
}
'gospodarować' => {
 'np(inst),np(nom)' => 7,
 'np(nom)' => 7,
 'np(acc),np(nom)' => 2,
 'np(acc),np(inst),np(nom)' => 1
}
'gotować' => {
 'np(acc),np(nom)' => 4,
 'np(nom)' => 3,
 'do+np(gen),np(nom),sie' => 3,
 'np(nom),sie' => 3,
 'do+np(gen),np(acc),np(nom),sie'
    => 1,
```

```
 'np(dat),np(nom),sie' => 1,
 'np(dat),np(nom)' => 1
}
'gościć' => {
 'np(nom)' => 24,
 'np(acc),np(nom)' => 14
}
'gratulować' => {
 'np(nom)' => 97,
 'np(dat),np(nom)' => 39
}
'grać' => {
 'np(nom)' => 267,
 'np(acc),np(nom)' => 21
}
'gromadzić' => {
 'np(nom)' => 10,
 'np(nom),sie' => 10,
 'np(acc),np(nom)' => 8,
 'np(acc),np(nom),sie' => 4,
 'na+np(acc),np(acc),np(nom)' => 3,
 'na+np(acc),np(nom)' => 1,
 'na+np(acc),np(nom),sie' => 1
}
'grozić' => {
 'np(dat),np(nom)' => 83,
 'np(inst),np(nom)' => 54,
 'np(nom)' => 38,
 'do+np(gen),np(dat),np(nom)' =>
    11,
 'do+np(gen),np(dat),np(nom),
    za+np(acc)' => 10,
 'do+np(gen),np(acc),np(dat),
    np(nom)' => 7,
 'np(dat),np(inst),np(nom)' => 6
}
'gubić' => {
 'np(nom),sie' => 12,
 'np(nom)' => 6,
 'np(acc),np(nom)' => 4,
 'np(acc),np(nom),sie' => 2
}
'gwarantować' => {
 'np(acc),np(nom)' => 55,
 'np(nom)' => 41,
 'np(nom),ZE' => 26,
 'np(acc),np(dat),np(nom)' => 8,
 'np(dat),np(nom)' => 6,
 'np(dat),np(nom),ZE' => 4
}
```

References

1. Arun, A., Keller, F.: Lexicalization in crosslinguistic probabilistic parsing: The case of French. In: Proceedings of the 43rd Annual Meeting of the Association for Computational Linguistics, Ann Arbor, MI, pp. 306–313 (2005)
2. Bańko, M. (ed.): Inny słownik języka polskiego. Wydawnictwo Naukowe PWN, Warsaw (2000)
3. Briscoe, T., Carroll, J.: Automatic extraction of subcategorization from corpora. In: Proceedings of the 5th Applied Natural Language Processing Conference, Washington, DC, pp. 356–363. ACL (1997)
4. Carroll, J., Minnen, G., Briscoe, T.: Can subcategorisation probabilities help a statistical parser? In: Proceedings of the Sixth Workshop on Very Large Corpora, Montreal, pp. 118–126 (1998)
5. Dębowski, Ł.: Valence extraction using the EM selection and co-occurrence matrices, 5 Dec. 2007. arXiv:0711.4475v2 [cs.CL] (2007)
6. Dębowski, Ł., Woliński, M.: Argument co-occurence matrix as a description of verb valence. In: Vetulani, Z. (ed.) Proceedings of the 3rd Language & Technology Conference, Poznań, Poland, pp. 260–264 (2007)
7. Dębowski, Ł., Woliński, M.: Nowe metody ekstrakcji walencji czasowników z tekstów w języku polskim. Referat wygłoszony na seminarium Zespołu Inżynierii Lingwistycznej IPI PAN, Warszawa, 22 października 2007 (2007)
8. Hajnicz, E., Murzynowski, G., Woliński, M.: ANOTATORNIA – lingwistyczna baza danych. In: Materiały V konferencji naukowej InfoBazy 2008, Systemy * Aplikacje * Usługi, Gdańsk, pp. 168–173. Centrum Informatyczne TASK, Politechnika Gdańska (2008)
9. Hare, M., McRae, K., Elman, J.: Sense and structure: Meaning as a determinant of verb subcategorization preferences. Journal of Memory and Language 48(2), 281–303 (2003)
10. Kawahara, D., Kaji, N., Kurohashi, S.: Japanese case structure analysis by unsupervised construction of a case frame dictionary. In: Proceedings of the 18th International Conference on Computational Linguistics (COLING 2000), Saarbrücken, pp. 432–438 (2000)
11. Korhonen, A.: Subcategorization Acquisition. Ph. D. dissertation, University of Cambridge (2002)
12. Korhonen, A.L.: Using semantically motivated estimates to help subcategorization acquisition. In: Proceedings of the Joint SIGDAT Conference on Empirical Methods in Natural Language Processing and Very Large Corpora, ACL (2000)
13. Lapata, M., Keller, F., Schulte im Walde, S.: Verb frame frequency as a predictor of verb bias. Journal of Psycholinguistic Research 30(4), 419–435 (2001)
14. Mędak, S.: Praktyczny słownik łączliwości składniowej czasowników polskich. Universitas, Cracow (2005)
15. Polański, K. (ed.): Słownik syntaktyczno-generatywny czasowników polskich. Zakład Narodowy im. Ossolińskich / Instytut Języka Polskiego PAN, Wrocław / Cracow (1992)
16. Przepiórkowski, A.: Long distance genitive of negation in Polish. Journal of Slavic Linguistics 8, 151–189 (2000)
17. Przepiórkowski, A.: On the computational usability of valence dictionaries for Polish. IPI PAN Research Report 971, Institute of Computer Science, Polish Academy of Sciences, Warsaw (2003)

18. Przepiórkowski, A.: The IPI PAN Corpus: Preliminary version. Institute of Computer Science. Polish Academy of Sciences, Warsaw (2004)
19. Przepiórkowski, A.: Towards the design of a syntactico-semantic lexicon for Polish. In: Kłopotek, M.A., Wierzchoń, S.T., Trojanowski, K. (eds.) Intelligent Information Processing and Web Mining. Advances in Soft Computing, pp. 237–246. Springer, Berlin (2004)
20. Przepiórkowski, A.: What to acquire from corpora in automatic valence acquisition. In: Koseska-Toszewa, V., Roszko, R. (eds.) Semantyka a konfrontacja językowa, vol. 3, pp. 25–41. Slawistyczny Ośrodek Wydawniczy PAN, Warsaw (2006)
21. Przepiórkowski, A.: Krótka instrukcja anotacji składniowej. Unpublished manuscript, Institute of Computer Science, Polish Academy of Sciences (2008)
22. Przepiórkowski, A.: Powierzchniowe przetwarzanie języka polskiego. Akademicka Oficyna Wydawnicza EXIT, Warsaw (2008)
23. Przepiórkowski, A., Buczyński, A.: ♠: Shallow Parsing and Disambiguation Engine. In: Vetulani, Z. (ed.) Proceedings of the 3rd Language & Technology Conference, Poznań, Poland, pp. 340–344 (2007)
24. Roland, D., Jurafsky, D.: How verb subcategorization frequencies are affected by corpus choice. In: Proceedings of COLING 1998, Montreal, pp. 1122–1128 (1998)
25. Schiehlen, M., Spranger, K.: Authomatic methods to supplement broad-coverage subcategorization lexicons. In: Proceedings of the Fourth International Conference on Language Resources and Evaluation, LREC 2004, Lisbon, pp. 29–32. ELRA (2004)
26. Schulte im Walde, S.: Evaluating verb subcategorisation frames learned by a German statistical grammar against manual definitions in the *Duden* dictionary. In: Proceedings of the 10th EURALEX International Congress (2002)
27. Świdziński, M.: Syntactic dictionary of Polish verbs. Version 3a. Unpublished manuscript, University of Warsaw (1998)
28. Ushioda, A., Evans, D.A., Gibson, T., Waibel, A.: The automatic acquisition of frequencies of verb subcategorization frames from tagged corpora. In: Boguraev, B.K., Pustejovsky, J. (eds.) SIGLEX ACL Workshop of Acquisition of Lexical Knowledge from Text, Columbus, OH, pp. 95–106 (1993)
29. Vetulani, Z. (ed.): Proceedings of the 3rd Language & Technology Conference, Poznań, Poland (2007)
30. Woliński, M.: Komputerowa weryfikacja gramatyki Świdzińskiego. Ph. D. dissertation, Institute of Computer Science, Polish Academy of Sciences, Warsaw (2004)
31. Woliński, M.: An efficient implementation of a large grammar of Polish. Archives of Control Sciences 15(3), 251–258 (2005)

Semantic Annotation of Verb Arguments in Shallow Parsed Polish Sentences by Means of the EM Selection Algorithm

Elżbieta Hajnicz

Institute of Computer Science, Polish Academy of Sciences
Elzbieta.Hajnicz@ipipan.waw.pl

Abstract. The ultimate goal of our work is to extend a syntactic valence dictionary of Polish verbs by adding some semantic information to verb arguments. This information consists of wordnet *semantic categories* of words. In order to provide *syntactic slots* of dictionary entries with lists of appropriate semantic categories of corresponding nouns, we need a treebank with all nouns semantically annotated with such categories, as both syntactic (i.e., argument structure) and semantic information is required.

We aim here at *Word Sense Disambiguation* (WSD). To solve this task for our specific application, we adapt EM selection algorithm elaborated for extraction of syntactic valence frames.

In the paper, the whole process of data processing is shown. The main focus is put on WSD task. Three versions of the EM selection algorithm are presented: the original one and its two modifications. Finally, the evaluation and comparison of the algorithms is performed.

Key words: corpus lingustics, wordnets, valence dictionary, word sense disambiguation, Polish

1 Introduction

A number of resources and tools necessary for Natural Language Processing (NLP) are already available for the Polish language: morphological analysers (or dictionaries) [19,45,35,4,46], deep parsers [29,34,44], and shallow parsers [32]. The aforementioned parsers often use elaborate syntactic valence dictionaries [28,33,31,40].

Recognising the syntactic structure of texts is insufficient to obtain satisfactory results in solving NLP tasks such as machine translation, information extraction, and question answering. Semantic information is indispensable. In practical applications focused on specific domains (e.g., medicine, finance, sport) such information is often gathered in ontologies. More general lexical semantic resources, such as wordnets [13,42] and FrameNet [2,14] are also being created.

The main goal of our work is to enrich the valence dictionary of Polish verbs by adding semantic information, represented by means of wordnet semantic categories of nouns. The plain syntactic valence dictionary is a collection of

M. Marciniak and A. Mykowiecka (Eds.): Bolc Festschrift, LNCS 5070, pp. 211–240, 2009.

predicates (here: verbs) provided with a set of verb frames. Verb frames consist of syntactic slots that represent phrases occurring in the corresponding position in a sentence. Thus, our goal is to provide syntactic slots (here: NPs/PPs) with a list of appropriate semantic categories for the corresponding nouns.

In order to automatically acquire semantic information for a syntactic valence dictionary, we need a large treebank where all nouns are semantically annotated. In this understanding, our problem intersects with the Word Sense Disambiguation (WSD) task. An extensive overview of WSD is presented in [1]. Methods used for WSD can be divided into supervised, which rely on a manually annotated subcorpus to train the algorithm [12,37,38], unsupervised, i.e., based on clustering words that occur in similar context [24,26,36], semi-supervised [47] and knowledge-based, applying electronic lexicons and lexical knowledge bases, such as wordnets [3,27]. Most of those techniques are focused on fairly fine-grained word senses, hence they are applied to a small set of words or they need a large corpus to operate on. Specifically for Slavic languages, the vast majority of WSD experiments have considered a multilingual environment [21]. They are based on parallel resourses (e.g., wordnets and/or corpora) which makes them inapplicable to our goal. To the best of our knowledge, the only exception constitutes research on WSD for Czech [23].

In order to create a semantic valence dictionary, we need sense annotations for words which are immediate arguments of a verb (heads of phrases). Therefore, we apply the syntactic information (the valence of the main verb in a clause) which we have at our disposal to solve the WSD task. In fact, such information is rarely used to perform this task [16]. Although [11], a frequently cited work, does use some syntactic information, the authors disambiguate only the semantic category of a verb (based on classes proposed in [25]) using the set of possible verb classes and the syntactic frame of a sentence. They are not interested in the semantic categories of verb arguments.

In [18] we described our initial experiment in this task. We used the corpus of 165 263 single-verb sentences for 99 verbs. Sentences were parsed by means of the "permissive" version of the *Świgra* parser [44,43], which did not use any valence dictionary, and hence accepted any configuration of arguments with a verb. In that paper we focused on analysis of linguistic phenomena appearing in data set and their influence on the various steps of analysis.

At present we describe the whole process of analysis, starting from plain text. A special attention is given to algorithms we propose for slot-based WSD.

The composition of this article is as follows. In section 2, we present the resources used in our approach. In section 3, preparatory steps for data processing are describeed. In section 4, we discuss three versions of the EM selection algorithm used for slot-based WSD. In section 5, we present manual annotation of sentences used for evaluation. Finally, in section 6, we evaluate the algorithms and compare them. All example sentences presented throughout the paper come from a corpus used in the experiment.

2   Data Resources

Beginning our work, we paid a lot of attention to the choice of verbs to be considered in the experimental semantic valence dictionary [17]. The set of 32 verbs was chosen manually, taking into account the need to maximise the variability of syntactic frames (in particular, diathesis alternations) on one hand, and polysemy of verbs within a single syntactic frame, on the other. Another important criterion for this choice was frequency.

For the purpose of algorithm evaluation, we also needed a manually annotated subcorpus, referred to as HANDKIPI; with corresponding semantic categories assigned to the semantic head of each phrase (cf. section 5). We have prepared a test set which contains 5634 sentences for 32 verbs. The set of verbs chosen for experiments is referred to as CHVSET.

2.1   Corpus

Our main resource was the IPI PAN Corpus of Polish written texts [30], referred to as KIPI. The texts are segmented into paragraphs and sentences and annotated with morphosyntactic tags. The 2nd edition contains 250 mln segments (roughly, words).

From this corpus, we selected a small subcorpus, referred to as SEMKIPI, containing 195 042 sentences. Selected sentences contained:

(a) one or more verbs from CHVSET,
(b) at most three verbs in all.

Thus, we limited our considerations to the list of 32 verbs from CHVSET. The reason was that we could evaluate the algorithm only for the verbs from this list (cf. section 5).

2.2   Valence Dictionary

Our experiment was done using an extensive valence dictionary, which was specially prepared for the task (cf. section 3 below for the parser construction). The main component of our dictionary was Świdziński's [40] valence dictionary, which contains 1064 verbs. This dictionary was supplemented with entries for some additional verbs, as explained below.

Firstly, we made sure that the used valence dictionary provided entries for all 32 verbs of the benchmark set given in [17]. Only 24 of them were included in Świdziński's dictionary literally, whereas for 4 verbs we managed to adjust the available entries of their respective aspectual counterparts.[1] Entries for the remaining 4 verbs were elaborated by analysing entries of the automatically created valence dictionary by Dębowski [6]. All entries of this part of the dictionary were carefully studied, modified and augmented.

---

[1] In Polish, aspect is accomplished by different verbs.

Similar additions were done for other verbs. We adjusted the entries of the missing 269 aspectual counterparts of verbs described by Świdziński's dictionary. Moreover, we used the valence dictionary extracted from the corpus data by [6,7], from which the entries of 955 verbs were added verbatim to increase the coverage of our dictionary on SEMKIPI.

Syntactic valence frames are list of slot, which can be parametrised. Parameters are separated by colons. A nominal phrase (NP) is encoded as np, its only parameter is its case. A prepositional-nominal phrase (PP; encoded as prepnp) has two parameters: a preposition and the case of its NP complement.

Świdziński's dictionary includes frames that are subframes of other frames. The idea was to list all non-elliptic frames. For instance, *kupić* has a frame with slots for two nominal elements in accusative and nominative case, i.e., np:acc np:nom, which is instantiated in the sentence *Tak, panie, ja kupiłem pański dom.* (*Yes, sir, I bought your house.*). But larger frames can be found in this dictionary as well, such as np:acc np:nom prepnp:od:gen (cf. sentence *Ja kupiłem pański dom od pańskiej żony.*; *I bought your house from your wife.*). From the point of view of automatic parsing, however, listing subframes in the valence dictionary only slows down sentence parsing, since the *Świgra* parser accepts all subframes of each frame listed, to automatically account for the phenomenon of ellipsis. Thus, we deleted all subframes from our dictionary.

Table 1. Number of syntactic frames for verbs from CHVSET

lemma	frames	slots	lemma	frames	slots	lemma	frames	slots
bronić	9	7	powiedzieć	6	6	rozpoczynać	5	5
interesować	2	3	powtarzać	5	7	rozpocząć	6	5
kończyć	10	8	powtórzyć	4	5	skończyć	11	8
kupić	10	13	proponować	3	5	spotkać	2	4
lubić	2	4	przechodzić	10	14	trzymać	14	11
minąć	5	5	przejść	18	17	uderzyć	8	8
mówić	7	7	przygotować	6	6	widzieć	10	7
odnieść	4	6	przygotowywać	6	6	zaczynać	8	7
odnosić	6	6	przyjmować	8	8	zacząć	8	7
pisać	8	9	przyjąć	9	9	zakończyć	8	7
postawić	6	8	robić	13	10			

The numbers of syntactic frames for verbs from CHVSET are listed in Table 1 (the *frames* columns). In the *slots* columns we specify the number of NPs/PPs that can fill arguments of a particular verb (in any frame). The maximal number of syntactic frames per verb in the whole dictionary is 25, their mean is 2.8 and their median is 2. In the dictionary of selected verbs, it is 18, 7.9 and 7, respectively. In Świdziński's dictionary the numbers are 25, 3.9 and 3. The maximal number of arguments per frame (including subject and reflexive marker *się*) is 5, the median is 3 and the mean varies from 2.5 in the whole dictionary to 2.9 in the dictionary of selected verbs.

Having a variety of syntactic frames was an important criterion for selecting verbs for our experiments, which is confirmed by the largest number of frames per verb in the dictionary of selected verbs. On the other hand, the aforementioned Dębowski's method of valence dictionary creation chooses most frequent, strongly supported frames, i.e., a small number of rather short ones. This explains the smallest number of frames per verb in the whole dictionary.

2.3 Słowosieć—Polish Wordnet

In order to prepare an initial sense annotation for nouns (to be later automatically disambiguated), we used the Polish WordNet [8,10,9], called *Słowosieć* (English acronym PLWN). *Słowosieć* is a network of lexical-semantic relations, an electronic thesaurus with a structure modelled on that of the Princeton WordNet and those constructed in the EuroWordNet project. Polish WordNet describes the meaning of a lexical unit of one or more words by placing this unit in a network which represent such relations as synonymy, hypernymy, meronymy, etc. For the present work we do not use the whole structure of the net, but the set of 26 predefined categories (see Table 2) which are situated at the top of the actual hierarchy. Using these categories, 6917 nouns (most frequent in the balanced subcorpus of the KIPI) were classified by the Polish WordNet group. One to five categories are assigned to each noun. The mean of categories per noun is 1.2.

Słowosieć does not include gerunds, which are lematised to verbs. Their implicit semantic categories are act, event, process and state. This enlarges the actual mean (and probably median) of categories assigned to the semantic head of NPs/PPs.

Table 2. Predefined set of general semantic categories in Polish WordNet

Nr name	Nr name	Nr name
01 Tops	10 feeling	19 possession
02 act	11 food	20 process
03 animal	12 group	21 quantity
04 artifact	13 location	22 relation
05 attribute	14 motive	23 shape
06 body	15 object	24 state
07 cognition	16 person	25 substance
08 communication	17 phenomenon	26 time
09 event	18 plant	

The linguists performing manual annotation of the test corpus HANDKIPI were allowed to extend the set of semantic categories of a particular noun (cf. section 5). They have added 893 categories to 726 nouns. As a result, the maximal number of semantic categories per noun has increased to 7, and their mean has increased to 1.4.

3   Data Preparation

Before we start WSD, we need to annotate sentences from SEMKIPI with their reduced parses and semantic categories of nouns.

3.1   Parsing

SEMKIPI was parsed with the *Świgra* parser [44,43] based on the metamorphosis grammar GFJP [39]. In contrast to the experiment discussed in [18], this time we used a version of the grammar provided with the valence dictionary presented in section 2.2. The parser assumes that the boundaries of sentences were already identified in the corpus, but it ignores the disambiguation of morphosyntactic annotation established by the tagger (i.e., it takes all tags produced by morphosyntactic analyser *Morfeusz* into account).

In order to reduce sparseness of data, for the present experiment we considered only phrases being arguments of a verb (i.e., the subject and complements included in its valence frames). This means that each obtained parse was reduced to its flat form representing only these phrases. Each phrase has the syntactic and the semantic head (cf. [31]). Usually, these heads are the same (i.e., they are heads of phrases determined by the grammar), but for instance, the syntactic head of a preposition phrase is the preposition whereas its semantic head is the noun (i.e., the head of the complement noun phrase). Observe that these slots are the most interesting ones for our further semantic analysis. It should be emphasised here that the parser recognises phrases as arguments of a verb only on the basis of their syntactic form. If a complement has the same syntactic form as an adjunct, two parses are obtained with one of the phrases treated as a complement, and the other one treated as an adjunct, only one of them being a proper interpretation of a sentence. Similar error-causing effects are caused by the syncretism of cases. As a result, we obtain *reduced parses* of a sentence composed of a verb and a set of slots.

Table 3 lists all types of slots (verb arguments) considered in the analysis. Only slots present in sentences qualified for final semantic analysis are listed.

The table shows that the distribution of frequency of prepositions is as Zipfian as of other words. As for NPs, the proportion of their frequencies is not surprising: nominative and accusative cases are the most frequent and genitive case is the least frequent. Observe that NPs are 4 times more frequent than PPs.

If a sentence is composed of more than one clause, its parses are decomposed into subparses, each of them transformed into a separate reduced parse, linked to the predicate of a clause.

3.2   Selection of a Single Parse

Świgra tends to produce large parse forests. The mean of parses per sentence is 11 418 041 672 whereas the median is 42. The maximal number of parses is 592 491 499 739 040 (with 707 reduced parses) obtained for the sentence *Kraków to jedna z pierwszych stolic duchowych Polski, Wawel to jest wzgórze wielkości*

Table 3. Argument types considered in the analysis and their frequencies

slot	freq.	slot	freq.	slot	freq.
nom	29722	na:loc	497	bez:gen	14
acc	23007	na:acc	484	przy:loc	11
o:loc	4341	dla:gen	381	u:gen	10
dat	3974	w:acc	227	po:loc	7
do:gen	3415	z:gen	143	pod:inst	4
inst	1810	przez:acc	88	za:inst	3
od:gen	1433	za:acc	75	o:acc	3
z:inst	1128	przed:inst	74	między:inst	2
w:loc	1001	nad:inst	33	wśród:gen	1
gen	532	NP total	59045	PP total	13337

naszego narodu, tu Go zostawimy wśród tych wszystkich wielkich duchów, które w ciągu całego tysiąclecia przeszły po tym wzgórzu. (Cracow is among the prime Polish spiritual capitals. The Wawel hill is the peak of our nation's grandeur. We will let Him here, amidst all these great spirits that crossed this hill during the whole millennium.). The number of reduced parses of a sentence is much smaller than the number of entire parse trees, the more so as we have only considered actual arguments of verbs (without adjuncts). Nevertheless, this number is still considerable (the mean of reduced parses per sentence is 19, the median is 7). The maximal number of reduced parses is 6085 (for 36 920 464 complete parses). It was obtained for the sentence *Tak więc możliwości obrony ze strony obrońcy nie doznawałyby znacznego uszczerbku w myśl tego, co w tym projekcie się proponuje. (Thus the defense capacity of the defendant would not suffer a significant loss from what is proposed in this draft.).* The first sentence has 4 predicates (including *jest* (*is*) and predicative *to* having similar function). The idiosyncrasy of *to* probably influences the number of parses. The second sentence has 2 predicates. Note the weak dependency between the number of complete and reduced parses.

Thus, for each sentence we obtain a forest of reduced parses. Observe that such a forest is heterogeneous, as it contains reduced parses of the main clause and its subordinate clauses together. The selection process of the appropriate parse (for each clause) consists of two steps: a purely statistical one that is preceded by one based on the heuristics described below.

First, *Morfeusz* contains an extremely rich dictionary of Polish words, some of them very rare (e.g., dialectical or archaic). Thus, we delete parses containing such words which are homonyms of other, more frequent words. The procedure is based on the frequency list of lexemes in the whole KIPI. Some very rare interpretations of words are explicitly listed.

Next, although Polish is a free word order language, some orders of words are very unusual (e.g., inverse orders). However, *Świgra* produces all proper parses of a sentence. In order to delete parses containing such rare constructs, we compare two parses and delete:

- the parse with an NP corresponding to a pair of consecutive ⟨PP, NP1⟩ in another parse;
- the parse with an NP corresponding to a pair of consecutive ⟨NP1, NP2⟩ in another parse, where NP1 is in genitive and NP, NP2 have the same head;
- the parse with AdjP corresponding to a pair of consecutive ⟨AdjP, NP2⟩ in another parse.

Additionally, for the sake of semantic annotation, we exchanged wh-pronouns being heads of NPs/PPs with nouns being heads of NP/PP of the main clause modified by the corresponding wh-clause. For instance, the clause *który tak mówi* (*who speaks like that*) from the sentence *Człowiek, który tak mówi, opisuje swoje wrażenie emocjonalne.* (*A man who speaks like that describes his emotional impressions.*) the subject wh-clause *który* is exchanged with the noun *człowiek* (*man*).

Finally, we chose only those reduced parses from the forest which were headed by verbs from CHVSET.

The resulting reduced parse forests were disambiguated by means of an EM algorithm referreded to as *EM selection algorithm*. The algorithm was proposed by Dębowski [6] for the task of creating a syntactic valence dictionary. It selects the most probable element of a list (here: a list of syntactic frames of a verb for a particular sentence) on the basis of information about the frequency of the occurence of these elements (frames) in the whole data set (of sentences). In section 4.1 we use this algorithm for choosing the most probable semantic frame from a list obtained for the fixed syntactic frame.

The algorithm consists of three steps. In the first step, it ignores sentences with more than 50 frames or having parses with some idiosyncratic words (such as interrogative and demonstrative pronouns *co* (*what*), *ten* (*this*)). Moreover, it separates parses of different clauses of a sentence. Thus, the resultant forests are homogeneous and linked to a particular predicative verb.

The second step consists in transforming reduced parses into syntactic valence frames they represent. The order of phrases in a sentence is neglected. Thus, the algorithm possesses no information about the frequency of syntactic frame instantiation in different orders, hence it cannot influence the probability of a frame for any particular sentence. This is the reason for using some heuristics for this phenomenon.

The number of frames for a particular clause could be smaller than the number of corresponding reduced parses, as two or more parses may correspond to the same frame. Note that this means that selecting a single frame does not entail selecting a single reduced parse. This phenomenon is exemplified in sentence (7); cf. section 4.

The third step carries out the actual EM selection algorithm. It operates on valence frames corresponding to particular reduced parses and selects the most probable one.

Let us show the whole process by examples. They contain lists of reduced parses of exemplary sentences. A reduced parse starts with a verb heading a clause, together with morphosyntactic tags separated by ':' sign (aff,neg show whether a sentence is affirmative or negative). A verb is acompanied by a list of phrases (np, prepnp, advp etc.). Phrase type is followed by its syntactic head, optionally its semantic head and other corresponding morphosyntactic tags. Numbers show boundaries of a phrase (or a whole clause) in a sentence.

(1) % 'Góralskie przysłowie mówi, że późna zima mocno trzyma.'
    (The highlander's proverb says that late winter holds "strong".)

```
% trees:   24
0-19 mówić aff:fin:sg:_:ter   []
0-19 mówić aff:fin:sg:_:ter   [0-1:np:góralski:pl:acc:m2:ter]
0-19 mówić aff:fin:sg:_:ter                                              +
     [0-1:np:góralski:pl:acc:m2:ter, 1-2:np:przysłowie:sg:nom:n:ter]
0-19 mówić aff:fin:sg:_:ter   [0-2:np:przysłowie:sg:acc:n:ter]
0-19 mówić aff:fin:sg:_:ter   [0-2:np:przysłowie:sg:nom:n:ter]
0-19 mówić aff:fin:sg:_:ter                                              +
     [0-2:np:przysłowie:sg:nom:n:ter, 3-19:cp:że:ctr]
0-19 mówić aff:fin:sg:_:ter   [1-2:np:przysłowie:sg:acc:n:ter]
0-19 mówić aff:fin:sg:_:ter   [1-2:np:przysłowie:sg:nom:n:ter]
0-19 mówić aff:fin:sg:_:ter
     [1-2:np:przysłowie:sg:nom:n:ter, 3-19:cp:że:ctr]
0-19 mówić aff:fin:sg:_:ter   [3-19:cp:że:ctr]
5-9  trzymać aff:fin:sg:_:ter [5-7:np:zima:sg:nom:f:ter]
5-9  trzymać aff:fin:sg:_:ter                                           +
     [5-7:np:zima:sg:nom:f:ter, 7-8:advp:mocno]
```

(2) % 'Prawie każdy bank przygotował na okres urlopowy
    specjalną linię kredytową dla klientów.'
    (Almost every bank had prepared a customised credit line
    for its clients for the holiday period.

```
% trees:   368
0-12 przygotować aff:fin:sg:m3:ter [0-3:np:bank:sg:nom:m3:ter,
        4-10:np:linia:sg:acc:f:ter, 10-12:prepnp:dla:klient:gen]
0-12 przygotować aff:fin:sg:m3:ter [0-3:np:bank:sg:nom:m3:ter,
                           4-12:prepnp:na:linia:acc]
0-12 przygotować aff:fin:sg:m3:ter [0-3:np:bank:sg:nom:m3:ter,
                           4-12:prepnp:na:okres:acc]
0-12 przygotować aff:fin:sg:m3:ter [0-3:np:bank:sg:nom:m3:ter,
        4-7:prepnp:na:okres:acc, 7-12:np:linia:sg:acc:f:ter]
```

(3) % 'Sąd uznał, że w tym procesie nie interesuje
    go sprawa moskiewskich pieniędzy.'
    (The court of law decided that during this process
    it is not interested in the case of Moscow money.)

```
% trees:   150
0-21 uznać aff:fin:sg:m3:ter      [0-1:np:sąd:sg:nom:m3:ter]

0-21 uznać aff:fin:sg:m:ter       []                                  ✕
0-21 uznać aff:fin:sg:m:ter       [0-1:np:sąd:sg:acc:m3:ter]
4-13 interesować neg:fin:sg:_:ter []
4-13 interesować neg:fin:sg:_:ter [10-11:np:sprawa:sg:nom:f:ter]
4-13 interesować neg:fin:sg:_:ter
     [10-11:np:sprawa:sg:nom:f:ter, 11-13:np:pieniądz:pl:acc:m3:ter]
4-13 interesować neg:fin:sg:_:ter [10-13:np:pieniądz:pl:acc:m3:ter]
4-13 interesować neg:fin:sg:_:ter [10-13:np:sprawa:sg:nom:f:ter]
4-13 interesować neg:fin:sg:_:ter [9-10:np:on:sg:acc:mn:ter]
4-13 interesować neg:fin:sg:_:ter                                     +
     [9-10:np:on:sg:acc:mn:ter, 10-11:np:sprawa:sg:nom:f:ter]
4-13 interesować neg:fin:sg:_:ter
     [9-10:np:on:sg:acc:mn:ter, 10-13:np:sprawa:sg:nom:f:ter]
4-13 interesować neg:fin:sg:_:ter [9-11:np:sprawa:sg:nom:f:ter]
4-13 interesować neg:fin:sg:_:ter                                     +
     [9-11:np:sprawa:sg:nom:f:ter, 11-13:np:pieniądz:pl:acc:m3:ter]
4-13 interesować neg:fin:sg:_:ter [9-13:np:pieniądz:pl:acc:m3:ter]
4-13 interesować neg:fin:sg:_:ter [9-13:np:sprawa:sg:nom:f:ter]       +
```

The sentence (1) has 24 trees for 2 verbs, both belonging to CHVSET, with 10 reduced parses for *mówić* (*to say*) and 2 for *trzymać* (*to hold*). The heuristic part of the process chooses 2 reduced parses for *mówić* and 1 for *trzymać* (marked by + symbol).

The sentence (2) has 368 trees for just one verb *przygotować* (*to prepare*). The number of reduced parses is quite large (37), so we listed only the 4 chosen by the heuristics.

The sentence (3) has 150 trees for 2 verbs. However, only *interesować* (*to interest*) belongs to CHVSET, with 12 reduced parses obtained for it. 3 of the parses were chosen by the heuristics.

In the preparation step, the EM selection algorithm finds syntactic valence frames corresponding to a particular reduced parse (for each verb separately). For the above examples we obtain:

```
ad (1)   – mówić:
            mówić np:acc np:nom
            mówić np:nom sentp:że                    +
         – trzymać:
            trzymać advp np:nom        +
ad (2)   przygotować
         przygotować np:acc np:nom prepnp:dla:gen    +
         przygotować np:acc np:nom prepnp:na:acc
ad (3)   interesować
         interesować np:acc np:nom      +
```

The frames selected by the EM selection algorithm are marked by the + symbol. For sentences (1) and (3) the selected frames are the correct ones. For sentence

(2) both source frames are actually subframes of an entire frame, so both choices are equally correct. Nevertheless, the evaluation of the process of frames selection on manually annotated data shows that only 43.5% of manually annotated sentences were given the same syntactic valence frame as during automatic process (see Table 7). This is the significantly worse result than the 75% reported in [6,7]. The most probable reason of the difference is the way the set of sentences was selected.

Finally, the reduced parses that match selected frames are found. The reduced parse chosen for sentence (2) is shown in (4).

Unfortunately, each of the applied tools impose specific conditions on the input sentences, which results in a reduction of the subcorpus. First, the present version of *Świgra* analyses only a subset of Polish syntactic constructions. Next, the EM selection algorithm discards sentences that do not satisfy the conditions mentioned above. Finally, all clauses containing no NPs/PPs (such as *Ogromnie się cieszę; [I] am very glad*) are neglected.[2] As a result, after taking into account these constraints, the number of sentences has decreased from 195 042 to 47 184. The reduction of the number of sentences during the whole process is presented in Table 4. The table contains the lemmata of verbs, their aspect (i for imperfect and p for perfect), English glosses, the number of sentences containing these verbs in the whole SEMKIPI, the number of sentences (per verb) parsed by *Świgra*, the number of sentences after the heuristic phase and after EM-based valence selection. The last column shows the number of sentences with NPs/PPs such that at least one of them has a semantic category (i.e., it is not a pronoun, an adjective interpreted as noun, etc.). The general statistics of sentences, their reduced parses and the phrases they include, are presented in Table 5.

3.3   Semantic Categories

In our current work we are interested in assigning semantic categories to nouns being semantic heads of NPs and PPs. Thus, we can ignore other syntactic slots.[3]

For this raeson, we provide a list of semantic categories for each semantic head of an NP or a PP in each reduced parse of every clause. The categories are taken from *Słowosieć*. Assuming that a sentence is not ambiguous and its syntactic frame has been properly selected, exactly one reduced parse is adequate and exactly one category is adequate for every NP/PP.[4] The resultant syntactic-semantic reduced parse for sentence (2) corresonding to the chosen syntactic frame (in <> brackets) is presented in (4).

---

[2] Such clauses contain no nouns and hence no semantic categories of nouns to disambiguate. Thus, they would have a single semantic frame to choose and they would not influence the algorithm's behaviour.

[3] Neglecting the existence of other syntactic slots means reducing the number of different valence frames, hence reducing the sparseness of data. However, this also means some loss of information.

[4] Unfortunately, this assumption is not always met.

(4) % 'Prawie każdy bank przygotował na okres urlopowy
     specjalną linię kredytową dla klientów.'
     <przygotować np:acc np:nom prepnp:dla:gen>
     0-12 przygotować aff:fin:sg:m3:ter
              [0-3:np:bank:sg:nom:m3:ter::   group location,
              4-10:np:linia:sg:acc:f:ter::   shape location relation,
              10-12:prepnp:dla:klient:gen::   person]

Table 4. The reduction of number of sentences per verb

lemma	asp.	gloss	source	parsed	prepared	valence selected	parses with sem.categs.
bronić	i	protect	2547	821	805	763	646
interesować	i	interest	2850	1113	1102	922	873
kończyć	i	finish	3982	1587	1551	1425	1332
kupić	p	buy	2293	544	527	477	452
lubić	i	like	3194	1519	1473	1358	939
minąć	p	pass	2206	719	668	520	499
mówić	i	speak	35792	10418	9993	8874	6072
odnieść	p	*polysemous*	4952	1260	1259	1210	1129
odnosić	i	*polysemous*	2626	869	869	775	748
pisać	i	write	3574	1305	1266	1174	958
postawić	p	put	3260	958	956	931	879
powiedzieć	p	say	28882	5286	4635	3912	1907
powtarzać	i	repeat	2821	759	637	555	355
powtórzyć	p	repeat	2154	391	374	310	224
proponować	i	propose	6351	3164	3142	2977	2194
przechodzić	i	*polysemous*	2592	1537	1531	1478	1442
przejść	p	*polysemous*	5032	1875	1861	1772	1659
przygotować	p	prepare	3735	1234	1228	1194	1147
przygotowywać	i	prepare	2470	1135	1130	1101	1052
przyjmować	i	*polysemous*	2554	1067	1036	937	868
przyjąć	p	*polysemous*	8155	3809	3743	3192	3091
robić	i	do/make	10215	3355	3221	2175	1646
rozpoczynać	i	start	1669	877	873	857	834
rozpocząć	p	start	8707	4096	4089	4032	3852
skończyć	p	finish	4213	1383	1363	1203	991
spotkać	p	meet	4174	1452	1451	1382	1166
trzymać	i	hold	1870	673	670	622	532
uderzyć	p	hit	1017	476	473	448	372
widzieć	i	see	10006	3890	3769	3382	2550
zaczynać	i	start	3642	1636	1623	1553	1126
zacząć	p	start	10872	4670	4661	4508	3059
zakończyć	p	finish	6635	2816	2801	2706	2590
total	–	–	195042	66694	64780	58725	47184
mean	–	–	6095	2084	2024	1835	1475
median	–	–	3642	1383	1363	1210	1126

Table 5. Simple statistics for sentences and their parses

(*Source number of sentences: 195 042*)	after syntactic analysis	after disambiguation of parses	sentences with NPs/PPs only
Number of sentences	74 904	58 723	47 194
Number of reduced parses	1 405 989	101 576	73 153
Number of empty parses	70 264	0	0
Mean of reduced parses per clause	18.772	1.730	1.550
Median of reduced parses per clause	7	1	1
Number of phrases	2 447 026	192 708	143 542
Number of NPs and PPs	2 049 558	159 312	122 941
Mean of phrases per parse	1.832	1.897	1.962
Median of phrases per parse	2	2	2
Mean of NPs and PPs per parse	1.534	1.568	1.681
Median of NPs and PPs per parse	2	2	2

4   EM Selection Algorithm for Semantic Category Disambiguation

Our goal here is the disambiguation of semantic categories of nouns being semantic heads of NPs and PPs. To solve this task, we adapted the EM selection algorithm (a version of a well known Expectation Maximisation algorithm) initially used by Dębowski in [6] to select the correct valence frame of a clause. The algorithm is not supervised, i.e., it does not need any training data. On the other hand, its behaviour is highly dependant on the size of the corpus.

We decided to work on semantic valence frames of a clause instead of entire parses. Thus, we combined sets of categories attached to each NP and PP from every reduced parse.

Next, we split the resultant syntactic-semantic valence frames in such a way that each NP/PP had only one category assigned. The disambiguation process consists in selecting the most probable frames. So, the sentence (2) has the semantic frame shown in (5) which after splitting transforms into the 6 frames presented in (6):

(5) % 'Prawie każdy bank przygotował na okres urlopowy
        specjalną linię kredytową dla klientów.'
    <przygotować np:acc:shape,location,relation np:nom:group,location
                    prepnp:dla:gen:person>

(6) acc: shape,   dla:gen: person, nom: group
    acc: shape,   dla:gen: person, nom: location
    acc: location, dla:gen: person, nom: group
    acc: location, dla:gen: person, nom: location
    acc: relation, dla:gen: person, nom: group
    acc: relation, dla:gen: person, nom: location

As we have noticed in section 3.2, there exist sentences (clauses) with more than one reduced parse matching the selected syntactic frame. The typical case of such situation occurs in simple sentences with the nominative subject and accusative object, when nouns that occupy these positions have syncretic nominative and accusative case. To eliminate such cases we impose special conditions on the splitting operation. We demonstrate how they work on the example (7). Each NP head in each reduced parse has 2 semantic categories assigned. So, their maximal number per NP/PP head is 2 as well. However, the number of semantic categories of each slot of the semantic frame (presented in <> brackets) is 4 and the sets of categories are equal. This suggests that they were combined from two reverse reduced parses, which is the reason for ignoring the sentence.

(7) % 'Zarząd widzi możliwość kontynuacji testów.'
   ([A/The] board of management notices the possibility of continuing tests.)
   <widzieć np:acc:act,group,state,top np:nom:act,group,state,top>
   0-5 widzieć aff:fin:sg:_:ter
              [0-1:np:zarząd:sg:acc:m3:ter::   act group,
               2-5:np:możliwość:sg:nom:f:ter::   state top]
   0-5 widzieć aff:fin:sg:_:ter
              [0-1:np:zarząd:sg:nom:m3:ter::   act group,
               2-5:np:możliwość:sg:acc:f:ter::   state top]

4.1   EM Selection Algorithm Applied to Whole Semantic Frames

Firstly, we tried using the EM selection algorithm in the same fashion as it was originally used by Dębowski in [6] to select the correct syntactic parse, namely, to treat each semantic frame as an atomic value. This approach will be called the *EM-whole* algorithm here. The scheme is as follows.

Let A_i be the set of alternative pairs $\langle v, f \rangle$ of verb v with frame f for the i-th clause with verb v in the corpus, $i = 1, 2, ..., M_v$, where M_v is the number of clauses containing verb v. Moreover, symbol $p_n(v, f)$ will stand for the effective probability, or frequency, of verb v with frame f in the n-th iteration. We set the initial equidistribution $p_1(v, f) = 1$ and iterated:

$$p_n(v, f|i) := \begin{cases} p_n(v, f)/\sum_{\langle v', f' \rangle \in A_i} p_n(v', f') & \text{for } \langle v, f \rangle \in A_i, \\ 0 & \text{else}, \end{cases}$$

$$p_{n+1}(v, f) := \frac{1}{M_v} \sum_{i=1}^{M_v} p_n(v, f|i).$$

The iteration was stopped at $n = 15$ and, for each clause i, we selected all pairs $\langle v, f \rangle$ that had the maximal frequency $p_n(v, f|i)$.

The algorithm is based on the observation that:

1. effective probability $p_n(v, f|i)$ that frame f of verb v is appropriate for clause i is proportional to the frequency of frame f in all clauses of verb v;

2. it is better to consider effective probability $p_n(v, f)$ of frame f in all clauses of verb v (normalised w.r.t. number of clauses M_v) counted proportionally to effective probability $p_n(v, f|i)$ of frame v in each clause i instead of simple frequency.

At the beginning, all semantic frames of a verb are equally probable. Thus, the algorithm estimates the probability of semantic frames of a clause on the basis of their overall probability for a verb predicating the clause.

4.2   EM Selection Algorithm for Independent Occurrence of Verb Arguments in a Frame

According to the experimental results of [18], the *EM-whole* algorithm very often selects more than one frame per clause. This is due to the sparseness of data: quite frequently each frame of a particular clause occurs only once in the data set. Thus, all interpretations of the clause remain equally probable during EM iterations.

In order to reduce the sparseness of data, we decided to modify the selection algorithm so that the occurrence of each NP/PP argument $a \in f$ of a semantic frame f be treated as independent of the occurrence of other arguments. In this version, called the *EM-indep* algorithm, we iterated:

$$p_n(v, f|i) := \begin{cases} \prod_{a \in f} p_n(v, a) / \sum_{\langle v', f' \rangle \in A_i} \prod_{a' \in f'} p_n(v', a') & \text{for } \langle v, f \rangle \in A_i, \\ 0 & \text{else}, \end{cases}$$

$$p_{n+1}(v, a) := \frac{1}{M_v} \sum_{i=1,2,\dots,M_v : a \in f \in A_i} p_n(v, f|i),$$

with the initialization $p_1(v, a) = 1$.

The above changes are the consequences of the following:

1. due to the assumption of independence, effective probability $p_n(v, f)$ is a product of $p_n(v, a)$ for $a \in f$;
2. we count effective probability (frequency) $p_n(v, a)$ of argument a considering all frames f including it.

Thus, this time the algorithm estimates the probability of semantic frames of a clause, basing on the overall effective probability of all arguments from this very frame.

4.3   Incremental Version of the EM Selection Algorithm

The assumption that the arguments of a verb in a clause are independent is an obvious simplification. Thus, either version of the algorithm presented above has its own merits and limitations. In [18] it was observed that the *EM-whole* algorithm tends to make no decisions, whereas the *EM-indep* algorithm makes more mistakes.

In order to combine the merits of both algorithms and to reduce their short-comings, we decided to use yet another solution, referred to as the *EM-incr* algorithm. For clauses having frames composed of at most two NP/PP slots, the *EM-incr* algorithm works the same way as the *EM-whole* algorithm. For clauses having larger frames, *EM-incr* incrementally applies the *EM-whole* for subframes composed of 2, 3 up to all slots. First, all 2-element syntactic sub-frames of a whole frame are considered. The larger subframes are established on the basis of the subframe chosen in the previous step. Each time one slot is added to each of the selected subframes, i.e., we obtain all its one-slot larger su-perframes. The EM procedure treats semantic frames corresponding to different syntactic frames independently (as *EM-whole* applied here). Nevertheless, we select the most probable frame(s) from the whole set. Thus, in the subsequent step we work on frames obtained from syntactic frames which semantic frames were chosen as most probable ones (for a particular sentence). Finally, we select the most probable frames (subframes composed of all slots).

Using this procedure, we have to disambiguate only the semantic categories of the last-added slot. The only exception is the first step, when the algorithm works on subframes containing two non-disambiguated slots. On the other hand, the number of clauses relevant to a particular subframe is larger than in the case of the *EM-whole* algorithm for frames of the same syntactic nature; the shorter frame is considered, the larger number of clauses. Both features should reduce the sparseness of data.

4.4   The Degree of Reduction of Semantic Categories

In order to disambiguate semantic categories, the algorithms have to reduce their number per noun.

In Table 6 we present the mean of the number of semantic categories assigned to occurrences of nouns, including baseline algorithms presented in section 6.2. For each data set (cf. section 5.1), the first column shows the mean for all nouns; in the second column pronouns were not counted. Median is always 1, even for source data. Initially, all 26 semantic categories are assigned to pronouns.

All the algorithms reduce the number of semantic categories per noun. The *MaxNoun* algorithm makes no decisions for pronouns and nouns from outside the set of manually annotated nouns (cf. section 5), which results in its poor behaviour.

Table 6. Results of reducing the number of semantic categories per noun

algorithm	AutoProc		ValMatch		ParseMatch	
source	3.423	1.852	3.434	1.851	3.435	1.816
EM-indep	1.002	1.002	1.002	1.002	1.003	1.003
EM-whole	1.077	1.026	1.080	1.027	1.080	1.027
EM-incr	1.120	1.026	1.126	1.027	1.122	1.028
MaxNoun	2.901	1.289	2.913	1.290	2.921	1.262
MaxSlot	1.002	1.002	1.002	1.002	1.002	1.002

5 Manually Annotated Data for Evaluation of Algorithms

In order to evaluate the algorithms, a small subcorpus of SEMKIPI was syntactically and semantically annotated by a group of linguists. 240 sentences for each verb from CHVSET were selected randomly from SEMKIPI.[5] The linguists performed three different tasks:

1. correction of morphosyntactic tagging (tagger errors),
2. division of sentences into phrases, i.e., pointing out their boundaries and syntactic and semantic heads,
3. assignment of a single PLWN semantic category to each noun in a sentence.

The correction of morphosyntactic tagging was limited to choosing one of the suggestions provided by the morphosyntactic analyser. In contrast, the linguists were allowed to add semantic categories of nouns (from the whole repertoire of categories presented in Table 2) that were not anticipated by PLWN authors for a particular noun.[6] Considered phrases should be semantically connected to a verb (its complements and adjuncts). Functional expressions were ignored. If any of the above requirements could not be satisfied (e.g., the sentence was grammatically incorrect), the linguists were allowed to reject it.

The process of annotation was performed by means of the program *Anotatornia* via the Internet [20]. Each sentence was annotated by two linguists. In the case of differences between their annotations, a process of negotiations were performed until they reached an agreement. If the negotiations took more than two iterations, the moderator judged the sentence.

The process of manual annotation is still in progress. Thus, we should keep in mind that until the process of manual annotation is finished, its results are not stable and could change. The statistics for hand-annotated sentences w.r.t. automatic processing is presented in Table 7. Since the linguists were allowed to reject sentences, the column *hand* shows the number of accepted ones. Their total number is 5634, which means 94%. Note that sentences in HANDKIPI do not need to have NP/PP arguments, as there were not such criteria for their selection. Nevertheless, as many as 91.2% of accepted sentences include at least one NP/PP with assigned semantic category (column *with sem. categs.*). HANDKIPI is simply a subset of SEMKIPI, and only 43.3% of manually annotated sentences were accepted by the automatic process (column *common*); most of them were lost during *Świgra* parsing. The reason is that linguists could annotate bizarre sentences which are rejected by the parser.[7] Worse still, only 43.5% of sentences belonging to both sets had the same syntactic valence frame chosen automatically and by the linguists (column *common valences*).

---

[5] More precisely, only single-verb sentences were chosen for this sampling.

[6] The reason for this decision is that the PLWN is not yet completed and our work bases on the preliminary classification.

[7] All percentages in the table are counted w.r.t. the *hand* column in order to show the reduction of data.

Table 7. The statistics for annotated sentences

lemma	hand	with sem. categs.	proc.	common	proc.	common valences	proc.
bronić	216	188	87.0	83	38.4	42	19.4
kończyć	229	195	85.2	95	41.5	70	30.6
kupić	231	221	95.7	63	27.3	11	4.8
lubić	229	196	85.6	69	30.1	36	15.7
minąć	221	218	98.6	57	25.8	21	9.5
mówić	221	189	85.5	56	25.3	10	4.5
odnieść	234	220	94.0	102	43.6	56	23.9
odnosić	234	223	95.3	83	35.5	51	21.8
pisać	223	192	86.1	67	30.0	14	6.3
postawić	228	224	98.2	100	43.9	36	15.8
powtarzać	202	168	83.2	70	34.7	30	14.9
powtórzyć	236	204	86.4	33	14.0	14	5.9
proponować	223	217	97.3	103	46.2	61	27.4
przechodzić	233	225	96.6	147	63.1	15	6.4
przejść	227	211	92.9	113	49.8	30	13.2
przygotować	235	223	94.9	96	40.9	39	16.6
przygotowywać	231	222	96.1	125	54.1	48	20.8
przyjmować	226	220	97.3	97	42.9	37	16.4
przyjąć	227	225	99.1	126	55.5	55	24.2
robić	228	160	70.2	65	28.5	29	12.7
rozpoczynać	224	218	97.3	118	52.7	64	28.6
rozpocząć	226	219	96.9	144	63.7	95	42.0
skończyć	229	198	86.5	63	27.5	36	15.7
spotkać	227	190	83.7	73	32.2	45	19.8
trzymać	194	171	88.1	71	36.6	20	10.3
total	5634	5137	91.2	2219	43.3	965	18.7

The percentage for sentences accepted both by linguists and the automatic process is quite irregular and varies from 14.0 (*powtórzyć, to repeat*) to 63.7 (*rozpocząć, to start*). Similarly, the percentage of sentences with the same frames ("common valences") varies from 4.5 (*mówić, to speak*) to 42.0 (*rozpocząć, to start*).[8] This influences the number of sentences used for the actual evaluation of the algorithm. We should remember that the results of the evaluation for verbs with the smallest percentages are less valuable.

5.1   Preparation of Data

One of the consequences of the ongoing work of manual annotation is is the fact that obligatory arguments are not differentiated from adjuncts. As a result,

---

[8] Counted w.r.t. *hand* column. The percentage of sentences with the same syntactic frame obtained by computer and manual processing counted w.r.t. *common* column varies from 10.5 (*przechodzić*, polysemous) to 73.7 (*kończyć*).

these sentences are connected with syntactic frames that have no counterparts in *Świgra*-parsed sentences (are their superframes). This would influence the performance of the algorithm for HANDKIPI sentences. In order to avoid such an influence, we have automatically selected maximal subframes corresponding to valence dictionary frames for each sentence. However, this procedure is error-prone.

As we have noticed above, only 43.3% of sentences from HANDKIPI underwent computer processing. Furthermore, the reduced parses obtained by the EM selection algorithm in the data preparation process (section 3) and through manual annotation differs for 66.5% sentences. This influences the data and hence the evaluation procedure.

Thus, we decided to evaluate all the algorithms on three sets of sentences. The first set (referred to as AutoProc) consists of purely automatically pre-processed sentences. Then the evaluation is performed on manually annotated sentences that were automatically parsed as well. The second set (referred to as ValMatch) was obtained by merging valence schemata of automatically pre-processed sentences and manually annotated ones. Then reduced parses were again found automatically. The third set (referred to as ParseMatch) consists of automatically preprocessed sentences extended with manually annotated ones. If a sentence belongs to both sets, the manually annotated version was chosen. Manually annotated semantic categories of nouns were deleted. Thus, evaluation performed on this set is independent from the preprocessing steps of the analysis, hence it is most reliable if we want to evaluate slot-based WSD step solely instead of the whole process.

6   Evaluation and Comparison of the Algorithms

Ultimately, we had three versions of the EM selction algorithm at our disposal. In order to know how they actually work we had to evaluate each of them.

We decided to evaluate the algorithms using the simplest measure, i.e., correctness, which means counting all scores for every evaluated item w.r.t. the number of items. We think that in the case when the manual annotation assigns a single semantic category to each noun, this measure is satisfactory. Traditionally, if semantic categories of an NP/PP head are equal in computer and manual annotation, the score is set to 1. However, in our experiments each algorithm can choose more than one semantic category, and then one of them can match manual annotation. In order to take into account such situations in the evaluation as partial success, we decided to set the score to $\frac{m-k}{m \cdot k}$, where m is the number of all semantic categories of a noun, and k is the number of categories chosen by the algorithm (for $k \geq 2$). This score increases for decreasing k and increasing m, hence for our data it obtains the maximal value 0.375 for $k = 2$ and $m = 8$ (for pronouns $m = 26$, hence we obtain maximal value 0.462). The fact that the mean of k is 1.003 for the *EM-indep* and *MaxSlot* algorithms and 1.12 for *EM-whole* and *EM-incr* shows that this way of coutning scores for multi-element choices does not significantly influence the resultant evaluation of a particular algorithm. On the other hand, the *MaxNoun* algorithm makes no decisions for pronouns and some nouns, which means $k = m$, i.e., we obtain 0 score. Thus, the average $k = 2.9$ does not entail a big influence on the evaluation of this algorithm as well.

The evaluation has been performed in two ways. Firstly, we compared each NP/PP slot separately. Secondly, we evaluated entire sentences, i.e., we checked whether the results agree for every NP/PP slot in a sentence. Then, scores for all NPs/PPs in the sentence were added.

6.1   Choice of the Number of Iterations

The number of iterations n of each EM algorithm had to be set experimentally. In this section we show the corresponding procedure.

The results of the evaluation for n set to 6, 10, 15 and 20 are summarised in Table 8. The evaluation was carried out in two ways: for entire sentences and for separate slots (cf. section 6), which is shown in the figure. For each algorithm, the best result of the evaluation is underlined.

Table 8.   Evaluation of the EM selection algorithms for various number of main iteration

data set	EM. alg.	whole sentences				separate NPs/PPs			
		6	10	15	20	6	10	15	20
AutoProc	whole	74.15	74.27	<u>74.37</u>	<u>74.37</u>	77.70	77.78	<u>77.89</u>	77.86
ValMatch	whole	71.06	<u>71.43</u>	71.42	71.42	76.78	<u>77.10</u>	77.09	77.05
ParseMatch	whole	80.45	80.52	<u>80.59</u>	80.55	84.43	84.51	<u>84.54</u>	84.50
AutoProc	indep	<u>76.13</u>	<u>76.13</u>	76.12	76.12	80.82	<u>80.86</u>	<u>80.86</u>	<u>80.86</u>
ValMatch	indep	<u>73.02</u>	<u>73.02</u>	72.98	72.98	<u>80.33</u>	<u>80.33</u>	80.31	80.30
ParseMatch	indep	80.97	<u>81.08</u>	81.03	81.03	86.16	<u>86.21</u>	86.18	86.18
AutoProc	incr	74.81	74.86	<u>75.08</u>	74.91	78.74	78.85	<u>79.02</u>	78.90
ValMatch	incr	71.63	71.78	<u>71.84</u>	71.70	77.87	78.02	<u>78.06</u>	78.03
ParseMatch	incr	80.61	<u>80.79</u>	80.72	80.66	84.60	<u>84.69</u>	84.54	84.46

First, observe that the results are very regular: the differences between them never exceed 0.4%. Next, there is an increase between 6 and 10 iterations for all cases except 2 equalities for whole sentences and 1 equality for separate slots. Surprisingly, in most cases 20 iterations execution of the algorithms show worse results than the best ones. The only exception is the *EM-whole* algorithm evaluated on whole sentences and *EM-indep* algorithm evaluated on separete NPs/PPs, both for AutoProc data set, where the result for 20 iterations equals the best one.

Thus, the reasonable choice is between 10 and 15 iterations. In this area, we can observe an increase in results as often as their decrease. The average increase is 0.08% and the average decrease is 0.04%. Decreases plague the *EM-indep*

algorithm, hence we have chosen 10 iterations for it. The most heterogeneous
is the *EM-incr* algorithm, with the maximum increase for the AutoProc data
set (0.22%, 0.17%) and the maximum decrease for ParseMatch data set (0.05%,
0.15%). We have chosen 15 iterations.

6.2   Baseline Algorithms

A comparison of the three versions of the algorithm does not suffice to make an
equitable evaluation. Since we cannot compare our results against any existing
WSD work for Polish, some baseline solutions are needed.

Word meanings exhibit Zipfian distribution. Hence a very popular choice
of a baseline algorithm for the WSD problem is *the most frequent sense of a
word* [15,1]. However, this heuristics is applicable only to those few languages
for which significantly large sense-tagged corpora are available. Currently, we
have no such resource at our disposal. Instead we can use the semantic category
that was most frequently assigned to a particular word in HANDKIPI (referred
to as the *MaxNoun* algorithm). However, this method is highly dependent on
this very corpus. Note that *MaxNoun* makes no decisions for nouns absent in
HANDKIPI, similarly as for pronouns.

In Table 9 we present some statistics for manually annotated data, counted
for all nouns and for nouns with more than one semantic category (referred to
as *multi-cat. noun*). *Frequency* is referred to as *freq.* and *semantic category*
is referred to as *sem.cat.* *Freq. with sem.cat.* means the number of times a
particular noun has a particular *sem.cat.* assigned; *freq of max. sem.cat.* means
how many times a particular noun has the most often *sem.cat.* assigned. Only
nouns present in HANDKIPI are considered. We have 899 nouns in HANDKIPI
with obvious Zipfian distribution, the most frequent is the pronoun *to* (*this*) with
596 occurrences. Observe that about 2/3 of occurrences of nouns in sentences are
polysemous nouns. Additionally, 426 (25%) semantic categories of nouns were
never assigned (0 frequency). The most important information is that 82.66%
of the occurrences of nouns are the most frequent semantic category occurrences
(74.67%, with mean 2.4 semantic categories, for polysemous nouns).

The percentage of semantic categories of a noun not assigned to any token,
and the percentage of tokens assigned to the most frequent semantic category
of a noun, is puzzling. Thus, the evaluation of the *MaxNoun* algorithm on
the same corpus for which the most frequent semantic categories were counted
is unreliable. The situation is similar to the case when an ML algorithm is
overtrained.

Therefore, we propose another baseline algorithm (referred to as the *MaxSlot*
algorithm), which does not use data from HANDKIPI, hence it is independent of
it. For each verb and each syntactic slot, it counts all occurrences of a particular
semantic category (for multi-category nouns, the corresponding fraction is considered) in a given set of sentences. Next, for each NP/PP, the most frequent
semantic category for this slot is chosen (from the categories of its semantic
head).

Table 9. Statistics for semantic categories of nouns in HANDKIPI

| | nouns | | | | nouns | |
	all	multi-cat.			all	multi-cat.
total nouns	899	592	max. noun freq.		596	201
total sem.cat.	1734	1427	mean noun freq.		9.23	9.60
max. sem.cat.	8	8	med. noun freq.		2	3
mean sem.cat.	1.93	2.41	max. freq. with sem.cat.		596	131
med. sem.cat.	2	2	mean freq. with sem.cat.		4.79	3.98
total noun freq.	8300	5681	med. freq. with sem.cat.		1	1
total freq. of			proc. freq. of			
max. sem.cat.	6861	4242	max. sem.cat.		82.66	74.67

Table 10. General results of evaluation of the algorithms

| | whole sentences | | | separate NPs/PPs | | |
algorithm	AutoProc	ValMatch	ParseMatch	AutoProc	ValMatch	ParseMatch
EM-indep	76.12	72.98	81.03	80.86	80.30	86.17
EM-whole	74.39	71.50	80.59	78.53	77.80	84.53
EM-incr	75.04	71.84	80.81	78.99	78.06	84.68
MaxNoun	72.71	71.42	84.90	71.57	71.56	86.26
MaxSlot	75.47	72.33	81.40	80.52	80.07	86.46

6.3   General Evaluation

Now we have 5 algorithms to evaluate on 3 sets of sentences. To begin, we want to discuss the general results for all the sets together. They are summarised in Table 10. Note that the results are very similar for all the algorithms. First of all the algorithms show the best results for the ParseMatch set of data, in which HANDKIPI is included without any changes. Hence, the evaluation is performed on the correct data, without influence of the preprocessing phases. However, we should keep in mind that the algorithms are executed on larger (for most verbs substantially larger) set of sentences. The syntactic frames of sentences from outside HANDKIPI were obtained through an automated process.

The only exception is the *MaxNoun* algorithm. Its behaviour depends only on a particular noun. What is more, it is based on the maximal frequency of nouns counted for HANDKIPI itself (cf. section 6.2). Therefore, good behaviour of this algorithm for this set is natural. Still, evaluated on separate NPs/PPs, the *MaxSlot* algorithm is even better.

All the algorithms obtain the worst results on the ValMatch set of sentences. Observe that this set of data is evaluated on the smallest set of sentences (cf. column *common valences* in Table 9), hence the results are less reliable. As we noticed in section 3.2, the original EM selection algorithm used to select a valence frame tends to choose the simplest frames. Such simple frames are usually adequate for simple sentences. Thus, probably most sentences for which hand and automatically selected frames match are simple. This could influence the EM algorithms based on co-occurrence of slots.

Both AutoProc and ValMatch data sets were best processed by the *EM-indep* algorithm, which obtains the best results within the EM algorithms group for ParseMatch data sets as well. The worst results are obtained by the *MaxNoun* algorithm. The most likely reason may be that the choice of nouns as heads of slots during the parsing process, particularly for frequent nouns, was incorrect or at least ambiguous.

The *EM-whole* algorithm is the worst within the EM algorithms group which is not surprising, because of data sparseness (cf. syntactic frames variety shown in Table 1). Table 5 shows that frames of most sentences include 1 or 2 arguments, which explains why the results of *EM-incr* are so similar to the results of EM-whole. The differences between these two algorithms would probably be more distinct for sets of sentences with longer frames.

However, the fact that the *EM-indep* algorithm behaviour is better than the *EM-incr* algorithm contradicts our intuition and assumptions based on data investigation (cf. [18]). This would suggest that semantic categories assigned to particular NP/PP heads are independent of each other! Observe that the *MaxSlot* algorithm based on the frequency of semantic categories counted independently for each slot of a verb, obtains very similar results.

Another possible explanation is that sparseness of data is still less crucial for the independent counting of arguments than for treating of frames as a whole.

6.4   Details of the Evaluation

Let us discuss the results of the evaluation in detail. Figure 1 shows evaluation of the algorithm performed on particular NPs/PPs w.r.t. frequency of verbs in AutoProc, ValMatch and ParseMatch sets of sentences, respectively. Figure 2 shows the evaluation of the algorithm performed on whole sentences in similar fashion. The frequency axis is logarithmic.

The frequency of verbs shows no distinct impact on the results of the algorithms; the percentage of matches changes rapidly from one set of sentences to another. The differences are greater for the evaluation performed on whole sentences.

On the other hand, all the algorithms behave similarly for both methods of evaluation and all sets of sentences. The only exception is the *MaxNoun* algorithm. Observe that it shows more regular results, especially for whole sentence evaluation and for the ParseMatch set. This is an obvious consequence of its sole dependence on nouns.

All the algorithms[9] show best results for verbs *minąć, rozpoczynać, rozpocząć* for all data sets and both methods of evaluation. Note that the frequency of these verbs differs greatly. Vast majority of sentences predicated with the less frequent verb *minąć (to pass)* have some period of time as a subject. Usualy this is the simplest syntactic frame np:nom, e.g., *Styczeń minął. (January*

---

[9] In what follows we ignore the *MaxNoun* algorithm.

(a) set of sentences AutoProc

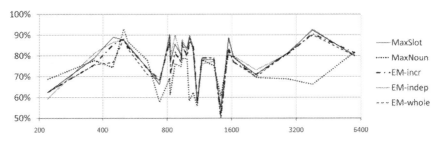

(b) set of sentences ValMatch

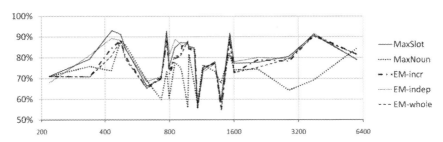

(c) set of sentences ParseMatch

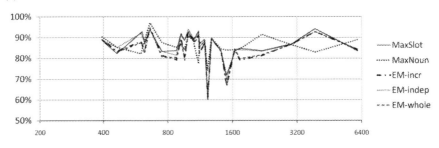

Fig. 1. Diagrams of evaluation of the algorithms on particular NPs/PPs

passed.) or *Godzina minęła bardzo szybko.* (*An hour passed very quickly.*). A similar situation exists for the frame `np:nom prepnp:od`, e.g., *Od tragedii minął rok.* (*A year has passed since the tragedy.*), where *tragedy* is denoted as an event. Other frames have lower frequency, but still they present an analogous semantic regularity. Verbs *rozpoczynać, rozpocząć* (*to start*) are sementically very regular as well, meaning usually that someone starts an action or an event, e.g., *Następnego dnia jej brat rozpoczął poszukiwania.* (*The next day her brother started the search).*), or that they start themselves *Rozpoczęły się prace modernizacyjne na olkuskim rynku.* (*Modernisation works have started on the Olkusz market place.*).

All the algorithms show the worst results for whole sentence evaluation, for *odnieść*, and these are really bad results: from 26% (ValMatch data set) to

(a) set of sentences AutoProc

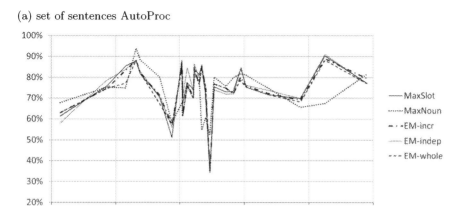

(b) set of sentences ValMatch

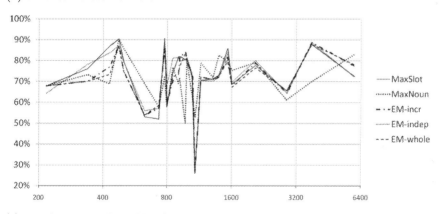

(c) set of sentences ParseMatch

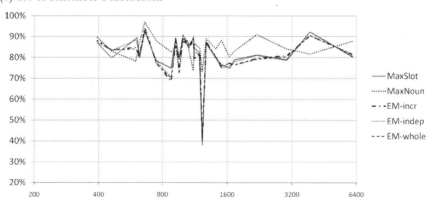

Fig. 2. Diagrams of evaluation of the algorithms on whole sentences

40.3% (SenseMatch data set; *EM-whole* and *EM-incr* algorithms). Observe that if we ignore this verb, the results would be much better and more regular. The syntactic frame np:acc np:nom was chosen for almost 2/3 of sentences predicated by this verb. The subject is usually, as one may expect, a person or a group. There are two types of objects that are most frequent. The first type is usually represented by two nouns:

- *zwycięstwo* (*a victory*), e.g., *Szwajcaria odniosła pierwsze zwycięstwo w eliminacjach.* (*Switzerland gained their first victory in the eliminations.*),
- *sukces* (*success*), e.g., *Premiera opery odniosła wielki sukces.* (*The premiere performance of the opera was a big success.*),

Both nouns *zwycięstwo, sukces* have the same semantic categories event, motive assigned in *Słowosieć*. The second type is usually represented by three nouns:

- *obrażenie*, (*injury*), *rana* (*wound*) e.g., *Obrażenia ciała odniósł pasażer.* (*The passenger received bodily injuries.*), *Odniósł on poważne rany gardła, ale przeżył.* (*He received serious wounds of the throat, but survived.*),
- *kontuzja* (*contusion*), e.g., *Sylwia odniosła w niedzielę kontuzję.*, (*Sylwia sustained a contusion on Sunday.*).

The noun *obrażenie* was categoriesed in *Słowosieć* as act, attribute or body and *rana* was categoriesed univocally as body. Contrary, *kontuzja* was categoriesed univocally as an event. This influenced the EM algorithms to treat *zwycięstwo* and *sukces* as events and *obrażenie* as body, which contradicts the decisions of annotators, who in such contexts denoted *zwycięstwo, sukces* as a motive and more controversially *obrażenie* as attribute.

The next verb with poor results is *odnosić*, the imperfective counterpart of *odnieść*. Other quite poor results for verbs *bronić, przyjmować* show differences between algorithms and data sets.

The worst result for separate NPs/PPs evaluation are observed for *odnieść, przechodzić* and *bronić* for the ValMatch set of sentences.

The EM algorithms, especially *EM-whole* and *EM-incr*, depend on the number of syntactic frames of verbs. On the other hand, *MaxSlot* and *EM-indep* depend rather on the number of slots being potential arguments of verbs. However, the results show no dependency to these numbers (cf. Table 1). The algorithms rather show sensitivity to semantic regularity withing a single schema. Such regularity could affect results both positively (*minąć*) and negatively (*odnieść*).

7   Conclusions

In the paper we presented the process of assigning wordnet-like semantic categories to nouns being semantic heads of NPs/PPs. The process starts with plain text (a set of sentences extracted from a corpus). Sentences are parsed using the *Świgra* parser which applied the morphosyntactic analyser *Morfeusz*. Parses obtained in this way are reduced to their flat form. Then, a single syntactic frame together with its corresponding reduced parses is selected for each sentence.

We focus on the last part of this process: providing arguments of syntactic verb frames with semantic categories of NPs/PPs filling them in corresponding reduced parses. For this we have adapted Dębowski's EM selection algorithm in three ways, referred to as *EM-whole, EM-indep* and *EM-incr* algorithms. In order to evaluate them, we consider two baseline algorithms: *MaxNoun* and *MaxSlot*.

During evaluation, all the algorithms demonstrate very similar results. In its present form, *MaxNoun* is adjusted to the SEMKIPI set of sentences, which seriously limits its application.

The best and very similar results of evaluation are obtained by the *EM-indep* algorithm and the simpler *MaxSlot* algorithm: 75–76% for the whole sentence evaluation and 86–86.5% for separate NPs/PPs evaluation.

The EM selection algorithm, as other statistical algorithms, tends to choose more frequent elements (here: semantic frames). Thus, more frequent frames are probably overrepresented and less frequent ones are underrepresented (sometimes, totally ignored). The totally irreleveant semantic frames are rather rare and hence prone to filtering.

As we have mentioned in the Introduction, our main goal is to extend the syntactic dictionary of Polish by semantic categories of verbs. We plan to count semantic frames obtained by means of the EM selection algorithm described in this paper, and then to cluster them w.r.t. a particular similarity function between categories. The results of all the algorithms described above are good enough for this purpose.

We also plan to perform the WSD task by means of the EM selection algorithm described in this paper, for NPs/PPs heads annotated with senses coming from the complete hypo/hiperonym hierarchy of *Słowosieć*.

Acknowledgements. This paper is a scientific work supported within the Ministry of Science and Education project No N N516 0165 33

We would like to thank Agnieszka Mykowiecka, Adam Przepiórkowski, Joanna Rabiega-Wiśniewska, Marcin Woliński and the other members of Linguistic Engeneering Group of ICS PAS for their comments on the subsequent drafts of the paper.

References

1. Agirre, E., Edmonds, P. (eds.): Word Sense Disambiguation. Algorithms and Applications. Text, Speach and Language Technology, vol. 33. Springer, Dordrecht (2006)
2. Baker, C.F., Fillmore, C.J., Lowe, J.B.: The Berkeley FrameNet Project. In: Proceedings of the COLING-ACL'98 Conference, Montreal, Canada, pp. 86–90 (1998)
3. Banerjee, S., Pedersen, T.: An adapted Lesk algorithm for word sense disambiguation using WordNet. In: Gelbukh, A. (ed.) CICLing 2002. LNCS, vol. 2276, pp. 136–145. Springer, Heidelberg (2002)

4. Broda, B., Piasecki, M., Radziszewski, A.: Towards a set of general purpose morphosyntactic tools for Polish. In: [22], pp. 441–450 (2008)
5. Proceedings of the 19th International Conference on Computational Linguistics (COLING-2002), New Brunswick, Canada (2002)
6. Dębowski, Ł.: Valence extraction using the EM selection and co-occurrence matrices. arXiv (2007)
7. Dębowski, Ł., Woliński, M.: Argument co-occurrence matrix as a description of verb valence. In: Vetulani, Z. (ed.) Proceedings of the 3rd Language & Technology Conference, Poznań, Poland, pp. 260–264 (2007)
8. Derwojedowa, M., Piasecki, M., Szpakowicz, S., Zawisławska, M.: Polish Word-Net on a shoestring. In: Data Structures for Linguistic Resources and Applications: Proceedings of the GLDV 2007 Biannual Conference of the Society for Computational Linguistics and Language Technology, Universität Tübingen, Tübingen, Germany, pp. 169–178 (2007)
9. Derwojedowa, M., Szpakowicz, S., Zawisławska, M., Piasecki, M.: Lexical units as the centrepiece of a wordnet. In: [22] (2008)
10. Derwojedowa, M., Piasecki, M., Szpakowicz, S., Zawisławska, M., Broda, B.: Words, concepts and relations in the construction of Polish WordNet. In: Tanacs, A., Csendes, D., Vincze, V., Fellbaum, C., Vossen, P. (eds.) Proceedings of the Global WordNet Conference, Seged, Hungary, pp. 162–177 (2008)
11. Dorr, B.J., Jones, D.: Role of word sense disambiguation in lexical acquisition: Predicting semantics from syntactic cues. In: Proceedings of the 16th International Conference on Computational Linguistics (COLING-1996), Copenhagen, Denmark, pp. 322–327 (1996)
12. Escudero, G., Arquez, L.M., Rigau, G.: Naive bayes and exemplar-based approaches to word sense disambiguation revisited. In: Proceedings of the 14th European Conference on Artificial Intelligence (ECAI'00), Budapest, Hungary, pp. 421–425 (2003)
13. Fellbaum, C. (ed.): WordNet — An Electronic Lexical Database. MIT Press, Cambridge (1998)
14. Fillmore, C.J., Johnson, C.R., Petruck, M.R.: Background to FrameNet. International Journal of Lexicography 16(3), 235–250 (2003)
15. Gale, W., Church, K., Yarowsky, D.: Estimating upper and lower bounds on the performance of word-sense disambiguation programs. In: Proceedings of the 30th Annual Meeting of the Association for Computational Linguistics (ACL'92), Newark, DL, pp. 249–256 (1992)
16. Gaustad, T.: Linguistic knowledge and word sense disambiguation. PhD thesis, Rijksuniversiteit Groningen, Groningen (2004)
17. Hajnicz, E.: Dobór czasowników do badań przy tworzeniu słownika semantycznego czasowników polskich. Technical Report 1003, Institute of Computer Science, Polish Academy of Sciences, Warsaw (2007)
18. Hajnicz, E.: Towards extending syntactic valence dictionary for Polish with semantic categories. In: Lingustic Investigation into Formal Description of Slavic Languages, Peter Lang, Leipzig (2008)
19. Hajnicz, E., Kupść, A.: Przegląd analizatorów morfologicznych dla języka polskiego. Technical Report 937, Institute of Computer Science, Polish Academy of Sciences, Warsaw (2001)
20. Hajnicz, E., Murzynowski, G., Woliński, M.: Anotatornia — lingwistyczna baza danych. InfoBazy2008 (2008), http://www.infobazy.gda.pl/
21. Ion, R., Tufiş, D.: Multilingual word sense disambiguation using aligned wordnets. Romanian Journal of Information Science and Technology 7(1–2), 183–200 (2004)

22. Kłopotek, M.A., Przepiórkowski, A., Wierzchoń, S.T. (eds.): Proceedings of the Intelligent Information Systems XVI (IIS'08). Challenging Problems in Science: Computer Science, Zakopane, Poland. Academic Publishing House Exit (2008)
23. Král, R.: Three approaches to word sense disambiguation for Czech. In: Matoušek, V., Mautner, P., Mouček, R., Tauser, K. (eds.) TSD 2001. LNCS (LNAI), vol. 2166, pp. 174–179. Springer, Heidelberg (2001)
24. Landauer, T.K., Foltz, P.W., Laham, D.: Introduction to latent semantic analysis. Discourse Processes 25, 259–284 (1998)
25. Levin, B.: English verb classes and alternation: a preliminary investigation. University of Chicago Press, Chicago (1993)
26. Lin, D., Pantel, P.: Concept discovery from texts. In: [5], pp. 577–583 (2002)
27. McCarthy, D., Carroll, J.: Disambiguating nouns, verbs and adjectives using automatically acquired selectional preferences. Computational Linguistics 29(4), 639–654 (2003)
28. Mędak, S.: Praktyczny Słownik Łączliwości Składniowej Czasowników Polskich. Universitas, Cracow (2005)
29. Obrębski, T.: Automatyczna analiza składniowa języka polskiego z wykorzystaniem gramatyki zależnościowej. PhD thesis, Institute of Computer Science, Polish Academy of Sciences, Warsaw (2002)
30. Przepiórkowski, A.: The IPI PAN corpus. Preliminary version. Institute of Computer Science, Polish Academy of Sciences, Warsaw (2004)
31. Przepiórkowski, A.: What to acquire from corpora in automatic valence acquisition. In: Koseska-Toszewa, V., Roszko, R. (eds.) Semantyka a konfrontacja językowa, vol. 3 (2006)
32. Przepiórkowski, A., Buczyński, A.: ♠: Shallow parsing and disambiguation engine. In: Vetulani, Z. (ed.) Proceedings of the 3rd Language & Technology Conference, Poznań, Poland, pp. 340–344 (2007)
33. Przepiórkowski, A., Fast, J.: Baseline experiments in the extraction of Polish valence frames. In: Kłopotek, M.A., Wierzchoń, S.T., Trojanowski, K. (eds.) Proceedings of the Intelligent Information Systems New Trends in Intelligent Information Processing and Web Mining IIS:IIPWM'05, Gdańsk, Poland. Advances in Soft Computing, pp. 511–520. Springer, Heidelberg (2005)
34. Przepiórkowski, A., Kupść, A., Marciniak, M., Mykowiecka, A.: Formalny opis języka polskiego. Teoria i implementacja. Academic Publishing House Exit, Warsaw (2002)
35. Rabiega-Wiśniewska, J.: Podstawy lingwistyczne automatycznego analizatora morfologicznego Amor. Poradnik Językowy 10, 59–78 (2004)
36. Schütze, H.: Automatic word sense discrimination. Computational Linguistics 24(1), 97–123 (1998)
37. Stevenson, M., Wilks, Y.: The interaction of knowledge sources in word sense disambiguation. Computational Linguistics 27(3), 321–349 (2001)
38. Suárez, A., Palomar, M.: A maximum entropy-based word sense disambiguation system. In: Proceedings of the 19th International Conference on Computational Linguistics (COLING-2002), New Brunswick, Canada, pp. 960–966 (2002)
39. Świdziński, M.: Gramatyka formalna języka polskiego. Rozprawy Uniwersytetu Warszawskiego. Wydawnictwa Uniwersytetu Warszawskiego, Warsaw (1992)
40. Świdziński, M.: Syntactic Dictionary of Polish Verbs. Uniwersytet Warszawski / Universiteit van Amsterdam (1994)
41. Vetulani, Z. (ed.): Proceedings of the 3rd Language & Technology Conference, Poznań, Poland (2007)

42. Vossen, P. (ed.): EuroWordNet: a multilingual database with lexical semantic network. Kluwer Academic Publishers, Dordrecht (1998)
43. Woliński, M.: An efficient implementation of a large grammar of Polish. In: Vetulani, Z. (ed.) Proceedings of the 2nd Language & Technology Conference, Poznań, Poland, pp. 343–347 (2005)
44. Woliński, M.: Komputerowa weryfikacja gramatyki Świdzińskiego. PhD thesis, Institute of Computer Science, Polish Academy of Sciences, Warsaw (2004)
45. Woliński, M.: Morfeusz — a practical tool for the morphological analysis of Polish. In: Kłopotek, M.A., Wierzchoń, S.T., Trojanowski, K. (eds.) Proceedings of the Intelligent Information Systems New Trends in Intelligent Information Processing and Web Mining IIS:IIPWM'06, Ustroń, Poland. Advances in Soft Computing, pp. 503–512. Springer, Heidelberg (2006)
46. Wołosz, R.: Efektywna metoda analizy i syntezy morfologicznej w języku polskim. Academic Publishing House Exit, Warsaw (2005)
47. Yarowsky, D.: Unsupervised word sense disambiguation rivaling supervised methods. In: Proceedings of the 33rd Annual Meeting of the Association for Computational Linguistics (ACL'95), Cambridge, MA, pp. 189–196 (1995)

Adjectives: Constructions vs. Valence

Anna Kupść[*]

Université de Bordeaux/ERSSàB and SIGNES
Université Michel de Montaigne, UFR des Lettres
Domaine Universitaire, 33607 Pessac, France
akupsc@u-bordeaux3.fr
Institute of Computer Science, Polish Academy of Sciences
ul. J.K. Ordona 21, 01-237 Warszawa, Poland

Abstract. The paper approaches adjectives in French and Polish from two perspectives: a linguistic description and an automatic text analysis. In particular, we aim at specifying adjective valence and distinguish it from components with which they occasionally occur in various syntactic constructions. Then, we apply linguistic knowledge to annotated data and automatically extract valence lexicons for adjectives. For French, a richly annotated treebank is available whereas the Polish corpus we use currently contains only morphosyntactic information. The paper focuses on results obtained for French as valence extraction for Polish requires additional data processing.

Key words: adjectives, syntactic valence, treebank, adjectival constructions, French, Polish, valence extraction

1 Introduction

Valence describes the combinatory potential of a predicate. For example, the English verb *read* requires two elements (two NPs) to express its meaning 'someone reads something'. The valence is often considered on two levels: semantic and syntactic. In this paper, we will be concerned with the latter: the surface realization of semantic (or logical) arguments is expressed by the *syntactic* valence, also called the (syntactic) 'argument structure' or the 'subcategorization frame'. It specifies the grammatical category (a type of phrase) and morphosyntactic properties of each argument. Additionally, arguments can be associated with their prototypical grammatical functions. For example, in English, the verb *read* selects two NPs, the subject and the direct object (a complement),

---

[*] Various versions of my work on French have been publicly presented on several occasions: *Journées au Vert* of group SIGNES, May 2008, the Intelligent Information Systems Conference, Zakopane, June 2008 and TALC seminar in Nancy, October 2008. I wish to thank their audiences and reviewers for comments and suggestions. Additionally, I benefited from crucial discussions with Anne Abeillé and Jesse Tseng which gave me insights on the intricacies of French adjectives. Finally, thanks to Adam Przepiórkowski several imprecise statements, especially concerning the Polish part, have been corrected.

M. Marciniak and A. Mykowiecka (Eds.): Bolc Festschrift, LNCS 5070, pp. 241–269, 2009.

whereas in German, which has morphological cases, the two arguments of *lesen* are additionally distinguished by different case values: NP[nom] (the subject) and NP[acc] (the complement). In linguistics, the concept of valence has been initially applied to verbs, [60], but then it has been adopted to other categories as well, e.g., nouns or adjectives. In this paper we focus on the syntactic valence of adjectives in two typologically dissimilar languages, French and Polish (a Romance and a West Slavic language), which also allows for a comparative study.

We consider syntactic valence as a lexically specified 'canonical' list of syntactic arguments, which provides a basis for various grammatical realizations. The passive and active voice are a classical example of different realizations of the same valence: the respective syntactic functions and their surface forms change but the semantic arguments are preserved. Mismatches between valence arguments and their overt realization can go even further: for instance, the PP-agent in passive voice can be left out on the surface. On the contrary, in extraposed phrases (for instance in French or English), the impersonal subject is added to surface arguments. Such different realizations of the syntactic valence will be called (syntactic) constructions here (cf. active vs. passive construction).

Adjectives turn out to be particularly prone to participating in various syntactic constructions (e.g., comparatives) which interfere with their inherent valence properties. Our goal in this paper is to identify such constructions and relate them to the syntactic (canonical) valence. Additionally, we aim at automatically detecting these constructions in an annotated corpus and creating valence lexicons for the two languages discussed here.

Subcategorization lexicons, i.e., resources which store information about the syntactic combinatory potential of a predicate, play a crucial role in various NLP applications, related both to parsing, e.g., [7], [8], [54], and generation, e.g., [10], [22]. For French and Polish, just like for many other languages, such resources have been mostly developed for verbs, applying diverse methods ranging from time-consuming but detail-oriented work of human experts, cf. [19,21,36] (French), [43,3] (Polish), to various recent automatic techniques: [6,9,18,13,52,31] (French), [16,11,34,47] (Polish). We can mention two European research and development initiatives, concerning French (among others), which as a side effect involved creating valence lexicons: EAGLES (GENLEX, [37]) and LE-PAROLE ([51]). For Polish, two government-funded research projects that involved creating valence lexicons of verbs are described in [56] and [47].

Valence lexicons for other types of French or Polish predicates are scarce. [20] contains information on subcategorization frames of French nouns, adjectives and adverbs but it has not been adapted for automatic text processing, whereas a syntactic lexicon of French prepositions, which can be used for NLP purposes, has been only recently created (PREPLEX, [17]). As far as Polish is concerned, a detailed linguistic description of syntactic properties of adjectives, including a list of items which take complements, has been presented in [58], whereas a theoretical approach has been discussed in [61]. Recently, a semantic classification of Polish adjectives oriented towards a practical NLP application (namely, English–Polish machine translation) has been proposed in [25]. Regrettably, the resulting syntactico-semantic lexicon is not publicly available. In this paper, we describe our experiments on creating syntactic lexicons of French and Polish adjectives, adjusted to NLP applications.

The organization of the paper is as follows. Section 2 provides a description of syntactic properties of French and Polish adjectives, especially with respect to their valence, summed up with a brief cross-linguistic comparison. Section 3 takes a practical stance and presents extraction techniques applied. The French corpus we used is much more sophisticated than the Polish one, hence our work on French is much more advanced. We focus on providing details of creating the French lexicon since work on Polish is currently at a preliminary stage. Section 4 contains final remarks and concludes the paper.

2 Properties of Adjectives

In traditional linguistics, in order to specify a predicate's valence, a strict distinction between obligatory (complement) and optional (adjunct) components is often made. It has been repeatedly noticed, mainly with respect to verbal valence, that the borderline is far from clear, e.g., [23], [5], [45], and should be considered gradual rather than binary, [39], [15], [14]. The valence of adjectives is particularly tricky as, in most cases, the syntactic realization of arguments is optional; for instance, [41] mentions only a few adjectives, such as *enclin* 'inclined', *exempt* 'exempted' or *désireux* 'desirous', among those which "do not make sense without a complement". This implies that the strongest 'obligatoriness' criterion, often used to identify complements of verbs, is practically inapplicable to adjectives. Moreover, syntactic realization of adjective arguments is more variable than with verbs: several equivalent syntactic realizations of one semantic argument are possible (at least in Polish, [58]), which makes the specification of required components even more challenging. These difficulties probably made [25] abandon the term 'valence' altogether (with respect to adjectives) and replace it with a statistically-oriented notion, the collocation. Despite the problems with recognizing their complements, there is clear evidence that adjectives do have subjects, cf. sec. 2.3–2.4. Surprisingly enough, the subject is usually ignored with respect to adjective valence, especially in Polish literature. In this paper, we shall maintain the term valence.

2.1 Morphological Properties

Both in French and in Polish, adjectives are inflected for number (singular and plural) and gender. In addition to masculine and feminine, as in French, Polish has the neuter gender.[1] Polish has seven morphological cases[2] and adjectives

---

[1] In fact, a more fine-grained specification of up to 9 genders can be proposed for Polish, e.g., [35,53].

[2] Nominative (NOM), Genitive (GEN), Dative (DAT), Accusative (ACC), Instrumental (INST), Locative (LOC) and Vocative (VOC).

are also inflected for case. In NPs, adjectives agree with nouns in number and gender; in Polish, additionally in case.

Adjectives in both languages are gradable. In French, apart from rare exceptions, the comparison of adjectives is analytical and a separate adverb, e.g., *plus* 'more' or *moins* 'less', is used to obtain the comparative form; in superlative forms, the adverb is additionally preceded by a definite article. In Polish, the formation of comparative forms is rather complex and depends on the morphophonological structure of the adjective. Generally speaking, the analytic comparative is used with longer or derived adjectives, whereas the suffixed comparative is formed with other adjectives. In the latter case, the analytical comparative is sometimes possible as an option, e.g., *zielony* 'green': *zieleńszy* (suffixed) vs. *bardziej zielony* (analytical) 'greener'. The superlative is always obtained by adding the prefix *naj-* to the comparative form: either directly to the suffixed form, *najzieleńszy* 'the greenest', or to the adverb in the analytical form, *najbardziej zielony*.

In the paper, participles which carry inflectional marks, e.g., number or gender, will be considered only to the extent they behave as adjectives. This means, for instance, that the French past participle (*participe passé*), is excluded from our consideration if it appears in an auxiliary construction (*Pierre a écrit une lettre* 'Pierre has written a letter') but will be taken into account if it modifies a noun (*une lettre écrite* 'a written letter').

2.2   Types of Arguments

In French, complements of adjectives can be realized by three syntactic categories: prepositional phrases (PP), subordinate clauses (Ssub) or infinitival verb phrases (VPinf), illustrated in (1).[3]

(1)   sûr [$_{PP}$ de sa réussite] / [$_{Ssub}$ qu'il   réussira]   /[$_{VPinf}$ de réussir]
      sure   of his success   that he will succeed   to succeed

      'sure of his success / that he will succeed / to succeed'

In Polish, complements can be additionally realized by nominal phrases (NP), [58,25].

(2)   pewny [$_{NP}$ swojego sukcesu] / [$_{Ssub}$ że   ona odniesie   sukces]
      sure   self's   success   that she will return success

      'sure of his success / that she will succeed'

(3)   skłonny [$_{PP}$ do wyjazdu] / [$_{VPinf}$ wyjechać]
      inclined   to departure   leave

      'inclined towards departure / to leave'

Therefore, our quest for adjectives' complements will be restricted to these types of phrases.

---

[3] Of course, not all adjectives have three types of complements.

2.3   Attributive vs. Predicative

Adjectives have two main uses: attributive and predicative. In the former, the adjective modifies a noun, (4), whereas in the latter, it forms a predicate with the copula, (5), or is a secondary predicate in the sentence, (7)–(8).[4] In (4)–(5), the same adjective appears in both uses. This is not always the case. The majority of adjectives can appear as nominal modifiers, whereas predicative uses are restricted to the so-called non-relational adjectives, e.g., [50,58]. This distinction will not be considered here.

(4)   **Miły**   chłopak   pomógł głodnej   dziewczynie.
       nice-NOM.M.SG boy-NOM.M.SG helped   hungry-DAT.F.SG girl-DAT.F.SG
       'A nice boy helped the hungry girl.'

(5)   Chłopak   był **miły**.
       boy-NOM.M.SG was nice-NOM.M.SG
       'The boy was nice.'

As mentioned above, in French and Polish, an adjective agrees with the noun in number and gender. In Polish, an attributive adjective shares also the case with the noun it modifies, (4). For predicative adjectives, the case may be different. In copular constructions, a predicative adjective refers to the subject (subject complement, fr. *attribut du sujet*) and is most often used in nominative in Polish, (5), but is possible in instrumental as well, see [28,42]:

(6)   Chłopak   zaczął się starać być **miły**   / **miłym**.
       boy-NOM.M.SG started RM try   be nice-NOM.M.SG   nice-INST.M.SG
       'The boy started trying to be nice.'

In general, as observed by [45], two different case assignments are possible for predicative adjectives: the adjective either agrees in case with the NP it predicates of or is assigned the instrumental case. The two options are illustrated in (7) for an adjectival secondary predicate (an object complement, fr. *attribut de l'objet*) and in (8) for an adjectival resultative adjunct.

(7)   Dziewczyna zapamiętała go   **miłego**   / **miłym**.
       girl   remembered him-ACC.M.SG nice-ACC.M.SG nice-INST.M.SG
       'The girl has remembered him nice'.

(8)   Chłopak   wrócił   do domu **głodny**   / **?głodnym**.
       boy-NOM.M.SG returned to home hungry-NOM.M.SG hungry-INST.M.SG
       'The boy returned home hungry'.

A predicative adjective agrees in number and gender with the noun which is predicated of, i.e., from the valence point of view, it is its subject. An attributive adjective also agrees with the noun it modifies and, from the semantic point of

---

[4] We illustrate the two uses with Polish examples as richer morphology allows us to present a more complete (yet more complex) picture.

view, the noun is its argument. As the majority of attributive adjectives can be used predicatively as well, for simplicity, the agreeing argument of an adjective which is either predicated of or modified will be referred to as the subject here.

2.4   Extraposition

In French, certain predicative adjectival constructions allow for the extraposition of a subordinate, (9), or an infinitival clause, (10). In the resulting impersonal construction,[5] (9a) and (10a), the extraposed clause is in fact the subject of the predicate, as shown in (9b) and (10b).[6] The extraposed subject should be thus distinguished from a true complement which cannot become the subject, (11) and (12). With respect to adjective valence, in the former construction, the propositional argument is the subject, whereas in the latter it is a complement of the adjective.

(9)   a. Il est agréable [$_S$ que Marie vienne].
           it is   pleasant   that Mary comes
           'It is pleasant that Mary comes.'

       b. [$_S$ Que Marie vienne] est agréable.
           'That Mary comes is pleasant.'

(10)   a. Il est agréable [$_{VPinf}$ de sortir].
           it is   pleasant   to go out
           'It is pleasant to go out.'

       b. [$_{VPinf}$ De sortir] est agréable.
           'Going out is pleasant.'

(11)   a. Paul est heureux [$_S$ que Marie vienne].
           Paul is happy   that Mary comes
           'Paul is happy that Mary comes.'

       b. *[$_S$ Que Marie vienne] est heureux (de Paul).

(12)   a. Paul est capable [$_{VPinf}$ de sortir tous les jours].
           Paul is capable   to go out every the day
           'Paul is capable of going out every day.'

       b. *[$_{VPinf}$ (De) sortir tous les jours] est capable (de Paul).

In Polish, some uses of predicative adjectives, (13a–b), resemble the French impersonal construction with extraposition. This surface similarity is misleading. Unlike French *il* 'it', the Polish pronoun *to* 'it/this' is always optional and should be considered a correlative, coreferential with the event expressed by Ssub rather than an impersonal subject pronoun. This is illustrated by the behavior of reflexive pronouns in (13c–d). Since Polish reflexives are subject-oriented, these

---

[5] The sentential subject is expressed by the impersonal invariable pronoun *il* 'it' or *ce* 'this'.

[6] See [40] for a study of discourse constraints on extraposition of sentential subjects (in English).

examples show also that Ssub is the subject of the predicative adjective,[7] just like in French.[8] Hence, Ssub in (13) is the subject of the predicative adjective, unlike in (14), where it is a complement.

(13) a. (To) jest jasne, [$_S$ że   Janek przyjedzie].
       it   is   clear COMP Janek will come
       'It is clear that Janek will come.'

      b. [$_S$ Że Janek przyjedzie] jest jasne.
         'That Janek will come is clear.'

      c. To$_i$, [$_S$ że   Janek przyjedzie]$_i$ jest jasne samo$_i$ w sobie$_i$.
         it   COMP Janek will come   is   clear self   in self
         'This, that Janek will come is clear in itself.'

      d. (To$_i$) jest jasne samo$_i$ w sobie$_i$, [$_S$ że Janek przyjedzie]$_i$.

(14) a. Dziecko jest szczęśliwe, [$_S$ że   Janek przyjedzie].
       child   is   happy   COMP Janek will come
       'The child is happy that Janek will come.'

      b. *[$_S$ Że   Janek przyjedzie] jest szczęśliwe (dla dziecka).
          COMP Janek will come   is   happy   for child

In Polish, an adjective can appear with an infinitive phrase as well but it is much less frequent than for a Ssub, as noted in [59], and sometimes considered slightly marked; (15b) comes from [59] but the grammatical judgement is ours. The use of the pronoun *to* with VPinf is excluded, (15a) and (15c).[9] The correlative becomes possible again if VPinf is introduced by the complementizer *żeby* '(in order) to', (15d). As coreference of reflexive pronouns shows, (15e), the infinitive phrase (with or without the complementizer) is the subject in (15). On the other hand, VPinf in (16) is a complement of the predicative adjective.

(15) a. Czasami   jest konieczne (*to) [$_{VPinf}$ pracować z   osobami,
       sometimes is   necessary it   work   with people
       których się nie lubi].
       who   RM NEG likes
       'Sometimes it is necessary to work with people one does not like.'

      b. ?[$_{VPinf}$ Pracować] jest konieczne.
           work   is   necessary
       'Working / To work is necessary.'

---

[7] Interesting details of the co-occurrence of the Polish *to* with Ssub are discussed in [59]. In particular, she mentions (fn.8, p.260) that in rare cases, the presence of the pronoun is **excluded** but she does not provide any examples.

[8] We are reluctant to consider (13a–b) as instances of a syntactic extraposition: the Polish data could be explained by a more flexible word order than in French. We leave the discussion here as this would lead us too far astray from the main topic.

[9] In (15a), *(*to)* indicates that the phrase is grammatical only if *to* is omitted and ungrammatical otherwise.

    c. *To [$_{VPinf}$ pracować] jest konieczne.

    d. Jest konieczne (to),   [$_S$ żeby   pracować].
       is   necessary it/this   COMP work
       'It is necessary to work.'

    e. [$_{S/VPinf}$?(Żeby) pracować]$_i$ jest konieczne samo$_i$ w sobie$_i$.
            COMP   work   is   necessary self   in self
       'Working / To work is necessary in itself.'

(16)   a. Dziecko jest zdolne [$_{VPinf}$ spać cały dzień].
       child   is   capable   sleep all   day
       'The child is capable of sleeping all day long.'

    b. *[$_{VPinf}$ Spać cały dzień] jest zdolne (dla dziecka).

2.5 Comparative Constructions

A comparison can be associated with adjectives of all degrees. As adding a comparison is possible with almost any adjective[10] and is always optional, none of comparative phrases belong to the adjective valence. We discuss them here in order to specify a set of PP, Ssub and VPinf phrases which should not be considered complements.

Positive. French positive adjectives can be compared by using *comme*[11] 'as' or *aussi / autant* 'as well':

(17)   Lise est jolie   comme Rose.
      Lise is   pretty as   Rose
      'Lise is as pretty as Rose.'

(18)   Lise est *(aussi / autant) jolie   que   Rose.
      Lise is   as well   pretty COMP Rose
      'Lise is as pretty as Rose.'

(19)   Lise est *(aussi / autant) jolie   que   sympathique.
      Lise is   as well   pretty COMP nice
      'Lise is as pretty as she is nice.'

Different constraints apply to the two constructions: *comme*, treated as a preposition, e.g., in [17], takes an NP complement, (17), whereas *aussi / autant* may compare the quality as well, (18)–(19). Note that the comparison part is introduced by *que* 'than' which should be distinguished from the homographic subordinate conjunction as what follows is not a subordinate clause. As mentioned above, neither PP[*comme*] nor the *que*-phrases will belong to the adjective valence.

    In Polish, positive comparative constructions can be introduced by *jak* 'as', often co-occurring with an 'equality' adverb *tak samo* 'the same', *równie*

---

[10] Only non-relational adjectives can be compared, e.g., [50,58].

[11] For its different uses, and a formal classification see [12].

'equally', or by *co* 'what', normally accompanied by the adverb *równie*; *co* is used to compare the quality, (20) vs. (21).

(20) Maria jest (tak samo / równie) piękna   jak marzenie / jej matka.
      Maria is   so   same   equally beautiful as   dream   her mother
      'Maria is as beautiful as a dream / as her mother.

(21) Maria jest *(równie) piękna   co   inteligentna.
      Maria is   equally   beautiful what intelligent
      'Maria is as beautiful as she is intelligent.'

In comparisons, *jak* has been considered a preposition or a conjunction, depending on the context, [57,26], while *co* can be considered a *wh*-word which introduces a subordinate (relative) clause or a preposition. Regardless of the actual category (PP vs. Ssub) or the syntactic structure of comparison phrases, they should be ignored with respect to the adjective valence.

Comparative. In French comparative constructions, the comparison is introduced by *que* 'than', (22), often used as a subordinate conjunction, even if the comparison does not involve a clause, (23).

(22) La réunion était *(plus) intéressante que je ne   pensais.
      the meeting was   more   interesting than I NEG thought
      'The meeting was more interesting than I thought.'

(23) La réunion était *(plus) intéressante que l'année   dernière.
      the meeting was   more   interesting than the year previous
      'The meeting was more interesting than last year.'

Again, the *que*-phrases should be distinguished from subcategorized for subordinate clauses as their presence is related to the comparative adverb, e g , *plus* 'more', *moins* 'less', rather than to the adjective: the corresponding sentences without the adverb are ungrammatical.

In Polish, a comparison in comparatives can be introduced in two main ways: by *niż* (or its variants: *niźli, aniżeli, niżby*) 'than' or by the preposition *od* 'from' followed by NP[gen], occasionally replaced by *nad* 'above' or *ponad* 'beyond' with NP[acc], (25), [58].

(24) On jest młodszy niż ty   / od   ciebie.
      he is   younger than you-NOM   from you-GEN
      'He is younger than you.'

(25) ważniejszy   nad   to wszystko jest dla mnie ...
      more important above it all   is   for me
      'more important than all that is for me ...'

(26) On jest młodszy niż Piotr przypuszczał.
      he is   younger than Piotr expected
      'He is younger than Piotr thought.'

In addition to different case assignment properties, (24), only *niż* may be used for comparing clauses (its variant *niżby* is used exclusively in such contexts), (26). Although *od / (po)nad* are unanimously considered prepositions, analyses of grammatical status of *niż* swing from a preposition, [57,27], to a conjunction, [4]. Regardless of its status, neither PP[*od*] nor *niż*-phrase is part of the adjective valence in comparatives.[12]

Superlative. In French superlative constructions, a PP headed by *de* 'of' or *parmi / (d')entre* 'among', can be used to specify the "range" (or "scope") of the comparison, (27) or (28). As in other comparative constructions, such PPs are not part of the valence frame of the adjective.

(27) récession la plus forte   [$_{PP}$ des dix derniers mois]
      recession the more strong   of ten last   months
      'the strongest recession of the last ten months'

(28) la plus sévère récession [$_{PP}$ parmi les Douze]
      the most severe recession   among the Twelve
      'the most severe recession among the Twelve'

In Polish, the superlative form can combine with a PP as well. [58] lists the following prepositions: *z*+NP[gen] 'from', *wśród/(s)pośród*+NP[gen], *między*+NP[inst] 'among', (29). None of them is part of the adjective valence in superlative constructions.[13]

(29) a. najwyższy ze szczytów
         highest   of peaks
         'the highest of the peaks'

      b. najpewniejszy kandydat z   wytypowanych osób
         most certain   candidate from selected   persons
         'the most certain candidate of the selected people'

2.6   Intensifier Constructions

Normally, adverbs are optional with adjectives as they are modifiers. Intensifier adverbs have a peculiar property: they become obligatory if the adjective is accompanied by an 'intensity degree' phrase. In (30), the VPinf[*pour*] is not associated with the adjective *fabuleuse* 'fabulous' but with the intensity adverb *trop* 'too' since its omission renders the sentence ungrammatical. In French, the 'intensity degree' phrase is expressed by VPinf[*pour*] requiring the adverbs *trop* 'too', *assez* 'enough', *bien* 'well', listed in [41], but also *suffisament* 'sufficiently' which is not mentioned there. Other adverbs, e.g., *si*, *tellement* 'so (much)', *à ce point* 'to this point', [41], are correlated with the 'intensity degree' phrase which is introduced by *que* 'that', (31) from [41].

---

[12] PP[*od*] can be a complement of an adjective: *zależny* 'dependent', *wolny* 'free' etc., [58].

[13] Again, some of these prepositions can introduce a complement otherwise: *dumny z* 'proud of', *znany wśród* 'known among', etc., [58].

(30)   Cette histoire est *(trop) fabuleuse [$_{VPinf}$ pour être vraie].
this   story   is   too   fabulous   for   be   true
'This story is too fabulous to be true.'

(31)   Antoine est *(si) inquiet [$_{Ssub}$ qu'il   n'   ose   rien   demander].
Antoine is   so   worried   that   he   NEG   dare   nothing ask
'Antoine is so worried that he doesn't dare asking (about) anything'.

Similar constructions can be found in Polish. For example, [55] mentions the combination of the adverb *tak* 'so' and the subordinate such as in (32) but, to the best of our knowledge, the correlation of the two elements has not been noticed so far.

(32)   Chłopak jest *(tak) przystojny, [$_{Ssub}$ że   wszyscy   jej   zazdroszczą].
boy   is   so   handsome   that   everybody   her   envy
'The boy is so handsome that everybody envies her.'

(33)   Ta   opowieść jest *(zbyt) piękna   [$_{Ssub}$ by   była prawdziwa] /
this story   is   too   beautiful   COMP was   true
[$_{Ssub}$ by w nią uwierzyć].
COMP in   it believe
'This story is too beautiful to be true / to believe in it.'

In Polish, *tak* 'so', *taki* 'such' or *do tego stopnia* 'to this point', mostly appear with a finite subordinate clause, introduced by *że* 'that' or *jakby* 'as if'. The adverbs *zbyt* 'too', *dość* 'enough', *dostatecznie, wystarczająco* 'sufficiently' are followed by the complementizer *by, żeby, aby* '(in order) to', which introduces either a subordinate or an infinitive clause, (33). Therefore, the partition of types of clauses correlated with each type of adverbs is similar to French. As presence of correlates is triggered by the adverb rather than by the adjective, correlates in neither language are valence arguments of the adjective.

2.7   *Tough* Adjectives

The so-called *tough*-constructions can be found in French: an adjective, such as *facile* 'easy' in (34a), selects a verbal complement with an extracted phrase. In French, the extracted NP has to be realized much more locally than in English:[14] roughly, it can be the subject in a copular construction, (34a), or a modified NP in attributive uses of the adjective (*les erreurs facile à comprende* 'the errors easy to understand'). In (34b), *facile* combines with a saturated VPinf but in the impersonal construction the complementizer is different and VPinf is in fact the extraposed subject, see (34c), sec. 2.4. Therefore, *facile* either takes a gapped VPinf[à] as a complement (and the extracted NP as the subject) or a saturated VPinf as the subject (and no complement). As discussed in [24,2],

---

[14] Boundedness constraints on French *tough*-constructions are discussed in [2].

there is no systematic correspondence between the two constructions since they do not involve the same set of adjectives.[15]

(34) a. [$_{NP}$ Ces   erreurs]$_i$ sont faciles [$_{VPinf}$ à comprendre _ _ $_i$].
         these mistakes are   easy   to understand
         'These mistakes are easy to understand.'

      b. Il est facile [$_{VPinf}$ **de** comprendre ces   erreurs].
         it is easy   to understand these mistakes
         'It is easy to understand these errors.'

      c. [$_{VPinf}$ (De) comprendre ces   erreurs] est facile.
                to   understand these mistakes is easy
         'Understanding / To understand these errors is easy.'

In Polish, *tough*-adjectives combine with a PP: a preposition (usually *do* 'to/for' or *w* 'in') is followed by a gapped phrase headed by a gerund (35) or a deverbal noun (36). The extracted element is identified with the modified NP, (35a), or with the subject of a copular construction, (36a). Note that the overtly realized NP is only coindexed with the gap as the two have different case values: nominative vs. genitive. If the NP is realized locally as the complement of a gerund, (35b), or a noun (36b), no preposition is needed and the gerund/noun phrase becomes the subject of the adjective. Morphologically, Polish gerunds are a subtype of nouns and the subject of the adjective is an NP rather than a VPinf as in French. Hence, Polish *tough*-adjectives either are 'normal' adjectives with an NP subject (and no complement), or take a gapped PP complement (and the extracted NP as the subject).

(35) a. [$_{NP}$ warunki]$_i$ niemożliwe [$_{PP}$ do przyjęcia _ _ $_i$]
         conditions-NOM impossible   to accepting _ _ $_{gen}$
         'the conditions impossible to accept'

      b. [$_{NP}$ Przyjęcie warunków]   jest niemożliwe.
         accepting conditions-GEN is   impossible
         'Accepting the conditions is impossible.'

(36) a. [$_{NP}$ Ten samochód]$_i$ jest trudny [$_{PP}$ w obsłudze _ _ $_i$].
         this car-NOM   is   difficult   in service   _ _ $_{gen}$
         'This car is hard in use.'

---

[15] *Tough* adjectives should be distinguished from control adjectives where no element is extracted from the VPinf complement but the sentential subject is coreferential with that of the infinitive:

(1) Jean$_i$ est lent [$_{VPinf}$ _ _ $_i$ à comprendre ses erreurs].
     Jean is slow   to understand his mistakes
     'Jean is slow to understand his mistakes.'

(Another example with the complementizer *de* is (12a).)

b. trudna [$_{NP}$ obsługa samochodu]
difficult   service car-GEN

'a difficult use of the car'

2.8 Restructurization

In French copular constructions, a complex NP subject (37) sometimes allows for a restructurization of its components, which results in an alternative realization of the restructured element as a dependent of the predicative adjective, (38), cf. [38]:

(37) [$_{NP}$ La forme du vase] est étonnante.
the form of vase is   surprising

'The shape of the vase is surprising.'

(38) [$_{NP}$ Le vase] est étonnant [$_{PP}$ de forme].
the vase is   surprising   of form

'The vase is surprising by its shape.'

The PP following the adjective is not its complement since real complements of adjectives do not accept such transformations:

(39) Léa est tremblante [$_{PP}$ d' émotion].
Léa is   shaking   of emotion

'Léa is trembling with emotion.'

(40) *[$_{PP}$ L' émotion de Léa] est tremblante.

Although [38] provides a number of grammatical tests which permit one to distinguish a restructured PP from a complement of the adjective, a detailed specification when restructurization is possible turns out to be extremely difficult. The author proposes complex syntactico-semantic characteristics of the observed regularities but, as she admits herself, they are not "unfallible rules". Therefore, we will not consider the opposition (37) vs. (38) to be systematic.

The Polish equivalents of (37) and (38) are given below. Because the conditions which determine when such transformations are possible are far from clear, we will not discuss these constructions any further here.

(41) [$_{NP}$ Kształt wazy] jest zadziwiający.
shape   vase   is   surprising

'The shape of the vase is surprising.'

(42) [$_{NP}$ Waza] jest zadziwiająca [$_{PP}$ pod względem kształtu].
vase is   surprising   with respect to shape

'The vase is surprising with respect to its shape.'

2.9 French Clitics

In French, pronominal clitics can be attached only to a verb but they can replace arguments of the verb's dependents as well.[16]

(43)   Jean reste   fidèle   [$_{PP}$ à ses amis].   ⇒ Jean   leur reste   fidèle.
       Jean remains faithful   to   his   friends Jean to-them remains

       faithful
       'Jean remains faithful to his friends.' ⇒ 'Jean remains faithful to them.'

(44)   Jean est sûr [$_{PP}$ de sa réussite].   ⇒ Jean en   est sûr.
       Jean is sure   of his success   Jean of-it is sure
       'Jean is sure of his success.' ⇒ 'Jean is sure of it.'

(45)   Jean est attentif [$_{PP}$ à la   situation en Géorgie]. ⇒ Jean y   est
       Jean is attentive   to the situation in Georgia   Jean CL is
       attentif.
       attentive
       'Jean is attentive to the situation in Georgia.' ⇒ 'Jean is attentive to
       it/there.'

Not only complements but also certain adjuncts, especially locative, can be pronominalized, e.g., (45) is ambiguous since y 'to it/there' can refer either to the complement PP ('to it') or to the location ('there', i.e., 'in Georgia'). Clitics *lui/leur* 'him/her/them' and (dative) *me/te/vous/nous* 'me/you/us' are more reliable as they most often replace a human PP[à] complement but they can be used as well as an ethical dative adjunct: *Range-moi ta chambre* 'Clean up your room for me/for my sake'. Therefore, the presence of clitics does not ensure a clear-cut distinction between adjuncts and complements.

2.10 Summary

In French as in Polish, components of comparative, intensifier and restructurization constructions are not valence arguments of an adjective. In attributive uses, the modified NP is the subject of the adjective. In predicative uses, the argument predicated by the adjective is its subject: NP, Ssub or VPinf subject in copular constructions and in phrases containing resultative adjectival adjuncts; an extraposed Ssub or VPinf; an NP or Ssub complement of a verb which selects an adjectival object complement. Conditions on extraposition are different in the two languages as no impersonal construction is involved in Polish. *Tough* adjectives take a gapped phrase (a VPinf[à] in French, a PP[do/w] in Polish) as a complement and the extracted NP as the subject; alternatively, the adjective becomes complementless and, in French, the saturated VPinf turns into the

---

[16] Polish pronominal clitics have quite different properties which are irrelevant with respect to the behaviour of arguments of adjectives and will not be discussed here.

subject, whereas in Polish, the subject is a saturated NP. The subject restructurization is not systematic but it does not affect the adjective valence in either language. In French, pronominal clitics by themselves do not reliably indicate valence arguments of an adjective.

In the next section, we will use these conclusions to determine the adjective valence in practice.

3 Valence Extraction

The approach to valence extraction we have adopted explores corpus annotations. In particular, we use linguistic knowledge (properties of adjectives described in sec. 2) to guide us in this process.

As mentioned in the introduction, this section is focused mainly on French (sec. 3.1–3.3) as extraction of adjective valence for Polish is not very advanced yet. A brief presentation of work done on Polish so far is given in sec. 3.4.

3.1 Adjectives in the Treebank

For French, we used the treebank of Paris7, [1], a journalistic corpus based on articles from *Le Monde* (1989–1993), a French daily newspaper. The corpus contains about one million words with rich grammatical information which has been validated by human experts. The procedure we applied to obtain adjective valence aims in particular at specifying the function of each argument on the valence list and separating the subject from complements.[17]

The treebank contains three levels of annotation: 1) morphosyntactic: category, lemma and morphological features are associated with every (simple or compound) word, 2) syntactic: all major syntactic constituents, including adjective phrases (AP), are indicated in the corpus, and 3) functional: direct verb dependents have their grammatical functions assigned.

The treatment of adjectives in the corpus is not uniform. In NPs, simple prenominal attributive adjectives are not considered APs and have only morphosyntactic (part of speech) tags, (46). Even if several simple adjectives precede a noun, they are not grouped as an AP, (47). On the other hand, if a prenominal adjective is modified by an adverb or has a complement, it does form an AP with its dependent, (48). All postnominal or predicative adjectives are annotated as APs, even if the adjective appears alone, (50) and (51). Additionally, since predicative adjectives are direct dependents of the verb, they are assigned a grammatical function: ATS (fr. *attribut du sujet*) or ATO (fr. *attribut de l'objet*), which indicates a predicate related to the subject, (50), or to the object, (51), respectively. Finally, certain adjectival constructions are not considered to be headed by adjectives: for example, the superlative construction in (49) is annotated as an NP (although the noun is embedded within PP) and no AP occurs. Note that the superlative form in (48) is treated as an AP (but the entire phrase is labelled NP).

---

[17] In [29], we considered only complements.

(46) $[_{NP}$ la $[_A$ moindre] $[_N$ réforme]]
       the   slightest   reform

'the slightest reform'

(47) $[_{NP}$ les $[_A$ dix] $[_A$ derniers] $[_N$ mois]]
       the   ten   last   months

'the last ten months'

(48) $[_{NP}$ la $[_{AP} [_{Adv}$ plus] $[_A$ grande]] $[_N$ discrétion]]
       the   most   big   discretion

'the utmost discretion'

(49) $[_{NP}$ le $[_{Adv}$ plus] $[_A$ froid] $[_{PP}$ des $[_{NP} [_A$ dix] $[_A$ derniers] $[_N$ mois]]]]
       the   most   cold   of   ten   last   months

'the coldest of the last ten months'

(50) $[_{NP}$ Cette $[_N$ comparison] $[_{AP} [_A$ préliminaire]]] semble $[_{AP} [_A$ valable]].
      SUJ this   comparison   preliminary   seems ATS   valid

'This preliminary comparison seems valid.'

(51) $[_{NP}$ Jean] trouve $[_{AP} [_A$ triste]] $[_{Ssub}$ que $[_S$ Marie parte]].
      SUJ Jean finds ATO   sad   OBJ that   Marie leaves

'Jean finds it sad that Marie is leaving.'

In order to extract the adjective valence from the corpus, we focused on AP constituents which potentially contain dependents. For other adjectives, e.g., in simple (prenominal) attributive uses (47) or in superlative constructions (49), we assume that their valence list contains the NP subject alone. It is a desirable result as in neither case a different subcategorized element is present: simple adjectives, (46)–(47), do not have complements, whereas the PP in (49) is not part of the adjective valence but is inherent to the construction.

The category adjective, as specified by the corpus annotation schema, comprises four subtypes: numerals (e.g., *trois* 'three', *deuxième* 'second'), quantifiers (*plusieurs* 'several'), interrogative adjectival pronouns (*quel* 'which') and qualitative adjectives (*chaotique* 'chaotic', *adorable* 'adorable', *possible* 'possible', etc.). Below we examine more carefully qualitative adjectives more carefully as only this category can take complements.

3.2 Identifying Arguments

As mentioned in sec. 2.2, French adjectives admit three types of complements: Ssub, VPinf or PP, whereas NP can only be the subject. Of course, not every constituent of this type is an argument of the adjective. As the treebank contains rich annotations but its size is not very big, we relied on linguistic knowledge rather than on the corpus statistics in order to select real arguments of adjectives. In particular, we attempted to identify the adjectival environments discussed in sec. 2 and separate the valence arguments from components specific to constructions.

Ssub. In the corpus, all phrases introduced by *que* 'that/than' are annotated as a subordinate clause (Ssub): a complement of a verb (51) or an adjective (52), an extraposed clause (53), a comparison in positive and comparative constructions (even if it does not involve a true subordinate clause (54)), or the 'intensity degree' phrase in intensifier constructions (55). We keep track of the mood (indicative vs. subjunctive) of the subordinate clause as it is related to the semantic classification of adjectives discussed in [33]. Corpus annotations do not always consider a Ssub argument a part of AP (in (51) and (53) Ssub is the subject of the adjective), whereas Ssub internal to an AP does not necessarily belong to the valence list, (54)–(55). Therefore, we cannot rely solely on phrase boundaries to identify arguments of an adjective.

(52) $[_{NP}$ Paul$]$ $[_V$ est$]$ $[_{AP}$ heureux $[_{Ssub}$ que Marie vienne$]]$.
      SUJ Paul   is   ATS happy   that Mary comes
      'Paul is happy that Mary is coming.'

(53) $[_V$ Il est$]$ $[_{AP}$ agréable$]$ $[_{Ssub}$ que Marie vienne$]$.
      SUJ it is   ATS pleasant OBJ that Mary comes
      'It is pleasant that Mary is coming.'

(54) La réunion était $[_{AP}$ plus intéressante $[_{Ssub}$ que l'année dernière$]]$.
      the meeting was   ATS more interesting   than the year previous
      'The meeting was more interesting than last year.

(55) Antoine est $[_{AP}$ si inquiet $[_{Ssub}$ qu'il   ne   dort plus$]]$.
      Antoine is   ATS so worried   that he NEG sleeps no more
      'Antoine is so worried that he doesn't sleep anymore'.

In comparative and intensifier constructions, we added adverbs to the list of elements which are recognized within an AP: if an adjective is accompanied by a Ssub (i.e., a constituent annotated in the corpus as Ssub) and a comparative (*aussi/autant* 'as much as', *plus* 'more', *moins* 'less', etc., sec. 2.5) or an intensity adverb (e.g., *si, tellement* 'so much, so', *à ce point* 'to this point', sec. 2.6), the Ssub is not an argument of the adjective; both the adverb and the Ssub are removed from the list of dependents.

As for the extraposed clause in impersonal predicative constructions, it does not form a constituent with the predicative AP in the corpus, (53). Moreover, Ssub and AP are treated as direct dependents of the verb and are assigned grammatical functions: OBJ (the object) and ATS, respectively. We use the following notation to specify the function and its corresponding category: OBJ:Ssub, ATS:AP. The impersonal pronoun (*ce* or *il*) is the subject cliticized to the verb and its function is indicated as well (SUJ). The whole construction can be specified as the sequence: SUJ:*ce/il* ATS:AP OBJ:Ssub. As discussed in sec. 2.4, the extraposed clause is the subject of the adjective. If the construction is found in the corpus, OBJ:Ssub is identified with the subject of the adjective. Similar transformations have been adopted for specifying the subject of an adjective predicating of the object, ATO, as in (51): the object of the verb (NP, Ssub, or VPinf) is recognized as the subject of the adjective, sec. 2.3.

In personal copular constructions, the subject of the copula is an NP and the subordinate (52) or infinitive (12) clause is annotated as a dependent of the adjective. There are no functional annotations within AP and the status of the dependent still needs to be established. The subordinate clause is treated as a complement, unless a comparative or an intensity adverb is present. As no such element is found in (52), Ssub is considered a complement of the adjective; its subject is identified with the subject of the sentence.

VPinf. In impersonal predicative constructions, the extraposed VPinf subject of an adjective, sec. 2.4, is recognized analogously to the extraposed Ssub discussed above. In intensifier constructions, the adverbs *trop* 'too', *assez* 'rather', and others mentioned in sec. 2.6, help us exclude VPinf[*pour*] from the adjective valence (similarly to adverbs correlated with Ssub in these constructions).

Extracted elements are not indicated in the corpus. Hence, *tough* adjectives cannot be directly distinguished from adjectives which take a saturated VPinf[*à*] complement, see fn. 15. To some extent, this distinction could be made by inspecting the internal structure of VPinf in the corpus: if the infinitive has an NP complement (i.e., OBJ is found in VPinf), the adjective does not belong to the *tough*-class as the NP complement of VPinf[*à*] must be extracted (i.e., it is not present). Of course, this condition is not sufficient to separate *tough* and control adjectives as the absence of OBJ:NP in VPinf can simply mean that the verb does not require it at all as, for instance, an intransitive verb. To improve this method, a subcategorization lexicon of verbs should be used to check the valence of the infinitive.

PP. In order to identify PP complements, we retained the lexical preposition which introduces the phrase. As discussed in sec. 2.5, PPs in comparative constructions are not part of the adjective valence. Their treatment in the corpus is not uniform. In superlative constructions, (49) or (27)–(28), the "range" PP is separated from AP, whereas in the positive comparative construction in (56), PP[*comme*] forms an AP with the (postnominal) adjective.

(56)   une pierre [$_{AP}$ noire [$_{PP}$ comme l'enfer]]
      a   stone   black   as   hell
      'a stone as black as hell'

For PPs which do appear within an AP, we used PREPLEX, [17], a lexicon which specifies argumental and non-argumental prepositions, i.e., prepositions which may or may not introduce arguments in French. Although the lexicon has been created for PP arguments of verbs rather than adjectives, we assume that the prepositions appropriate for the latter are a subclass of the former.

PREPLEX contains 49 argumental prepositions, both simple (mono-word) and complex (multi-word). Actually, all of them have a double function and can be used in non-argumental PPs as well. Thus, if a PP is headed by a preposition listed as non-argumental, we can exclude this PP from the valence frame but argumental prepositions do not reliably indicate complements. For adjectives, we added one more non-argumental preposition, *comme* 'as', since it is used only in comparative constructions, i.e., with non-argumental PPs. Other "comparative" prepositions, e.g., *parmi* 'among' or *de* 'of', are retained as they can introduce real complements as well, for example: *connu parmi les artistes* 'known among the artists' or *satisfait des résultats* 'satisfied with the results'. Among PP dependents of adjectives, we found several complex prepositions (expressions tagged as prepositions in the corpus) which are not listed in PREPLEX: *à la tête de* 'leading/at the head of', *à la limite de* 'at the borderline of', *à la suite de* 'as a consequence of', *au profit de* 'at the benefit of', *du fait de* 'from the fact of', *par l'intermédiaire de* 'by means of'. They were all considered non-argumental for adjectives.

As mentioned in sec. 2.9, in copular constructions, an argument of the adjective can be realized as a clitic on the verb. In general, not only arguments but also certain adjuncts can be cliticized, thus clitics by themselves cannot reliably identify arguments. However, as the treebank contains functional annotations for dependents of a verb, if an argument of an adjective is realized as a clitic on the verb, it will be assigned a grammatical function:

(57) $[_V$ Je n'   y   suis] $[_{Adv}$ pas] $[_{AP}$ favorable].
      SUJ/A-OBJ I   NEG CL am   not ATS in favour
      'I'm not in favour of it.'

Thus, we rely on the functional annotations present in the corpus. In copular constructions, we disjoin the clitic from the verb and interpret it as an argument of the predicative adjective (ATS) rather than of the copula. In (57), the clitic *y* will be replaced by PP[*à*], as indicated by the A-OBJ function associated with it in the corpus.[18] More problematic are the cases where *y* is associated with the P-OBJ function (we found 8 such examples in the corpus), as the form of the corresponding preposition cannot be recovered: P-OBJ indicates any prepositional complement different from PP[*à*] and PP[*de*]. Instead of imposing an arbitrary form on the preposition, we assume that it can surface as an element of a semantically specified class rather than as a particular single form. We indicate such PPs as PP[*loc*] since P-OBJ often replaces a locative argument.

If an adjective does not have any arguments specified, we assume that it has no complements but we put an NP subject on its valence list.

3.3   Results

The adjective valence is represented as a list of syntactic categories (arguments), each associated with a syntactic function: SUJ (the subject, realized as an NP, VPinf or Ssub), OBJ (a Ssub complement, in indicative, SsubI, or subjunctive, SsubS, mode, with the complementizer), P-OBJ (a PP, with a lexical or semantic value of the preposition, or a VPinf complement with the complementizer), and cl (for clitics which have not been mapped to 'canonical' arguments). The function is specified first, followed by the syntactic category of the argument.

---

[18] Subject pronouns are considered clitics attached to a verb, which explains the two functions tagged on the copula in (57).

If a frame contains several arguments, they are separated by the vertical bar (|). For example, the frame SUJ:NP indicates that the adjective has only the nominal subject. We call this realization *basic*.

There are 2153 different qualitative adjectives (types), or 16410 occurrences (tokens) in the corpus. The vast majority of extracted adjectives, almost 86% of all types (1849 adjectives; 11116 occurrences, or 67.7% of all tokens), appear only with the basic frame in the corpus. After applying the transformations described in sec. 3.2, we obtained 304 adjectives (5294 occurrences) which had a different frame in addition to or instead of a nominal subject. Among those, three quarters (78%) of adjectives (238 types; 4369 tokens, or 82.5% of their occurrences) have been found with a basic frame. Having subtracted these basic uses, this leaves us with 925 "interesting cases" (5.6% of all adjective tokens) of adjectives which have a different frame in the corpus. We will examine them more closely below.

The method described in the previous section discovered 41 valence patterns (including the basic one) in the corpus. Figure 1 presents the distribution of frame types (with frequency counts) of the 304 adjectives mentioned above. (Only the frequency of the basic frame is given for all 2153 qualitative adjectives in the corpus.) Among those adjectives, most of them (241) appear with more than one frame; there are about 1.16 frame types per adjective. (As the majority of all qualitative adjectives appears with a single (basic) frame, the number of frame types per adjective in general is close to one: 1.02.) The distribution of multi-frame adjectives is as follows: 1 adjective (*difficile* 'difficult') has 9 frames, 4 adjectives (*nécessaire* 'necessary', *facile* 'easy', *indispensable* 'indispensable', *rare* 'rare') appear with 6 frames, another 4 (*présent* 'present', *nombreux* 'numerous', *élevé* 'raised', *possible* 'possible') select 5 frames, 14 have 4 frames, 34—3 frames, and 186—2. There are 61 adjectives which have a single non-basic frame (a frame different than the nominal subject alone).

As mentioned above, the basic (SUJ:NP) frame is prevailing. This use is also most frequent with the majority of 238 adjectives (out of 304) which have another frame: only 24 of these adjectives use more often a different frame than the basic. Their list, with the ratio of basic to other realizations, is given in (58).

(58)   absurde (1/3), analogue (1/3), capable (7/25), comparable (3/7), compatible (2/5), conforme (1/9), conscient (1/13), constitutif (1/3), content (1/3), créancier (2/6), digne (1/3), dépourvu (1/4), enclin (1/3), inférieur (7/45), inscrit (1/3), insensible (1/3), originaire (1/3), proche (11/38), prêt (5/30), soucieux (2/15), spécialiste (1/4), supérieur (20/61), âgé (4/10), égal (2/12)

Had we done a diachronic study, more frequent uses of non-basic frames could indicate obligatory complements or a shift in use. This is the case of *capable* 'capable' whose complement only recently became optional, cf. [41]. On the other hand, the presence of some adjectives on this list at all is due to annotation errors: in the text, neither *enclin* 'keen', *originaire* 'originating', nor *conforme* 'complying/conform' appear with a nominal subject alone.

FRAME	freq.	#adjs
SUJ:NP (basic)	15485	2087
SUJ:NP|P-OBJ:PP[à]	278	81
SUJ:NP|P-OBJ:PP[de]	204	94
SUJ:NP|P-OBJ:VPinf[de]	83	44
SUJ:VPinf[de]	66	29
SUJ:NP|P-OBJ:VPinf[à]	53	16
SUJ:NP|P-OBJ:PP[pour]	35	29
SUJ:NP|P-OBJ:PP[en]	30	23
SUJ:NP|P-OBJ:VPinf[pour]	24	6
SUJ:NP|P-OBJ:PP[dans]	22	14
SUJ:SsubI[que]	18	11
SUJ:NP|OBJ:Ssub[que]	18	4
SUJ:NP|P-OBJ:PP[par]	13	12
SUJ:NP|OBJ:SsubI[que]	12	3
SUJ:NP|P-OBJ:PP[sur]	11	11
SUJ:NP|P-OBJ:PP[avec]	9	6
SUJ:NP|P-OBJ:PP[loc]	8	8
SUJ:NP|P-OBJ:PP[entre]	5	3
SUJ:SsubS[que]	6	5
SUJ:NP|P-OBJ:PP[chez]	4	3
SUJ:NP|P-OBJ:PP[depuis]	3	3
SUJ:VPinf[de]|P-OBJ:PP[à]	3	3
SUJ:NP|P-OBJ:PP[après]	2	2

SINGLETON FRAMES (freq. 1)
SUJ:NP|P-OBJ:PP[à]|P-OBJ:VPinf[sans]
SUJ:NP|P-OBJ:PP[à]|P-OBJ:VPinf[de]
SUJ:NP|P-OBJ:PP[de]|P-OBJ:PP[à]
SUJ:NP|P-OBJ:PP[dans]|cl:me/OBJ
SUJ:Ssub[que]|P-OBJ:PP[en]
SUJ:VPinf[de]|P-OBJ:PP[pour]
SUJ:NP|P-OBJ:PP[vis-à-vis de]
SUJ:NP|P-OBJ:PP[devant]
SUJ:NP|P-OBJ:PP[face à]
SUJ:NP|P-OBJ:PP[jusqu'à]
SUJ:NP|P-OBJ:PP[envers]
SUJ:NP|P-OBJ:PP[sous]
SUJ:NP|P-OBJ:PP[selon]
SUJ:NP|OBJ:VPinf
SUJ:NP|OBJ:Sint
SUJ:Ssub[que]
SUJ:VPinf[à]

Fig. 1. The extracted frames and their frequency counts

There are 61 adjectives which never occurred with the subject alone. A few examples are given in (59).

(59)   accessoire, accompagné, adhérent, admis, agrégé, allergique, amateur, apte, aride, attenant, avare, concessionnaire, condamné, coupable, coutumier, destructeur, distant, désireux, exempt, fier, fixé, incapable, interdit, semblable

Although it might seem that these adjectives have an obligatory complement, as for instance *exempt* 'exempted' in (59), in many cases the complement can be optional but the corresponding frame was not found due to insufficient data, e.g., *coupable* 'guilty', *distant* 'far/distant', *fier* 'proud' or *inhérent* 'inherent'.

Fig. 1 indicates that the most frequent complements are PPs introduced by *à* or *de*. The same forms are also used with infinitives, either as a complement or the subject; there are 29 adjectives which take a VPinf[*de*] subject, (60), i.e., they can occur in impersonal constructions. Additionally, three of them can appear with a PP complement headed by *pour* (*difficile* 'difficult') or *à* (*difficile, facile* 'easy', *possible* 'possible'). PP with the preposition *pour* or *en* is also a frequent complement (it appears with 29 and 23 adjectives, resp.), also in the text (35 and 30 occurrences). Among the six adjectives with the VPinf[*pour*] complement, we consider four correctly recognized (*indispensable* 'indispensable', *insuffisant* 'insufficient', *suffisant* 'sufficient', *nécessaire* 'necessary'), whereas for the other two (*énorme* 'enormous' and *étroit* 'narrow'), the infinitive complement has been confused with the *pour*-clause of the intensifier construction, sec. 2.6.

There are 16 adjectives which take a sentential (Ssub) subject: 11 appear with an indicative (SsubI), cf. (61), 5 with a subjunctive (SsubS) clause, cf. (62), whereas the mode of the subject of one adjective (*normal*) has not been specified in the corpus (the frame SUJ:Ssub in Fig. 1). [33] proposes a semantic classification of adjectives which is reflected also in their syntactic properties. Roughly, the following combinations of clausal arguments (a complement and/or the subject) are possible, which specifies three classes of adjectives: 1) an indicative and/or infinitive clause, 2) a subjunctive and/or infinitive clause, 3) only an infinitive clause. This classification is confirmed in our data. For example, the adjective *certain* 'sure', on the list in (61), can select also an indicative complement, while *indispensable* 'essential' in (62) admits an infinitival subject as well (60). An apparent counterexample is *fréquent* 'frequent' which has been found with (the subject in) either mode. After verification in the corpus, the use of subjunctive is due to the adverb *peu* 'not much, little' which is semi-negative. As mentioned in [33], an indicative argument of the adjective may change the mode to subjunctive if the sentence is negated.

(60)   **SUJ:VPinf[de]**: absurde, acceptable, anormal, difficile, déconcertant, désireux, exact, facile, fâcheux, important, impossible, impératif, indispensable, interdit, intéressant, inutile, nécessaire, pertinent, possible, prudent, préférable, prématuré, rare, ruineux, soucieux, superflu, susceptible, utile, venu

(61) **SUJ:SsubI[que]**: acquis, certain, clair, fréquent, indéniable, inévitable, probable, surprenant, sûr, urgent, vrai

(62) **SUJ:SsubS[que]**: compréhensible, fréquent, indispensable, logique

There are 4 adjectives which seem to select a Ssub complement with no mode specified in the corpus (SUJ:NP|OBJ:Ssub[que]). They are: *autre* 'other', *tel* 'such', *meilleur* 'better' and *pire* 'worse'. In fact, Ssub specification, adopted in the corpus, is misleading as neither of them requires a subordinate clause. The last two are irregular comparative forms and could be stripped off the *que*-phrase on a par with comparative constructions. The other two resemble comparatives (hence they do not necessarily select a subordinate clause but a phrase introduced by *que*, see sec. 2.5) but their form is positive, so the similarity is not complete. For simplicity, we retain these adjectives/frames unchanged.

The frame SUJ:NP|P-OBJ:PP[loc] results from interpreting the clitic *y*, coupled with the P-OBJ function in the corpus, as a locative argument, see sec. 3.2. It has been found with 8 adjectives: *facile* 'easy', *fort* 'strong', *hermétique* 'hermetic', *nombreux* 'numerous', *rare* 'rare', *réduit* 'reduced', *technique* 'technical', *élevé* 'raised'. Not all of these adjectives really select a PP complement; hence functional annotations of clitic arguments in copular constructions in the corpus should be regarded with caution.

As for the frames which appeared only once, they contain a lot of noise and major problems have been discussed in [29]. Similarly, different forms of a PP complement require verification. For most adjectives, several frames result from a different form of the preposition (or the complementizer) used in the same type of frame. For example, *présent* 'present' appears with 4 different prepositions: *dans* 'in', *à* 'at', *en* 'in' and *sur* 'on'. Although different syntactic realizations of a semantic argument are possible and quite common with adjectives, [58], in order to consider them complements their status has to be carefully verified.

3.4 First Polish Lessons

The corpus we use for Polish [46] contains morphological information but no syntactic or functional annotations are available. For initial experiments, we used a small (about half million words) subcorpus which contains morphological annotations verified and disambiguated by human experts. This subcorpus served as the basis for the *Frequency dictionary of contemporary Polish* [32] and comprises same-size text samples of five genres: popular science, news dispatches, editorials and longer articles, artistic prose and drama.

Morphosyntactic annotations in the corpus specify the lemma, the syntactic category and morphological features of each segment (which roughly corresponds to a word), according to a flexemic positional tagset described in [49]. The corpus has been initially processed by a morphological analyser and, in the balanced subcorpus we used, multiple tags have been manually disambiguated.

In order to obtain syntactic information, additional processing needs to be involved. For detecting basic syntactic structures related to adjectives, we decided to employ Spejd, [48], a shallow processing platform which has already been applied to the same corpus. As a starting point, we have taken the shallow grammar developed for extracting verbal valence from this corpus, [47].[19] The grammar is divided into several sections dealing with abbreviations, compounds, various named entities (e.g., dates, person and organization names), correcting morphosyntactic tags (e.g., adding or deleting interpretations) and, finally, recognizing syntactic groups, [44]. The original grammar has been focused on verb dependents, hence there are only a few rules which identify adjectives, usually as part of a nominal group rather than as separate groups. Our goal is to modify this grammar in order to recognize adjectives and their dependents.

Although this work is at a preliminary stage, several problems have already been revealed. First, no finer classification of adjectives is provided and all elements which are morphologically adjectives are assigned the same tag. For instance, ordinal numerals (*sześćdziesiąty* 'sixtieth', *ósmy* 'eighth', etc.), demonstrative (*ten* 'this', *tamten* 'that') and relative pronouns (*który* 'which') as well as qualitative adjectives (e.g., *planowy* 'planned' or *wysoki* 'high') are uniformly annotated as adjectives. From the valence point of view, only the latter are truly interesting and should be distinguished from the others. For closed-class elements, i.e., pronouns, additional tags can be added, whereas for numerals, we plan to extend rules identifying numbers and dates.

Second, in our search of adjective complements, we rely on syntactic specification of potential arguments presented in [58,25], sec. 2.2. Since certain NPs (in dative, genitive or instrumental) can be complements of adjectives in Polish, special attention has to be paid to the case value of adjectives. As discussed in sec. 2.3, an attributive adjective agrees in case (as well as number and gender) with a modified noun. Thus, if the two elements agree in case, the adjective is most likely a modifier rather than a governor of the noun, e.g., *miłego chłopaka* 'a nice boy (GEN)'. However, the adjective itself may be assigned the same case as its complement, e.g., *(chłopaka) żądnego sławy* '(a boy-GEN.M) desirous-GEN.M of fame-GEN.F'. In this example, the adjective *żądnego* can be still correctly recognized as the governor of *sławy* due to the gender mismatch (masculine vs. feminine). However, if there is no gender or number distinction between the same-case adjective and noun, rules cannot differentiate between a modifier and a predicate role of the adjective. As this happens very rarely, for simplicity, we assume that an adjective agreeing with a noun in case, number and gender is always considered a modifier.

Another, more serious problem, is related to Polish word order. Various discontinuous structures are allowed in Polish, see for instance (63).

(63)   w najszerszym   tego   słowa   znaczeniu
        in widest-LOC.M this-GEN.M word-GEN.M meaning-LOC.M

        'in the broadest sense of this word'

In (63), the attributive adjective, *najszerszym*, is separated from the modified noun, *znaczeniu*, by the genitive complement of the noun, *tego słowa*. The

---

[19] The grammar is available at: http://nlp.ipipan.waw.pl/PPJP/

complement of the noun is adjacent to the adjective and, because of the genitive case, can be misinterpreted here as a complement of the adjective. There is no easy solution to this problem. Even if we correctly regroup (via case, number and gender agreement) the main noun ('meaning') with the adjective ('broadest'), we cannot in general exclude that the genitive phrase is in fact a complement of the adjective. The final decision could be made based on heuristics (NP[gen] is often a complement of a noun) or on corpus frequencies (how often the adjective appears with the genitive noun in the corpus). We are more in favour of the second alternative but it might require running experiments on the entire corpus as the size of the subcorpus used for initial experiments is probably insufficient.

Finally, it is obvious that the existing grammar requires various modifications with respect to rules which recognize syntactic groups. Not only does the focus have to be shifted to adjectives but also nominal groups should be more accurately identified. More complex constructions, such as comparatives, extraposition or predicative uses of adjectives in copular constructions, will involve developing specific rules in order to capture these less local relations. Work on extracting valence frames of Polish adjectives is still in progress and a detailed description of adopted adjustments and obtained results will be reported separately.

4   Summary and Future Work

The paper discussed adjective valence in two languages, French and Polish, from linguistic and NLP perspectives. In the first part, we described the syntactic properties of adjectives, focusing on separating valence arguments from components which appear only in specific constructions. The second part has been devoted to applying this knowledge to real data and automatically creating a valence lexicon for adjectives. The corpus available for French contains much richer grammatical information than the Polish resource, which gives us a head start over the work required to build the Polish lexicon. For French, we discovered 41 frames associated with 2153 adjectives (16410 occurrences) in the corpus. All of these frames are found with only 304 adjectives as the remaining adjectives appear with the basic (the NP subject) frame alone. The extraction method relies on linguistic knowledge and corpus annotations, supplemented with PREPLEX, a lexicon of prepositions we used to distinguish argumental and non-argumental PPs in APs.

After inspecting the results, it seems that their quality depends on the argument type. The adopted approach targets mainly subordinate and infinitive clauses since mostly these two phrase types regularly participate in adjectival constructions. Therefore frames which include Ssub or VPinf arguments, as the subject or a complement, can be considered quite reliable. Moreover, the obtained results appear to comply with the semantic classification of adjectives proposed in [33]. Frames with PP complements are more problematic. They have been selected mainly according to corpus annotations and the argumental vs. non-argumental role attributed to each preposition in PREPLEX. This classification is insufficient because every argumental preposition can introduce a non-argumental phrase as well, depending on the context. Also the conversion of clitic arguments into PP complements is not perfect.

In order to improve the results, a more thorough verification of PP dependents has to be made. In particular, as there are few linguistic tests or generalizations which could be applied here, we plan to check the use and frequency of PPs in a bigger corpus and employ statistical techniques to reveal their correlation with adjectives, similarly to [15,14]. As for VPinf complements, we intend to incorporate a syntactic lexicon of verbs in order to be able to identify *tough* adjectives. In order to provide a more objective estimation of the quality of the results, a quantitative evaluation is foreseen.[20] Finally, our work on extracting the Polish lexicon has barely started and this part of the project is still in progress.

The French lexicon is freely available from the site:
http://erssab.u-bordeaux3.fr/spip.php?article150

References

1. Abeillé, A., Clément, L., Toussenel, F.: Building a treebank for French. In: Treebanks, Kluwer Academic Publishers, Dordrecht (2003)
2. Abeillé, A., Godard, D., Miller, P., Sag, I.A.: French bounded dependencies. In: Balari, S., Dini, L. (eds.) Romance in HPSG. CSLI Lecture Notes, vol. 75, pp. 1–54. CSLI Publications, Stanford (1998)
3. Bańko, M. (ed.): Inny słownik języka polskiego. Państwowe Wydawnictwo Naukowe, Warsaw (2000)
4. Bondaruk, A.: The status of *niż* in Polish comparative clauses. In: Stalmaszczyk, P. (ed.) Projections and Mapping: Studies in Syntax. PASE Studies and Monographs, vol. 5, pp. 13–26. Folium, Lublin (1998)
5. Bouma, G., Malouf, R., Sag, I.A.: Satisfying constraints on extraction and adjunction. Natural Language and Linguistic Theory 19(1), 1–65 (2001)
6. Bourigault, D., Frérot, C.: Acquisition et évaluation sur corpus de propriétés de sous-catégorisation syntaxique. In: Actes des 12èmes journées sur le Traitement Automatique des Langues Naturelles (2005)
7. Briscoe, T., Carroll, J.: Generalised probabilistic LR parsing for unification-based grammars. Computational linguistics (1993)
8. Carroll, J., Fang, A.: The automatique acquisition of verb subcategorisations and their impact on the performance of an HPSG parser. In: Proceedings of the 1st International Conference on Natural Language Processing, Sanya City, China (2004)
9. Chesley, P., Salmon-Alt, S.: Le filtrage probabiliste dans l'extraction automatique de cadres de sous-catgorisation. In: Journée ATALA sur l'interface lexique-grammaire, Paris (2005)
10. Danlos, L.: La génération automatique de textes. Masson (1985)
11. Dębowski, Ł.: Valence extraction using EM selection and co-occurrence matrices. Technical report, arXiv (2007)

---

[20] A preliminary quantitative evaluation of Treelex with respect to another valence lexicon of French adjectives is presented in [30].

12. Desmets, M.: Les typages de la phrase en HPSG: le cas des phrases en comme. PhD thesis, Université Paris X, Nanterre (2001)
13. van den Eynde, K., Mertens, P.: La valence: l'approche pronominale et son application au lexique verbal. French Language Studies 13, 63–104 (2003)
14. Fabre, C., Bourigault, D.: Exploiter des corpus annotés syntaxiquement pour observer le continuum entre arguments et circonstants. Journal of French Language Studies 18(1), 87–102 (2008)
15. Fabre, C., Frérot, C.: Groupes prépositionnels arguments ou circonstants: vers un repérage automatique en corpus. In: TALN Proceedings, Nancy, pp. 215–224 (2002)
16. Fast, J., Przepiórkowski, A.: Automatic extraction of Polish verb subcategorization: An evaluation of common statistics. In: Vetulani, Z., ed.: Proceedings of the 2nd Language & Technology Conference, Poznań, Poland, pp. 191–195 (2005)
17. Fort, K., Guillaume, B.: Preplex: a lexicon of French prepositions for parsing. In: ACL SIGSE07 (2007)
18. Gardent, C., Guillaume, B., Perrier, G., Falk, I.: Extraction d'information de sous-catégorisation à partir du lexique-grammaire de Maurice Gross. In: TALN 2006 (2006)
19. Gross, M.: Méthodes en syntaxe. Hermann (1975)
20. Gross, M.: Lexicon–grammar: the representation of compound words. In: Proceedings of the 11th coference on Computational linguistics, Morristown, NJ, USA, Association for Computational Linguistics, pp. 1–6 (1986)
21. Guillet, A., Leclère, C.: La structure des phrases simples en français. Droz, Genève (1992)
22. Han, C., Yoon, J., Kim, N., Palmer, M.: A Feature-Based Lexicalized Tree Adjoining Grammar for Korean. Technical report, IRCS (2000)
23. Hukari, T.E., Levine, R.D.: Adjunct extraction. Journal of Linguistics 31, 195–226 (1995)
24. Huot, H.: Constructions Infinitives du français: le subordonnant de. Droz, Genève (1981)
25. Jassem, K.: Semantic classification of adjectives on the basis of their syntactic features in Polish and English. Machine Translation 17, 19–41 (2002)
26. Kallas, K.: Syntaktyczna charakterystyka wielofunkcyjnego. Polonica XII, 127–143 (1986)
27. Kallas, K.: O konstrukcjach z przyimkiem niż. In: Grochowski, M. (ed.) Wyrażenia funkcyjne w systemie i tekście, pp. 99–110. Wydawnictwo Uniwersytetu Mikołaja Kopernika, Toruń (1995)
28. Klemensiewicz, Z.: Orzecznik przy formach osobowych słowa być. Prace Filologiczne 11, 123–181 (1927)
29. Kupść, A.: Adjectives in TreeLex. In: Kłopotek, M., Przepiórkowski, A., Wierzchoń, S., Trojanowski, K. (eds.) Intelligent Information Systems, Akademicka Oficyna Wydawnicza EXIT, pp. 287–296 (2008)
30. Kupść, A.: TreeLex meets Adjectival Tables. Poster presented at RANLP, September 14–16, Borovets, Bulgaria (2009)
31. Kupść, A., Abeillé, A.: Growing TreeLex. In: Gelbukh, A. (ed.) CICLing 2008. LNCS, vol. 4919, pp. 28–39. Springer, Heidelberg (2008)
32. Kurcz, I., Lewicki, A., Sambor, J., Szafran, K., Woronczak, J.: Słownik frekwencyjny polszczyzny współczesnej. Wydawnictwo Instytutu Języka Polskiego PAN, Kraków (1990)
33. Léger, C.: La complémentation de type phrastique des adjectifs en français. PhD thesis, Univérsité du Québec à Montréal (2006)

268   Anna Kupść

34. Łukasz Dębowski, M.W.: Argument co-occurrence matrix as a description of verb valence. In: Vetulani, Z. (ed.) Proceedings of the 3nd Language & Technology Conference, Poznań, Poland, pp. 260–264 (2007)
35. Mańczak, W.: Ile jest rodzajów w polskim? Język Polski XXXVI(2), 116–121 (1956)
36. Mel'cuk, I., Arbatchewsky-Jumarie, N., Clas, A.: Dictionnaire explicatif et combinatoire du français contemporain. Recherches lexico-sémantiques, vol. I, II, III, IV. Les Presses de l'Université de Montréal (1984, 1988, 1992, 1999)
37. Menon, B., Modiano, N.: Eagles: Lexicon Architecture. Technical Report EAG-CLWG-LEXARCH/B, EAGLES (1993)
38. Meydan, M.: La restructuration du GN sujet dans les phrases adjectivales à substantif approprié. Langages 133, 59–80 (1999)
39. Miller, P.H.: Compléments et circonstants: une distribution syntaxique ou sémantique? In: Proceedings of the 37th SAES Conference, Nice, 1997 (1997)
40. Miller, P.H.: Discourse constraints on (non)extraposition from subject in English. Linguistics 39(4), 683–701 (2001)
41. Noailly, M.: L'adjectif en français. Ophrys (1999)
42. Pisarkowa, K.: Predykatywność określeń w polskim zdaniu. Zakład Narodowy im. Ossolińskich (1965)
43. Polański, K. (ed.): Słownik syntaktyczno-generatywny czasowników polskich. Zakład Narodowy im. Ossolińskich / Instytut Języka Polskiego PAN, Wrocław / Kraków (1980–1992)
44. Przepiórkowski, A.: Towards a partial grammar of Polish for valence extraction. In: Proceedings of Grammar and Corpora 2007, Liblice, Czech Republic (to appear)
45. Przepiórkowski, A.: Case Assignment and the Complement-Adjunct Dichotomy: A Non-Configurational Constraint-Based Approach. PhD thesis, Universität Tübingen (1999)
46. Przepiórkowski, A.: The IPI PAN Corpus: Preliminary version. Institute of Computer Science, Polish Academy of Sciences, Warsaw (2004)
47. Przepiórkowski, A.: Przetwarzanie powierzchniowe języka polskiego. Akademicka Oficyna Wydawnicza EXIT (2008)
48. Przepiórkowski, A., Buczyński, A.: ♠: Shallow Parsing and Disambiguation Engine. In: Vetulani, Z. (ed.) Proceedings of the 3rd Language & Technology Conference, Poznań, Poland, pp. 340–344 (2007)
49. Przepiórkowski, A., Woliński, M.: A flexemic tagset for Polish. In: Proceedings of Morphological Processing of Slavic Languages, EACL 2003 (2003)
50. Riegel, M., Rioul, R., Pellat, J.C.: Grammaire méthodique du français. 3 edn. Presses universitaires de France (2004)
51. Ruimy, N., Corazzari, O., Elisabetta, G., Spanu, A., Calzolari, N., Zampolli, A.: The European LE-PAROLE Project and the Italian Lexical Instantiation. In: Proceedings of ALLC/ACH, Lajos Kossuth University, Debrecen, Hungary, July 5-10, 1998, pp. 149–153 (1998)
52. Sagot, B., Clément, L., de La Clergerie, E.V., Boullier, P.: The Lefff 2 syntactic lexicon for French: architecture, acquisition, use. In: Actes de LREC 06, Gènes, Italie (2006)
53. Saloni, Z.: Kategoria rodzaju we współczesnym języku polskim. In: Laskowski, R. (ed.) Kategorie gramatyczne grup imiennych we współczesnym języku polskim. Prace Instytutu Języka Polskiego, vol. 14, pp. 43–78. Ossolineum, Wrocław (1976)
54. Surdeanu, M., Harabagiu, S., Williams, J., Aarseth, P.: Using predicate-argument structures for information extraction (2003)

55. Świdziński, M.: Gramatyka formalna języka polskiego. Rozprawy Uniwersytetu Warszawskiego, vol. 349. Wydawnictwa Uniwersytetu Warszawskiego, Warsaw (1992)
56. Świdziński, M.: Własności składniowe wypowiedników polskich. Dom Wydawniczy Elipsa, Warsaw (1996)
57. Szupryczyńska, M.: Związki składniowe form stopnia wyższego polskiego przymiotnika. Polonica V, 115–137 (1979)
58. Szupryczyńska, M.: Opis składniowy polskiego przymiotnika. Uniwersytet im. M. Kopernika, Toruń (1980)
59. Szupryczyńska, M.: O zdaniach typu *Jest jasne, że . . . .* In: Gruszczyński, W., Andrejewicz, U., Bańko, M., Kopcińska, D. (eds.) Nie bez znaczenia . . . Prace ofiarowane Profesorowi Zygmuntowi Saloniemu z okazji jubileuszu 15000 dni pracy naukowej, pp. 255–268. Wydawnictwo Uniwersytetu Białostockiego, Białystok (2001)
60. Tesnière, L.: Éléments de Syntaxe Structurale. Klincksieck, Paris (1959)
61. Węgrzynek, K.: Składnia przymiotnika polskiego w ujęciu generatywno-transformacyjnym. Polska Akademia Nauk, Instytut Języka Polskiego (1995)

Part III

Applications

User-Centered Design for a Voice Portal

Krzysztof Marasek, Łukasz Brocki, Danijel Koržinek,
Krzysztof Szklanny, and Ryszard Gubrynowicz

Polish-Japanese Institute of Information Technology, Warsaw, Poland
{kmarasek,lucas,danijel,kszklanny,rgubryn}@pjwstk.edu.pl

Abstract. After a brief overview of voice portal technology, with special attention paid to Polish, we discuss some aspects of user-centered design and its influence on usability of the proposed solution. We describe the issues of voice portal preparation on the example of Warsaw city transportation hotline and main components of the voice portal. The system is effective and supports users in their needs: about 30% of users complete their requests through the automated system without talking to human operators.

Key words: voice portal, speech technology, user centered design, usability

1 Introduction

Voice portals give access to information through spoken commands and voice responses. Until the late 80s the voice portal was an information service provided by the telephone company and staffed by human agents. The advent of computer telephony allows more sophisticated services and the agent acts as an intermediary between the caller and the Internet. Automated voice portals appeared in the first years of this century. A combination of emerging technologies and/or substantial improvements in others, like automated speech recognition, naturally sounding speech synthesis, Internet technologies, application servers, database technologies, give the caller the possibility to use interactive voice services of various kinds, ranging from local and travel information up to stock and bank trading.

2 User-Centered Design and Its Application to Speech-Based Services

The preparation of services for naive users of computer technology has been a research topic for many years, not only resulting in rules or guidelines for proper human-computer interaction (HCI) but also methodologies of user interface design. User-centered design (UCD, ISO 13407: Human-Centered Design Process) places the end-user in the middle of the application preparation process for visually based HCI. Speech as a mode of interaction calls for different aspects of user behavior than visually-based technologies, however several authors

M. Marciniak and A. Mykowiecka (Eds.): Bolc Festschrift, LNCS 5070, pp. 273–293, 2009.

agree on the use of UCD as the main methodology for the preparation of voice portals [15].

Usability is a term describing the ease and clarity of human computer interaction for a given computer program or web page. For voice portals usability is not easy to define, even if there is common agreement what the term "usability" means in this context. The classical notion of this term in case of HCI takes into account four main dimensions [14]:

- **efficiency**: the extent to which a system supports user performance, can the task be accomplished;
- **effectiveness**: if it takes less time to accomplish a particular task;
- **learn ability**: easier to learn;
- **satisfaction**: more satisfying to use.

Usability, because of its often subjective character, is not easy to measure, even if several methods of usability evaluation exist [6] and design methods (in case of visual interfaces) are well known. For a better understanding of the application of HCI design methods to the voice user interface (VUI), first some differences between graphical user interfaces (GUI) and spoken interfaces have to be pointed out:

1. Time. Visual information is usually static and lets the reader take as much time as necessary to read it, understand it, or read it again and again. Application dialogs may let the caller repeat an utterance or ask the application to repeat, but at least a part of dialog has to be memorized by a user. Thus, the dialog and prompts (menus) need to have an appropriately simple structure, e.g. in one dialog step the number of options presented to the user has to be limited. It is typically suggested to limit the number of options to 5–7 depending on the existence of barge-in feature. This causes the users to focus on a much narrower context.

2. Sequential access. GUI are able to present a lot of information in parallel. A user can scan hundreds of items to quickly get to the desired information. Speech is not so effective: information is presented in sequence, users must carefully listen to various lists, dialog flow cues, and help prompts before they can proceed with an appropriate action. It is well known that most people can only remember between five and nine numbers for around twenty seconds after hearing them. Consequently, listening to long lists of choices is unreasonable, and purely hierarchical, menu driven applications are exhausting.

3. User Control. In the case of GUI, users can work at their own pace—in the case of VUI, users need to wait for instructions and then react. In a GUI, the available options are visible on the screen all the time which is not the case for the VUI.

4. Mental Model. The schemes employed in most web-based applications are familiar and consistent, which has given users a very clear mental model of how any GUI is likely to work. In contrast, a first time user of a VUI cannot predict the next dialog step without prior experience with the particular dialog system.

5. Errors. During interaction with graphical WIMP-style interface (Windows, Icons, Menus, Pointers) user errors may occur, but the correct interpretation of user interaction is generally assured. Despite the progress of Automatic Speech Recognition (ASR) technology, a proper hypothesis cannot be guaranteed: an appropriate acknowledge dialog is necessary.
6. Latency. If broadband a Internet connection is used, the latency of Web-based systems (at least to the first reactions) can be very limited. In a spoken dialog long pauses are very unnatural and may occur if the dialog system is not fast enough, breaking the naturalness of the voice communication.

Despite the problems mentioned above, user centered design methodology can be adopted to the preparation of voice portals. In this iterative design process (Fig. 1) four main design phases are applied.

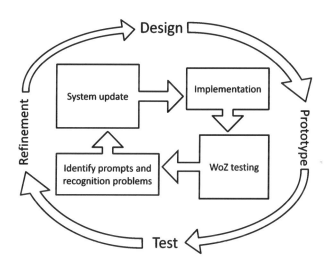

Fig. 1. Iterative voice portal design process (after [8]).

In the first phase of voice portal preparation, design, the call flow is defined, as are the initial scripts for the dialog between the application and the user (see 4.2). Here the data collected from the actual operators of services one wants to automate is especially useful. Additionally, interviews with potential users, analyses of their needs and profiles and usage scenarios are applied.

The second phase, prototyping, usually uses the Wizard of Oz (WoZ) testing technique. It is used to simulate the behavior of an automated system, by having a human agent play the role of the computer system. A call from a user is directed by a WoZ to the application based on a scenario (script) written according to the design phase. WoZ testing helps in understanding user behavior and gives a better view of potential user requests and expected application reactions (see 4.3).

The information collected in the design and prototype phases allows us to build the first version of the voice portal and initiate the testing phase. Here the main effort lies in tracking and analysis of the test calls to the automated system. During the testing phase, attention is concentrated on errors of the ASR, quality of grammars and dictionaries, Text-To-Speech (TTS) text generation, database interface, etc. An iterative refinement process is applied here as described in 4.3. One may also conduct surveys to get the callers' general feeling about the application during all preparation phases or just after the first deployment phase.

3   Voice Portal Technologies

Before going into details of the voice portal design a brief overview of enabling technologies is given. Generally, a voice portal consists of several main blocks (Fig. 2).

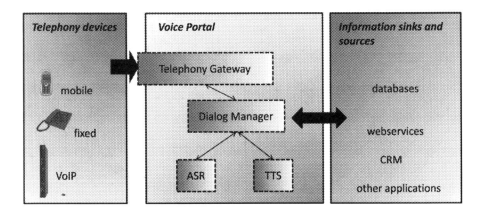

Fig. 2. Voice portal main elements.

It should be stressed that the voice portal technique goes beyond pure voice input and output, it needs a dialog manager with a spoken language understanding component, but also integration with existing telephone network, databases, etc. Below a brief description of components is given with a focus on the details of our solution.

3.1   Grapheme-Phoneme Conversion

Transcription from orthographic to phonetic form in the ASR or TTS systems is done automatically in our system using a rule-based system. Polish language is not extremely difficult to transcribe phonetically (at least for canonical pronunciation) and the basic rule-set consists of a manageable amount of regular expressions. The rules were created manually and tested on a large phonetic dictionary [11]. For non-native words, additional rules can be written to a file and read dynamically.

One peculiarity of the hybrid ASR system (see 3.2) is the need for multiple pronunciations in the dictionary. Since the acoustic model is trained to recognize phonemes, the phonetic transcription of words has to take into account any coarticulation effects that may or may not occur in any given utterance. Our grapheme-phoneme converter delivers multiple pronunciations accounting for phonetic effects typical for Polish (e.g. word final devoicing) and contributing to better results of the ASR.

3.2   Speech Recognition

Automatic speech recognition (ASR) is a process of converting an audio signal of recorded speech into its textual representation (Fig. 3). In the case of a voice portal an audio signal is derived from a telephone line, which due to limited bandwidth and often poor quality of transmission forms a challenge for the ASR engine. In Poland many types of telephone lines are in use (landline, cellular, VoIP), telephone traffic is high, which especially in the case of cellular calls, results in a lot of transmission breaks [12] and crackling noise. In the case of the presented system, typical speech signal parameterization was used to extract relevant features: signal energy, 12 MFCCs (Mel Frequency Cepstral Coeffcients, [1], their first and second derivatives computed in 10 ms tact (25 ms frame length, Hamming window) for telephone signals sampled at 8 kHz, with 16-bits resolution, and preemphasis.

There are many parameters that can be tweaked in the signal processing layer, and many different combinations were tested on representative corpora by our team. The parameterization described above performed best on our data, and results may vary for other experiments.

To improve the performance of the system, a special Voice Activity Detector (VAD) module is used. This module is similar to the acoustic model, but it is much faster and its only purpose is to recognize two events: speech and non-speech (ie. silence or static noise). The frames containing speech are forwarded to the computationally expensive acoustic model and non-speech frames are simply discarded.

Speech recognition typically involves statistical modeling of language and speech, thus the recognition system needs to be trained on a sufficiently large database for both the acoustic and language layers.

Acoustic Modeling. The purpose of acoustic modeling is to compute the phonetic probability distribution based on the acoustic observations from signal parameterization. This can be accomplished through several methods, the most popular being the Gaussian Mixture Model (GMM) [16]. For our project we chose a less popular method, based on Artificial Neural Networks (ANN) [3]. The practice of employing ANNs in Hidden Markov Model (HMM) based systems is often referred to as the "hybrid" method, because unlike GMM based models,

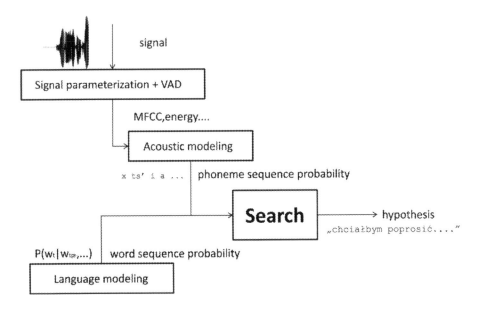

Fig. 3. Elements of speech recognition system.

neural networks are trained and tested separately from the rest of the system. This somewhat slows the whole learning process, but allows greater system flexibility.

Several types of ANNs have been tested in speech recognition systems. Even the simplest feed-forward networks perform reasonably well if given enough context [18]. To capture the context of a recorded utterance, the number of frames visible by the multilayer perceptron must be large enough—sometimes it is over ten frames. This approach is not always effective, because the increase in context length results in the rapid growth of the network, which, in turn, decreases the generalization and the speed of training. Another problem is that the length of the context the network can learn is constant, depending on the network topology. In speech recognition, the length of the context may depend on the phonemes, their neighborhood and their position in words or phrases. Therefore, the context length is not always known, which makes it hard to determine the topology of the network. To make things even more difficult, speech is often subject to non-linear time warp, which means that speech can be stretched and squashed in the time domain at different places. A solution to this problem cannot be achieved efficiently using the feed-forward network model.

Recurrent neural networks analyze only one frame at a time, however, the feedback loops in the network topology allow them to have short-term memory. There are two main methods for training recurrent neural networks, both based on gradient descent: Backpropagation Through Time (BPTT) [19] and Real Time Recurrent Learning (RTRL) [20]. Both of these have a major flaw: the backpropagated error either increases to infinity or diminishes to zero—it never remains stable. This makes the networks incapable of learning a context longer than a few frames. However, neural networks are used mostly for approximating the posterior probabilities of phonemes, leaving the higher level operations to statistical systems and therefore one doesn't really need a longer context.

Instead we chose to use a recurrent neural network called Long Short-Term Memory. LSTM networks (Fig. 4) consist of so called "memory blocks". Each block contains several gates which directly control special memory cells used to store information. The gates are simple sigmoidal units, whose activation function $F(x)$ returns values within the range of $(0, 1)$. The input to the memory cell is multiplied by the output of the input gate, so if the latter is near zero, it will stop any signal from reaching the memory cell.

Similarly, the output gate will block any signal from leaving the memory block. However, if the output values of these gates are near 1, the signal will flow almost unchanged. The forget gate's role is to reset the contents of the memory cell. If the value of this gate is zero, the contents of the memory cell will also be zeroed out. A good analogy to the LSTM memory block is a memory circuit which can execute read, write and reset operations. LSTM is a differentiable version of such a circuit.

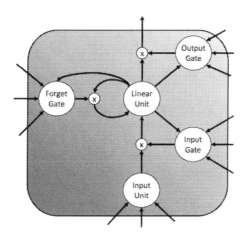

Fig. 4. Topology of a single memory block of the LSTM network.

The network topology in the described ASR was as follows: 39 inputs, 160 blocks, 2 cells per block. Each block had input, output and forget gates. There were 37 output units—one for each phoneme. The network was fully connected. All units were biased. The output of the memory cell was altered by a squashing function, which limits signal range to values $[-1, 1]$. Because the task of the network was classification, a softmax output layer with a cross-entropy error function was used. The error was calculated at each step. Synaptic weights were altered at each step as well. The learning rate was equal to 0.001 and the momentum was 0.95. The network was trained on a corpus consisting of recordings of 844 people who participated in the WoZ experiment [9] which corresponds to approx. 10 hours of speech. 100 speakers were chosen for the test set. The model was tested in a phoneme recognition task (37 phonemes) and achieved 64.66% accuracy. Please note, that in this task the search space was not constrained by any language model.

Word Decoding. Speech recognition is a spatio-temporal problem. Spatial classifiers, like ANNs and GMMs, are therefore usually insufficient to model speech. There were attempts to recognize whole words using ANNs [10], but the results were discouraging because of the amount of time needed to train each new word. That is why the classifier is used only to recognize phonemes. Their sequence is determined by a stochastic temporal model, such as the HMM.

HMM models temporal information using a probabilistic state machine approach. By modeling the probability of state occupancy at different points in time, and the probability of the transition from one state to the other, the goal is to find the most probable sequence of states given some observable data. This procedure is referred to as "decoding" and is a case of a search problem. A well known algorithm used to solve this problem is the Viterbi algorithm [1]. It finds the most probable sequence of "hidden" states (phonemes in the case of hybrid systems) given a transition model and a sequence of observations.

In our work, a time synchronous decoder is also used, however the Viterbi criterion is not applied when extending hypotheses. The algorithm sums up probabilities from different hypotheses sharing the same history and state. This approach is similar to full probability calculation. On the other hand the decoder uses beam thresholding to cut out all low probability hypotheses. Therefore, final probabilities are slightly different from the normal full-probability approach. The decoder uses search graphs which are built from base forms generated by the grapheme to phoneme converter (phonetizer). The phonetizer is custom built specifically for Polish language (see 3.1).

Language Modeling. The purpose of language modeling is to reduce the search space of the decoder by allowing only syntactically and semantically correct word sequences to be decoded. This is achieved by two methods: formal grammars and statistical language models.

Grammars define exactly which sequences are allowed. Algorithmically, they are represented by a Finite State Automaton, which is defined using a formal description often written in the Bachus-Naur Form (BNF) or some of its derivatives. In this work we used the ABNF format, which is a part of the VoiceXML standard (http://www.w3.org/TR/speech-grammar/). This grammar is designed by a human and sometimes partially generated by a program (e.g. to add data from a database or other external source) and compiled into its graph equivalent, which is directly used by the decoder.

Statistical language models offer greater flexibility than formal grammars, but at an increased uncertainty of the outcome. Instead of defining exactly which word sequences are allowed, they estimate the probability of a word sequence using a model trained on a large textual corpus. Usually only a short historical context is used, i.e. the model estimates only $P(w_t|w_{t-1})$ for bigram and $P(w_t|w_{t-1}, w_{t-2})$ for trigram language models. Even with such a small historical context, these models are very large, increasing polynomially with dictionary size. Because of this, the corpora used for training would have to be unfeasibly large to make the model statistically significant and many heuristics and tricks have to be used to make it work well [1].

Apart from modeling word sequences, an additional layer of abstraction can be built into the system. Concepts are groups of words in a phrase that represent a certain meaning. A person's first and last name can represent the concept of that person. A sequence of words describing an address can represent a concept of a destination, depending on its place in the dialog. Modeling of the concepts can often improve the overall accuracy of the dialog, even when the ASR doesn't perform well due to adverse conditions [5]. The goal of the LUNA project, of which this research was a part of, is to create a pipeline of methods to achieve effective concept models for use in dialog systems, such as the voice portal described in this paper.

3.3   Speech Synthesis

According to [7] speech synthesis is a process of generating speech from text and is often referred to as TTS. Speech synthesizers are classified by the manner in which they generate speech: rule-based (formant synthesis, articulatory) and concatenative synthesis. The most popular version of synthesis is currently an expanded model of concatenative synthesis—unit selection synthesis.

Concatenative speech synthesis generates speech by connecting series of acoustic elements of various length, occurring in natural speech (phones, diphones, triphones, syllables). A great advantage of this method is the small size of the database due to the limited number of acoustic units in speech. Speech synthesis with a small database performs faster and has lower system requirements.

Syllable concatenation gives pretty good results, but considering their number (e.g. in English—160,000, while there are only 40 phonemes) it is not a practical solution. A very common method is the concatenation of diphones, which delivers good quality at low cost of only ca. 1500 units for Polish.

Selection of acoustic units is one of the main problems of speech synthesis. Longer units allow for more natural sounding speech but are more expensive computationally. Additionally, in every case there remains the issue of the modification of prosody, i.e. durations of individual units and their intonation. This problem is solved in unit selection synthesis.

Cost Function. Instead of using a database that contains a single instance of every unit, a special corpus containing many instances of each unit and various types of acoustic units are used. Thanks to this, many custom concatenations are avoided and the generated speech sounds more natural. The most important factor in selecting proper units is the cost function. This function consists of the cost of selecting a given unit (target-cost) and the cost of concatenation (join-cost). The definition of the function is given by the following formula [1]:

$$d(\Theta, T) = \sum_{j=1}^{N} d_u(\Theta_j, T) + \sum_{j=i}^{N-1} d_t(\Theta_j, \Theta_{j+1}) \tag{1}$$

Where $d_u(\Theta_j, T)$ is the target cost and $d_t(\Theta_j, \Theta_{j+1})$ is the join cost of Θ_j, Θ_{j+1}. The optimal sequence of units can be given by:

$$\hat{\Theta} = \arg\min_{\Theta} d(\Theta, T) \tag{2}$$

The lower the value returned by the cost function, the more natural the resulting sentence is going to be. If all the join costs are equal, the sequence with the smallest number of units will have the lowest value. In practice, this entails the selection of the longest possible units. There is, however, a potential threat in choosing units in such manner. It is very unlikely that applying this strategy the generated speech will prosodically match with the sentence we want to synthesize. Therefore this is not the optimal solution.

The cost function which is language specific depends on many factors and the acoustic database has to be described accordingly. Various optimization techniques are used to speed up the search. The design of the cost function is a difficult task and there are not many publications on this subject. It is often the case that companies which create TTS systems treat their cost function as an important trade secret [2].

An evolutionary algorithm was used to estimate the cost function of our unit-selection speech synthesizer. Evolutionary algorithms search through the problem space looking for the best solutions. They are especially effective for complicated, non-linear problems. To estimate the parameters of the cost function, 20 professional linguists took part in the experiment. They had to chose the best sounding synthesis (using Mean Opinion Score of listeners, MOS) generated using an evolutionary optimized cost function based on 11 parameters (e.g. discontinuities in time and spectral domains, F0 mismatch, location in word or syllable) for specially a prepared representative sentence corpus [17].

The use of an evolutionary algorithm allowed us to solve the non-trivial problem of estimating the parameters of the cost function. The optimization and MOS tests show that these parameters can be optimized and improve the quality of synthesized speech. The drawback of this method is the time-consuming generation and grading of the individual utterances.

3.4   Dialog Manager

The dialog manager is a module that utilizes the ASR and TTS engines to control the flow of the dialog. In this project, it is implemented as a simple state machine, where each state executes a script written in a basic interpreted procedural language. Additional procedures can be added to expand the functionality of the system, e.g. provide access to additional information sources and sinks, interact with other programs and services, etc.

Some of the basic procedures include:

- flow-control—switch/case statements and state shifting methods;
- manipulating variables—methods to alter strings and various other data;
- speech—ASR and TTS methods;
- telephony—detecting DTMFs (touch-tone signals), routing and disconnecting calls;
- information—accessing database, downloading information from the Internet, etc.

State topology and scripts can be designed rapidly with a GUI tool (Fig. 5), that also serves as the script validator. Scripts can be altered and tested without the need of restarting the portal. Dialog manager has a built-in barge-in capability, which allows the user to interrupt the TTS signal. Barge-in capability can only be implemented when high quality hardware with echo cancellation is used. Otherwise there is a risk of ASR recognizing something from the TTS signal which is always present in the input.

3.5 Voice Portal Architecture

Voice Portal is a parallel network system consisting of 3 basic modules that communicate with each other over the local network (Fig. 2): Gateway, TTS server and ASR server. The parallel architecture allows easy scaling of the system. In case of a system that uses 4 telephone lines, all 3 programs can be installed on a single computer. In case of a larger project, e.g. for 120 lines, based on our experience, the system can be installed on 8 different machines: 1 gateway, 1 TTS server and 6 ASR servers, each ASR server conducting 20 conversations in parallel. Each of the mentioned servers is multithreaded and thus fully utilizes the multi-core capabilities of modern CPUs.

In the case of the voice portal installed at the Warsaw Transport Authority just one server is used and all programs are installed on it. The server has a four core processor with 4 GB of RAM.

Gateway. This is a program that is always installed on the computer with specialized telephony hardware. For our projects we used high quality Dialogic telephony voice boards. These cards have a separate Digital Signal Processing (DSP) module for each channel, which delivers a high quality signal free from noise and distortion often encountered in telephony environments. They support advanced echo cancellation important for barge-in and ASR .The gateway was designed to support both analog and digital telephone cards. It can use the analog cards to connect directly to the fixed-line telephone network or use digital cards to connect through e.g. T1/E1 lines. The purpose of the gateway is to transmit the audio signal from the telephony hardware to the ASR server and from the TTS server back to the telephone hardware. It is also responsible for

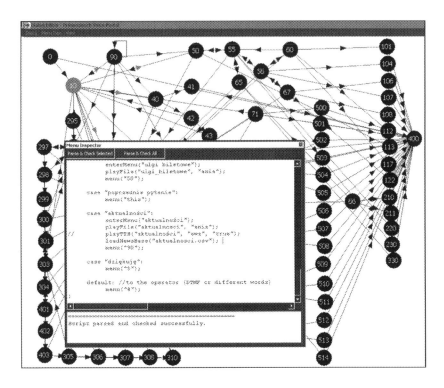

Fig. 5. GUI tool for creating dialogs. The main window contains a network of dialog states with transitions between them. The script editor is presented in the smaller window and it contains the script of a selected state in the graph.

controlling the telephone line, e.g.: connecting and disconnecting calls, sending DTMF and flash signals. Moreover, the gateway is responsible for managing the queue of people who are waiting to be connected to operators. Operators have indirect access to the gateway via a voice portal manager, which allows them to see how many concurrent connections there are and what is the duration of each call. They can also change the number of available operators.

ASR Server. This is the most important part of the voice portal. Its main task is to recognize speech. The integrated speech detector reduces the load from the computationally costly neural network component when the user is not speaking. The program architecture does not limit the number of dialogs to be run on the single server. This number is solely dependent on the computing power and the amount of memory of the machine.

Some off-line tests with ASR Server were performed to check it's word accuracy. The system was evaluated with 500 utterances recorded by the voice portal. Data was hand-transcribed and checked. 6 different grammar types were used in the test. The smallest grammar had only 4 possible choices (yes/no/no speech/garbage). The most complicated grammar had over 10,000 different words and it modeled all street names in Warsaw and surrounding towns (there are around 5,000 streets but many of them have more than one word in their name). Preliminary tests show 96,43% word accuracy rate. However, one must consider that all test data were of rather high quality and speakers were speaking clearly. Moreover in a real-world application of the voice portal, this accuracy may drop, because of really bad quality of voice transmission (even operators ask callers to repeat) and echo cancellation which may cut out some quiet parts of an utterance. The second significant problem is that some people do not answer the questions asked by the voice portal—they say something that was impossible to predict during the dialog creation process. In this case the best action for the ASR is to reply that the utterance was not recognized properly and the voice portal to ask the user to repeat.

TTS Server. This is a multithreaded program with an integrated speech synthesizer. Because the synthesizer can often be run in only a limited number of instances, the TTS server synchronizes all the requests so they are queued and executed in the optimal order. The synthetic speech can alternatively be saved as a file on the hard drive, so as to avoid re-synthesizing the same text. Moreover, the TTS server can also playback speech prerecorded by a professional speaker.

Voice Portal Manager. This is a web-based interface for monitoring and controlling different aspects of the Voice Portal. It is accessible from any browser that supports secure SSL connections. After logging in with a password, the user can control the whole voice portal, e.g.: shut down or restart the whole system, change the number of the human operators in the call center, change the database of the telephone routing service, change information synthesized by the portal, or listen to recorded messages. Depending on the implementation, different menus can be added or removed and the system can be integrated with other services available in the organization. Finally, the Voice Portal Manager collects and displays various statistics, e.g.: distribution of calls in time, length of calls or percentage-wise distribution of calls with regard to the subject or theme of the dialog.

4   Experiments and Results: UCD Methodology

We prepared the voice portal according to the methodology described in part 1. The system works at present at the Warsaw Transport Authority (Zarząd Transportu Miejskiego m.st. Warszawy—ZTM) which is the biggest city public transportation institution in Poland. It manages a call center that is accessible around the clock under the number +48-22-1-9484 and employs 10–20 operators working in different shifts. The call center provides information about departure times of city public transportation (buses, trams, metro and local railway), giving advice in choosing the best transport to reach a certain destination and other information pertaining to public transportation in Warsaw. The call center receives almost 30.000 calls per month with an average call duration of ca. 1 minute. Special attention was paid to the analysis of user requests and WoZ experiments.

4.1 Design and Prototype

Within the framework of the LUNA project (EC 6 FR IST 033549) a huge effort of speech data collection of real telephone dialogs has been completed: a corpus of human-to-human dialogs was collected at the Warsaw Transport Authority telephone information hotline and this data allow us to understand how the callers request information and how human operator complete the users' requests.

During data collection, we observed on-site call center operator activities to obtain a better insight into the center's work-flow, group dynamics and interactions between operators and callers. We found, that much of the required data is accessible via the ZTM web page, but some of it remains in an unstructured form used only by operators (text memos, detailed city maps with transportation lines).

On a sample of more than 500 information seeking dialogs five proximate topic classes were identified:

- information requests on the itinerary between given points in the city;
- timetable for a given stop and given line and the travel time from given stop to destination;
- information on line routes, type of bus, tram (e.g. is wheelchair access available or not);
- information on stops (the nearest from a given point in the city, stops for a given line, transfer stops, etc.);
- information on fare reductions and fare-free transportation for specified groups of citizens (children, youth, seniors, disabled persons, etc.).

For these categories a WoZ experiment has been prepared. In fig. 6 the architecture of the system is given [9].

The main parts of the system are the dialog manager and the TTS manager which coordinate the signal transmission through the telephony gateway. The WoZ operator chooses a response from a short list to simulate machine dialog steps. The WoZ simulation was done in its entirety for the domain of city transport fares reductions. For other domains, after a few steps of preliminary information collection from the user by the WoZ system, the operator would switch it off and continue the dialog with the caller.

The WoZ experiments gave us a better view into user requests. From 844 recorded calls, only 459 of them were retained and classified as belonging to one of the mentioned topics. Quite often, users (155) wanted to speak directly with the operator, ignoring the system or waiting silently for the operator to respond.

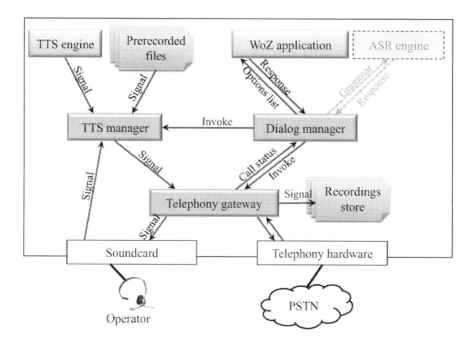

Fig. 6. Wizard-of-Oz system [9].

This showed, that the acceptance of automated service is quite low [9]. Moreover, many users were surprised that the system is automatic, so dialogs started with difficulty, contained long pauses, hesitations and WoZ prompts repetitions. It was a clear indication that the system voice should to be as natural as possible.

In our preliminary experiments we found that latency introduced by our TTS engine is unacceptable to the users of the service. Thus, we decided to switch to a commercial product offered by Loquendo, characterized by high speed and high quality of speech synthesis. Moreover, the most frequent and static prompts have been replaced by prerecorded speech. Careful analysis of the WoZ data also allowed to build better, more flexible grammars for the ASR module and to prepare more informative prompts spoken to the user.

The analysis of those dialogs shows that it is almost impossible to automate all classes of dialogs, mostly because of the complicated structure of the Warsaw public transport system, stops name conventions and imprecise users requests (see [13] for details). Thus we decided to choose the following dialog topics for automated services: departure times, ticket prices, fare reductions, news and complaints.

4.2 ZTM Voice Portal Features

After dialing in, the user connects automatically with the voice portal. It is possible to connect with the human operator at any moment by pressing any key on the phone keypad. If all operators are busy, the user is put into a queue and they are informed about their place in the queue. The user can always decide to leave the queue and to use the automatic features of the voice portal instead. These include: departure times, complaints, ticket prices, fare reductions and news. There are two more options: connect with operator and city routes, both of which connect directly to the human operator. Automated functions are described below.

Departure Times. The most elaborate feature of the voice portal in ZTM is providing departure times for buses and trams. The system recognizes about 4500 different names of bus and tram stops in Warsaw. It also recognizes all the transport line names, dates and times. After asking a few questions, the voice portal retrieves the times from a special database and synthesizes the response. The amount and type of questions the system has to ask depends on the chosen line and stop, and varies between 4–8 questions. The Warsaw transportation system is quite complex and the portal has to deal with a fair amount of disambiguation to be able to provide an accurate response every time. Also, the schedule is subject to slight changes very frequently (every 2–3 days) with not a lot of time in advance (sometimes only one day). This means that the system cannot provide reliable information too far into the future.

The dialog about departures starts (the user is warned before that the conversation is recorded) with the question whether the user wants the schedule for current or some other day. The schedules are different for weekends and holidays so this is a very important question. As follows from our research, 90% of users want the current schedule so it is reasonable to start from this question. If the user is interested in future schedules, the system asks for a specific date and accepts many different types of phrases: tomorrow, Monday, January 12th, etc.

The next question is about the route number, i.e. the number written on the side of the bus or tram. In Warsaw these are usually a number between 1–999 and sometimes combinations of letters and numbers, e.g. C-6. A list of n-best hypotheses returned by the ASR will sometimes be generated in this step due to acoustic similarity of some sequences in Polish, e.g. E-2 and N-2.

In the following question the system asks for the stop name. This is a difficult step, because the names often consist of several parts and each part has to be recognized individually or in connection with others (e.g. Central Railway Station 05). These names usually originate from streets and sites where the stops are located. The transport system of the city of Warsaw services a large area including the many small suburban localities. Therefore, a common name like "Szkolna" (i.e. School) street may occur several times. In such cases, the portal has to ask an additional question to disambiguate the stop. Finally, several stops in a single trip may have the same name, but are suffixed by an additional number. This infrequently merits an additional question from the portal.

Most of the stops have two opposite directions. If a stop has more than one direction for the chosen line, the system asks the user about the direction by giving a list of end destinations for the given line, e.g.: which direction are you interested in: towards the city center, towards Wilanów city district?

The final question is an approximate time of departure the user is interested in. Since most of the schedules contain up to 100 different departure times, it would be a waste of time to read them all. Instead the system reads the three closest of the given approximate time (1 earlier and 2 later) and allows the user to navigate the times by giving out commands: next departures, previous departures. The approximate time can also be given in many different ways, e.g.: 12 o'clock, half past eight, 7 in the morning, etc.

Ticket Prices. This feature gives the user information about the price of a ticket. There are up to 20 different ticket types available. The system asks the user what ticket he is interested in by asking a series of questions. There are a few options per dialog step and after a few steps the user is presented with a price for the chosen ticket.

Fare Reductions. Here, users can find out if they can benefit from a certain fare reduction. The guidelines for ticket reductions are governed by city officials and are quite complicated to understand for the average passenger. Similarly to the ticket prices, this dialog is organized in a tree-like fashion. The depth of the tree is at most 5 steps and each leaf contains comprehensive information about the chosen reduction.

News. This is just a database of synthesized news items updated manually by the ZTM staff. Each item consists of a title and some content. These may contain information about important changes to the infrastructure, lines, ticket prices, etc. The user is simply presented with a list of news topics (titles) and by picking the particular news item (by sequence number) its content is synthesized.

Complaints. The favorite feature of the voice portal among the human operators is the automatic recording of complaints. This function simply asks the user to provide all the information after a tone. This information is recorded and stored for further review. ZTM is an independent, government sponsored organization that governs and audits the different commercial transport carrier companies around the city. One of their main tasks is quality control and communication with the customers. Unfortunately, in times of heavy traffic, accidents or other unpredictable events, lots of harsh complaints from frustrated passengers pour in, which takes a big toll on the human operators and doesn't serve any purpose. When faced with an automatic complaint recorder, users are less likely to vent their anger and provide useful and constructive information.

4.3   Testing and Refinement

According to UCD methodology the next step in application preparation is its testing and refinement. 300 calls were analyzed in order to check how many people used automatic information. Around 30% of hotline clients used the voice portal and didn't connect to human operators at all. The most demanded automatic feature is information on departure times. The second most used automatic feature is filing complaints. From the first day of deployment all calls are monitored and usage statistics are collected (Fig. 7 and Fig. 8).

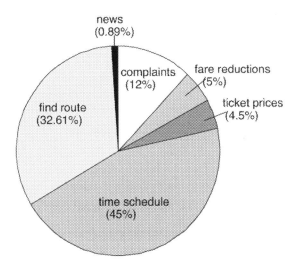

Fig. 7. Calls distribution at the ZTM voice portal (based on ca. 60000 calls).

We modified the dialog flow based on the analysis of recorded data, especially ASR error handling. The ASR performs very well, normal procedure is to acknowledge the user input by additional question and simple yes/no confirmation. However, if the system cannot recognize the user input for the third time, the user is prompted to press "0" key on the phone keyboard to redirect the call to the human operator; if the fourth attempt to recognize speech fails, the redirection is automatic.

On a side note, in previous versions the system asked the user to press any key if they wanted to connect directly to a human operator. Due to many mistakes we had to change that to "press 0"—callers were often confused to as what "any key" meant.

Also the prompt asking for a date of requested schedule has been modified: we observed that most callers asked for today schedule, so this is now the first option to choose.

Initially the cordial end of dialog was not being recognized (if the caller said "thank you, goodbye" system answered: "sorry, I did not understand"). Now the system reacts properly.

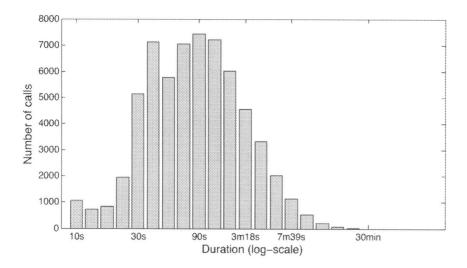

Fig. 8. Distribution of call durations (including waiting time) at the ZTM voice portal (based on ca. 60000 calls).

Finally, some users connected via old telephone equipment with rotary dialing instead of the standard touch-tone. This problem was solved by allowing the user to say "connect to operator" at any point in the dialog.

5 Conclusions

In Fig. 8 some important statistics are given. Most of the calls took ca. 90 seconds, which is an acceptable call duration (an average dialog with a human operator took on average 2 minutes, but included more complicated requests). The system is effective and supports users in their needs: about 30% of users complete their requests through the automated system. Moreover, because some of the calls can be fully automated, the voice portal successfully shortens the waiting time for connection to the human operator.

In the first days after the voice portal was deployed in ZTM hotline not many people were interested in using the automatic service. However, it was observed that with time more and more people started to use the fully automatic features. After three months several hundred people a day used automatic information and did not connect to the operators at all. It seems that the hotline divided customers into two groups. The first group (which consists mostly of pensioners and elderly people) always wants to speak with the operators. The second group (in which there are many students and younger people) uses automatic dialogs willingly, especially if the queue waiting time is significant.

It seems that many hotline clients are satisfied with the automatic service. The main advantage of such a system is that they never have to wait to get information that is available automatically. Most people are interested in departure times of buses and trams. It takes around 45 seconds for the human operator to provide proper information regarding this topic. The automated system needs twice the time. However, when one considers that people have to wait in queue sometimes for five or even ten minutes to get to the operator, the advantage of using the voice portal is obvious.

A more natural dialog can be achieved if its flow is not fully controlled by the computer system. We hope to apply the methods elaborated in the LUNA project towards more flexible, mixed-initiative dialog flow.

Acknowledgements. This work is partially supported by the LUNA project (EC 6 FR IST 033549). The voice portal technology described in this paper is the property of Primespeech (www.primespeech.pl).

References

1. Acero, A., Hsiao-Wuen, H., Xuedon, H.: Spoken Language Processing: A Guide to Theory, Algorithm, and System Development. Prentice-Hall, Englewood Cliffs (2001)
2. Black, A., Hunt, A.: Generating F0 contours from ToBI labels using linear regression. In: Proceedings of ICSLP '96, Philadelphia, USA, vol. 3, pp. 1385–1388 (1996)
3. Bourlard, H.A., Morgan, N.: Connectionist Speech Recognition: A Hybrid Approach. Kluwer Academic Publishers, Dordrecht (1993)
4. Byrne, B.: Turning GUIs into VUIs: Dialog Design Principles for Making Web Applications Accessible By Telephone. VoiceXML Review 1(6) (2001)
5. De Mori, R., Béchet, F., Hakkani-Tür, D., McTear, M., Riccardi, G., Tur, G.: Spoken Language Understanding for Conversational Systems. Signal Processing Magazine Special Issue on Spoken Language Technologies 25(3), 50–58 (2008)
6. Dix, A., Finlay, J., Abowd, G., Beale, R.: Human-Computer Interaction. Pearson/Prentice Hall, London (2004)
7. Dutoit, T.: An Introduction to Text-To-Speech Synthesis. Kluwer Academic Publishers, Dordrecht (1997)
8. IBM International Technical Support Organization, Speech User Interface Guide (2006)
9. Koržinek, D., Brocki, Ł., Gubrynowicz, R., Marasek, K.: Wizard of Oz Experiment for a Telephony Based City Transport Dialog System. In: Proceedings of the 16th Int. Conference Intelligent Information Systems, Zakopane, Poland, June 16–17, 2008 (in print)
10. Brocki, Ł., Koržinek, D., Marasek, K.: Recognizing Connected Digit Strings Using Neural Networks, Text Speech and Dialog 2006, Brno, Czech Republic, pp. 343–350 (2006)
11. Marasek, K., Gubrynowicz, R.: Multi-level Annotation in SpeeCon Polish Speech Database. In: Bolc, L., Michalewicz, Z., Nishida, T. (eds.) IMTCI 2004. LNCS (LNAI), vol. 3490, pp. 58–67. Springer, Heidelberg (2005)
12. Marasek, K., Gubrynowicz, R.: Design and Data Collection for Spoken Polish Dialogs Database, Language Resources and Evaluation Conference, Marrakech Morroco (2008)

13. Mykowiecka, A., Marasek, K., Marciniak, M., Rabiega-Wiśniewska, J., Gubrynow-icz, R.: Annotation of Polish spoken dialogs in LUNA Project, LTC'07, to appear in Springer LNAI (2009)
14. Nielsen, J.: Usability Engineering. Morgan Kaufmann, San Francisco (1994)
15. Polkosky, M.: What is speech usability anyway? Speech Technology Magazine (2005)
16. Redner, R.A., Walker, H.F.: Mixture densities, maximum likelihood and the EM algorithm. SIAM Review 26(2), 195–239 (1984)
17. Szklanny, K.: Cost function estimation in unit selection speech synthesis, Ph.D. Thesis, PJIIT Warszawa (in preparation, in Polish) (2009)
18. Waibel, A., Hanazawa, T., Hinton, G., Shikano, K., Lang, K.J.: Phoneme recognition using time-delay neural networks. In: Readings in speech recognition, Morgan Kaufmann, San Francisco (1990)
19. Werbos, P.J.: Backpropagation through time: what it does and how to do it. Proc. IEEE 78(10), 1550–1560 (1990)
20. Williams, R., Zipser, D.: A learning algorithm for continually running fully recurrent neural networks. Neural Computation 1(2), 270–280 (1989)
21. Yi, J.R.-W.: Corpus-Based Unit Selection for Natural-Sounding Speech Synthesis, Ph.D. Thesis, MIT Dept. of Electrical Engineering and Computer Science (2003)

Speech Understanding System SUSY—A New Version of the Speech Synthesis Program

Jerzy Cytowski

Faculty of Mathematics and Natural Sciences
College of Sciences
University of Cardinal Stefan Wyszyński
and Institute of Computer Science PAS
cytra@mimuw.edu.pl

Abstract. The method of digital synthesis of speech presented below was been worked out for Warsaw University in order to implement the acoustic output of an automated telephone information office. The paper presents a program which implements microphonemic synthesis of speech designed by a team of Professor Leonard Bolc in the mid-1970's. We had to build the program which would generate continuous speech with prosodic features similar to the natural language and understandable.

Key words: speech synthesis, microphonemic method

Those paper and synthesis program light to commemorate research works of Professor Leonard Bolc in speech processing.

1 Introduction

Speech synthesis is a very important branch of natural language processing. It is used for reading out textual material and describing figures. It is also an essential component of a bilingual computer, which helps in reducing the problems that one might face, while talking to someone who doesn't know the other's language. Synthetic speech has been used for a number of years as a final output stage in adaptive systems for visually impaired PC users. Visually impaired users make widely varying use of their PCs. At one end of the spectrum a casual user may read a book, or a 'phone bill for instance, with their home PC and special attachments such as an optical character recognizer. At the other end, the professional PC user, a programmer for instance, may use a talking PC for much of their working day.

Over the last 50 years since the invention of the modern computer, actually well before that, there have been several attempts to make a machine that would speak as humans do. In the mid-1970's a computer speech understanding system SUSY was designed [6]. Professor Leonard Bolc has done an excellent job of organizing the project and made important contribution to the advancement of theory of speech understanding systems and its application. The system SUSY consisted of a number of independent but closely cooperating modules.

M. Marciniak and A. Mykowiecka (Eds.): Bolc Festschrift, LNCS 5070, pp. 295–310, 2009.

The paper presents a microphonemic[1] approach to synthesis of speech. More precisely, the results of an experimental application of this method are presented. Speech synthesis concerns generation of speech from some kind of input, i.e. parametric representation of speech acoustics or speech production, or text [3]. The input to the synthesizer varies with following synthesis type [7]:

1. Articulation (rule-based) synthesis tries to model the human vocal organs as perfectly as possible. This type of synthesis involves a dynamic model of human articulators which makes it possible to simulate motions of the articulators during speech production. Another component is the glottis model which generates the excitation signal (i.e. random noise or a quasi periodic sequence of pulses). The synthesis also requires a method of generating and monitoring air pressure and velocity. For rule-based synthesis, the articulation control parameters are for example lip aperture, lip protrusion, tongue tip height, tongue tip position, tongue height, tongue position and velic aperture. Phonatory or excitation parameters are glottal aperture, cord tension and lung pressure. Articulation synthesis is very attractive for research on speech production and perception, but the quality of speech generated in this way is far from perfect. This is mainly due to computational and mathematical complexity of the underlying models and the insufficient knowledge concerning articulation processes involved in the production of natural speech. Articulation synthesis is potentially the most satisfying method to produce high-quality synthetic speech, but it is also one of the most difficult methods to implement.

2. Formant synthesis is based on the source-filter-model and knowledge concerning speech acoustics. It consists of two components: the generator of an excitation signal and formant filters that represent the resonances of the vocal tract. The former is considered as a source and the latter as a *filter* according to the *source-filter model* of speech production. Formant frequencies and bandwidths are modeled by means of a two-pole resonator. At least three formants are generally required to produce intelligible speech and up to five formants to produce high quality speech. Additional information concerning radiation characteristics of the mouth which influence formant frequencies are also necessary. There are two basic structures of formant synthesizer: parallel and cascade, but some kind of combination of these is usually used. Formant synthesis is very useful for research in the field of speech acoustics and phonetics, and for the study of speech perception. Speech generated by formant synthesizers is characterized by a metallic sound. However, this sort of speech synthesis can bring satisfying results if control parameters are hand-tuned, which is impossible in a fully automatic system. This method of synthesis is probably the most widely used in recent years.

---

[1] Microphonemic units are such segment of the signal which after their n-th repetition enable correct perception of the phonemes they represent. Number of repetitions n depends on the kind of phoneme, the fundamental frequency of model voice, the context and the stress. [6]

3. Concatenate (data-based) synthesis concerns the generation of speech from an input text. The speech results from concatenation of acoustic units stored in a database. The units are annotated for features referring to linguistic structure and the segment context in which they occur. One strategy to synthesize a new utterance is to perform a search over a large database and to select units that minimize a target cost and a concatenation cost accumulated over the entire utterance. One of the most important sapects in concatenate synthesis is to find the correct unit length. The selection is usually a trade-off between longer and shorter utterances. With longer utterances high naturalness, less concatenation points and good control of articulation are achieved, but size of database is increased. With shorter utterances less memory is needed, but the synthesis program becomes more difficult and complex. Features of the target utterance are determined on the basis of linguistic and contextual analyses which are carried out in NLP (*Natural Language Processing*) and DSP (*Digital Signal Processig*) components (Fig. 1) of a TTS (*Text- to-Speech*) system. The basic idea behind concatenate speech synthesis is that speech can be generated from a limited inventory of acoustic units and that their concatenation should account for articulation. Therefore systems based on phonemes and words are impractical—they do not account for articulation and as a result unnatural synthetic speech is obtained. Moreover, an inventory consisting of all words (or syllables) occurring in a given language is too large, requires too much memory and computationally too costly. These problems do not arise if microphonems are used as acoustic units. On the one hand they account for articulation, and on the other their inventories are not too large. However, in order to generate natural sounding synthetic speech, the modification of fundamental frequency and duration of the units is required.

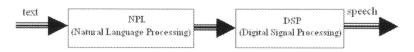

Fig. 1. Main component of a Text-To-Speech synthesis.

Concatenate synthesis produces the most natural-sounding synthesized speech and is divided into three subtypes of synthesis:

1. Unit selection, which uses large databases of recorded speech formed by different units such as individuals phones, syllables, morphemes, words, phrases and sentences; Diphone synthesis, whose minimal database contains all the diphones in a language, still used in research and which suffers from audible glitches;
2. Domain-specific synthesis, which concatenates prerecorded words and phrases to create complete utterances, which is used in a particular domain,

not for general purpose and needs additional complexity to be context-sensitive (e. g. when a consonant is not silent in that context);

3. Formant synthesis creates the synthesized speech using an acoustic model of a waveform with variable parameters like the fundamental frequency, the voicing and noise levels, it is very intelligible, but unnatural;

    a. articulatory synthesis uses computational techniques for synthesizing speech based on models of the human vocal tract and the articulation processes occurring there;

    b. HMM-based synthesis uses Hidden Markov Models to simultaneously model the frequency spectrum (vocal tract),the fundamental frequency and the duration (prosody) of speech;

    c. sinewave synthesis replaces the formants (main bands of energy) with pure tone whistles to synthesize the speech.

All of these technologies are used in a speech synthesizer, which can be implemented in software or hardware. To synthesize speech it is necessary to concatenate pieces of recorded speech that are stored in a database. It is still difficult to show the concatenation of these elements in the time domain. Systems for this tend to be natural (the output sounds like human speech) and intelligible (easy to understand the output) [1].

The most important attribute of speech synthesis is the *intelligibility* or *quality* of its output speech. Evaluations often consider just the quality of the best sentences and the percentage of sentences for which such quality is achieved, even though inferences from the quality of synthesized sound on one sentence to that of other sentences are not always reliable. The overall quality of speech synthesis is a tradeoff between the quality of the system using the best sentences and quality variability across sentences. This can be described for the different families of speech generation approaches as follows:

1. *Limited-domain waveform concatenation*: this approach can generate very high quality speech for a given limited domain using only a small number of prerecorded segments. This approach is used in most interactive voice response systems.

2. *Concatenate synthesis with no waveform modification*: Unlike the previous approach, these systems can synthesize speech from arbitrary text and can achieve good quality on a large set of sentences, but the quality can deteriorate drastically when poor concatenation occurs.

3. *Concatenate systems with waveform modification*: in these systems the waveform can be modified to allow for a better prosody match which gives these systems more flexibility. However, the synthetic prosody can hurt the overall quality which makes this approach questionable for second language learning acquisition in which very high quality synthesis is critical.

4. *Rule-based systems*: These systems tend to give a more uniform sound quality across different sentences, but at the cost of an overall lower quality compared to the above described systems.

Table 1 summarizes the tradeoff between speech quality, domain limitations, and resource requirements for the different synthesis approaches [2].

Table 1. Speech quality and domain limitations

Approach	Speech Quality	Free text generation
Limited-domain waveform concatenation	HIGH quality for limited domain (prerecorded sentences)	NO, can not provide speech output for arbitrary text
Concatenate synthesis without waveform modification	HIGH for known parts GOOD for free parts	YES, can provide output for free text BUT quality can deteriorate when poor concatenation occurs
Concatenate synthesis with waveform modification (also microphonemic method)	MID, prosody can hurt good quality	YES, can provide output for free text, more flexibility through synthetic prosody
Rule-based approach	LOW, compared to the other approaches	YES, uniform sound quality across different sentences

2   Main Principles of the Microphonemic Method

A typical text-to-speech system consists of three main parts, which are text analysis, prosody generation and speech synthesis. The text analysis part understands the text and determines the sound of each word. The prosody generation part generates parameters that control the variability of the speech. The speech synthesis part generates the speech utterance based on the pronunciation and prosody requirements.

Normal concatenation synthesis works by keeping a small set of units in system. During synthesis, a unit is selected and then modified using signal processing techniques according to prosody features. Synthesis in this way can generate speech with relatively high quality. However, the synthetic speech is more or less distorted due to the signal processing process.

Usually standard elements stored in the memory of the computer are words, syllables and, sometimes, speech sounds. The basic idea of the microphonemic method is to use some elementary segments of the speech signal which are shorter than speech sounds, but contain sufficient information about the latter to enable correct perception. An important feature of the method is that the segments can be modified during the synthesis process, so to obtain properly stressed and intoned synthetic utterances. This considerably improves the quality of the synthetic speech.

In our software implementation of the method microphonemic segments method a dictionary of prototypes is collected. The digital form of the synthesized utterance is generated by the program and saved in wave file. The process is controlled by so-called control instructions. These instructions can be prepared by an operator, or generated automatically by the program.

The phonetic transcriptions pipeline has the task of finding the morphemes corresponding to the data from the data stream and converting it into sections that can be pronounced as a whole. This task is done in three stages — normalization, morphemization and prosodization. This pipeline is optimized for integer operations that are more scalar in nature. The pipeline implements various partitioning algorithms that deal with the partitioning of the data stream into morphemes. The pipeline deals with the normalization step which is purely a database dependent step for which we have a separate database. But because such situations occur very infrequently in the text, we have not made any special arrangements for this and we have made allowance for this latency.

Morphemization is again a database intensive step that benefits from having a separate lookup table in local storage. This reduces latency and provides an excellent speedup. Prosodization is the final step into this pipeline that deals with the grouping of morphemes in a speakable unit which is finally converted into the signal.

3   Inserting a Text Into the Program

In the case of the automatic generation of the control instructions, an orthographic text is used as the input. This text can be delivered from different sources. In the automated telephone information system SUSY the text was generated by the system itself. Each piece of the text placed between two punctuation marks is treated by the program as a simple sentence. This is also true of single words separated by commas and single words ending with a period or any other punctuation mark. The program changes the text into a string of SAMPA symbols and generates the parameters modifying their amplitudes, pitch frequencies and time periods in order to produce the required prosodic features. Automation of the control instructions generation process of enables one to utilize the system in practice and generate any text [4] (Table 2 and Table 3).

An example of the string of SAMPA symbols for text is shown after [8].

Table 2. SAMPA symbols of vowels [9].

SAMPA symbol	Orthographic spelling	Transcription
i	PIT	/pit/
I	typ	/tIp/
e	test	/test/
a	pat	/pat/
o	pot	/pot/
u	puk	/puk/
e~	gęś	/ge~s'/
o~	wąs	/wa~s/

Table 3. SAMPA symbols of consonants [9].

SAMPA symbol	Orthographic spelling	Transcription
p	pik	/pik/
b	bit	/bit/
t	test	/test/
d	dyn	/dIn/
k	kit	/kit /
g	gen	/gen /
f	fan	/fan/
v	wilk	/vilk /
s	syk	/sIk/
z	zbir	/zbir/
S	szyk	/SIk/
Z	żyto	/ZIto/
s'	świt	/s'vit/
z'	źle	/z'le/
x	hymn	/xImn/
ts	cyrk	/tsIrk/
dz	dzwon	/dzvon/
tS	czyn	/tSIn/
dZ	dżem	/dZem/
ts'	ćma	/ts'ma/
dz'	dźwig	/dz'vik/
m	mysz	/mIS/
n	nasz	/naS/
n'	koń	/kon'/
N	gong	/goNg/
l	luk	/luk/
r	ryk	/rIk/
w	łyk	/wIk/
j	jak	/jak/

Do niedawna wytwarzanie sygnału było domeną systemów natural-
nych, to znaczy narządów artykulacyjnych człowieka. Obecnie oprócz
naturalnych źródeł sygnału mowy rozważać trzeba także jego wytwarzanie
przez systemy techniczne: syntezatory mowy i generatory sygnałów
mowopodobnych.

do n'edavna vItfaZan'e sIgnawu movI bIwo domeno~ sIstemuf
naturalnIx to znatSI naZonduf artIkulatsIinIx tSwovieka obecn'e oprutS
naturalIx z'rudew sIgnawu movI rozvaZats' tSSeba tagZe jego
vItfaZan'e pSes sIstemaI texn'itSne: sIntezatorI movI i generatorI
sIgnawuf movopodobnIx.

In the Polish language the translation of the orthographic text into phonetic form
is not so complicated as, for example, in the English language, so the algorithm

is rather simple. The program used the following additional information about the context of particular phonemes:

- vowels 'y', 'e', 'a', 'o' → phonemes /I, e, a, o/
- letters 'u' and 'ò' → phoneme /u/
- letter 'i' following consonants → /i/
- letter 'i' following vowels → /j/ after /m/, /f/, /v/, /l/, /r/
- letters 'ii' after /m/, /f/, /v/, /l/, /r/ and letters 'ch' → phonemes /ji/
- letters 'si' → /s'/
- letters 'ci' → /ts'/
- letters 'zi' → /z'/
- letters 'dzi' → /dz'/
- letters 'ni' → /n'/
- letters 'e', 'a' →
    - phonemes /e~/, /a~/ on the end of word
    - phonemes /em/, /om/ before 'p', 'b'
    - phonemes /en/, /on/ before 't', 'd', 'c'
    - phonemes /en'/, /on'/ before 'c', 'dz'
    - phonemes /eN/, /oN/ before 'k', 'g'
    - phonemes /e/, /o/ before 'l', 'ł'

4   Description of the Microphonemic Units

We distinguished 11 classes of phonemes:

1. /i/, /I/, /e/, /a/, /o/, /u/, /l/, /w/, /m/, /n/, /N/
2. /e~/, /o~/
3. /f/, /s/, /S/, /s'/, /x/
4. /v/, /z/, /Z/, /z'/
5. /p/, /t/, /k/
6. /b/, /d/, /g/
7. /j/
8. /ts/, /tS/, /ts'/
9. /dz/, /dZ/, /dz'/
10. /r/
11. empty

These phonemes are generated in a number of ways using microphonemic units. We used a set of words, which is shown in Table 4, for the preparation of microphonemic units.

Quasiperiodic phonemes are built by time repetition of these units with appropriate coefficients of amplification and intonation (example of synthetic phoneme /a/ in Fig. 2). All of the quasiperiodic phonemes can interfere with both proceedings and following phonemes thus generating transients having lengths of several pitch periods.

Pictures of microphonemic units of quasiperiodic classes are shown in Fig. 3.

Table 4. Set of words which was used for preparation of microphonemic units.

handel	ziemia	zanik	fala	ręka	hiena	węgiel	berło
stopa	bukiet	wąsy	mama	jazda	fotel	dżudo	mika
tragarz	wyłom	dzwonek	baza	biuro	gilza	dzięcioł	
rosa	Łomża	więzień	gong	giewont	podpis	szata	
banan	sieczka	kość	żaba	praca	chwila	groźba	

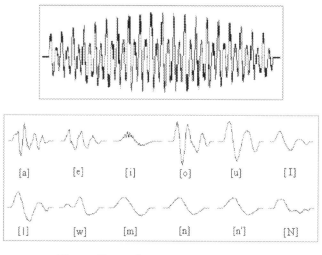

Fig. 3. Units of quasiperiodic phonemes.

Fig. 4. Unit of „basic".

The unit „basic" is a sinusoidal-like signal is shown in Fig. 4.

The microphonemic units of voiceless fricatives are shown in Fig. 5. They can interfere with both preceding and following phonemes. The length of interference is unrestricted. The phonemes are formed by repetition of their units. The first sample of each unit has the value of zero.

The generation of voiced fricatives is a little more complicated, and consists in superposition the sinusoidal-like unit marked „basic" (Fig. 4) with the unit of the correct voiceless fricative. This superposition is conducted in accordance with the following relation:

$$V_i = (B_{i \bmod N} + F_{i \bmod M})/2$$
$$i \in [0, pN]$$

where:

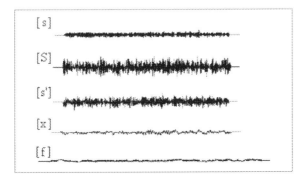

Fig. 5. Units of voiceless fricatives.

B_i—value of the i-th sample of the basic unit
F_i—value of the i-th sample of the voiceless fricative unit
V_i—value of the i-th sample of the synthetic voiced fricative
N—length of the basic unit
M—length of the voiceless fricative unit
p—length of the phoneme (in terms of pitch period).

An example of the synthetic voiced fricative /z/ is shown in Fig. 6.

Fig. 6. Synthetic phoneme /z/.

The voiceless stops [p], [t]. [k] consist of several empty units and one of the microphonemic units [p], [t], [k] which represents the plosion. Each of the three units mentioned above has a different length, as shown in Fig. 7. Their characteristic feature is that they must not interfere with adjoining phonemes.

The voiced stop consonant /g/ does not have its own unit (like /b/ and /d/). The first part of this consonant is modeled by time repetition of the basic unit using a changing coefficient of amplification. The plosion, like in in the case of

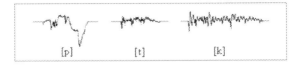

Fig. 7. Units of voiceless stops.

voiceless stops, is added to the end of the basic sequence (Fig. 8). Voiced stops can interfere with proceeding phonemes over any length. Interference with the following phoneme is inadvisable.

Fig. 8. Synthetic phoneme /g/.

The consonant /j/ does not have its own units and is modeled by time repetition of the microphonemic unit /i/ using a changing coefficient of amplification. In effect, a suitable envelope of the signal is obtained (Fig. 9).

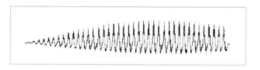

Fig. 9. Synthetic phoneme /j/.

The generation of the voiced stop-fricative /ts/ is shown in Fig. 10. The first part of this fricative is a zone of silence which is modeled by repetition of empty units. Next we add one /t/ unit and the unit /s/.

Fig. 10. Synthetic phoneme /ts/.

The generation of the voiced stop-fricative /d/z/ is still more complicated. The first part of this fricative is modeled by time repetition of the basic unit using a changing coefficient of amplification. The plosion /t/ is then superposed onto the last period of the basic sequence. The third part of the phoneme is generated as a superposition of the same basic unit and fricative unit /S/ (Fig. 11).

Fig. 11. Synthetic phoneme /dz/. The phoneme /r/ is treated as a single microphonemic unit (Fig. 12). There are some situations in which the [r]-unit is repeated in the synthetic text. The [r]-unit does not interfere with its adjoining phonemes.

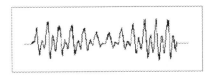

Fig. 12. Synthetic phoneme /r/. The total number of microphonemic units is smaller than the number of phonemes (Table 5).

Table 5. List of phonemes, microphonemic units and components of synthetic phoneme.

1	/i/	[i] [i]		21	/z/	- [z,s]
2	/I/	[I] [I]		22	/z'/	- [z,s']
3	/e/	[e] [e]		23	/Z/	- [z, S]
4	/o/	[o] [o]		24	/p/	[p] [,p]
5	/a/	[a] [a]		25	/t/	[t] [,t]
6	/u/	[u] [u]		26	/k/	[k] [,k]
7	/l/	[l] [l]		27	/b/	- [z,p]
8	/w/	[w] [w]		28	/d/	- [z,t]
9	/m/	[m] [m]		29	/g/	- [z,g]
10	/n/	[n] [n]		30	/j/	- [i]
11	/n'/	[n.] [n']		31	/ta/	- [, t,s]
12	/N/	[N] [N]		32	/tS/	- [,t,S]
13	/e~/	- [e,N]		33	/ts'/	- [,t,s']
14	/a~/	- [o,N]		34	/dz/	- [z,t,z,s]
15	/f/	[f] [f]		35	/dz'/	- z,t,z,s']
16	/s/	[s] [s]		36	/dZ/	- [z,t,z,S]
17	/S/	[S] [S]		37	/r/	[r] [r]
18	/s'/	[s'] [s']		38	pauza	[ ] [ ]
19	/x/	[x] [x]		39	podkład	[z] nie istnieje
20	/v/	- [z, f]				

5   Generation of Transients

In continuous speech, transients appear between successive phonemes. While rapid transitions are common in natural speech, spectral mismatch between concatenated phonemes dramatically lowers the quality of synthetic speech. Most current synthesis systems try to avoid spectral mismatch by careful selection of concatenation utterances. However, this approach can only be successful if the database consists of closely fitting utterances that have to be synthesised. Although studies of the Polish language have shown that transients are not necessary to the proper understanding of a text. In our program we generated suitably constructed signals between successive phonemes in order to improve the quality of the synthetic speech. It is assumed that transients are the result of the continuous transition of vocal tract parameters between the two states corresponding to the adjoining phonemes. Further it assumed that a satisfactory approximation of this phonemes is over a length of several pitch periods. This interference is produced by superposing successive samples of adjoining phonemes using a decreasing coefficient of amplification for the preceding phoneme and an increasing coefficient of amplification for the following one. This interference can be described by the relation:

$$C_i = \frac{L-i}{L} A_{i \bmod N} + \frac{i}{L} B_{i \bmod M}$$

where:

A_i—value of i-th sample of the preceding unit

B_i—value of i-th sample of the following unit

C_i—value of i-th sample of the transient

N—length of the preceding unit

M—length of the following unit

L—length of the transient

If both units belong to quasiperiodic phonemes, than N should equal M. An example of microphonemic unit interference is shown in Fig. 13.

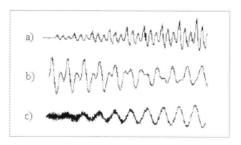

Fig. 13. Interference of microphonemic units:
a) empty and /e/
b) /l/ and /n/
c) /s/ and /m/

6   Control Instructions

The main part of program is the synthesis instructions which convert a given sequence of control instructions into a digital sequence representing the required speech signal. The control instructions contain complete information regarding the method of generating the required utterance These modifications and operations are represented by three parameters

$$\alpha\, P\, T$$

where:

α—microphonemic unit
P—coefficient of duplication
T—coefficient of interference

An example of control instructions is shown in Table 6 and in Fig. 14.

Table 6. Control instructions for example phonemes.

Phoneme symbol	Control Instructions
a	a 40 0
ts'	_ 10 0 t 1 0 s' 1 0
g	z 20 0 k 1 0
k	_ z 20 0 k 1 0
v	z 0 15 f 0 0
ts	_ 10 0 t 1 0 S 1 0
dz'	z 20 0 t 1 0 z 0 4 s' 0 0

7   Conclusion

Text-To-Speech can enable the reading of computer display information for the visually challenged person, or may simply be used to augment the reading of a text message. Current TTS applications include voice-enabled e-mail and spoken prompts in voice response systems. TTS is often used with voice recognition

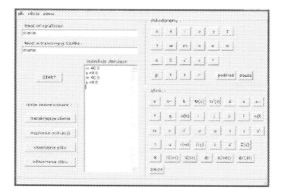

Fig. 14. Print screen of the program. String of control instructions automatic generated for the word „mama" [5].

programs. It is very useful both for language and speech research. The program of microphonemic synthesis of speech enables one to convert any orthographic Polish text into a speech signal. In the mid-1970's one of the last sentences of paper [5] was: „The execution time of the synthesis programs is 30 times longer than the duration of the synthesized text..". Our program, of course, is generating speech in real time. For the Polish language the quality of the synthetic speech generated by the program is satisfactory.

The program can be used to generate synthetic speech for other languages, but such an application requires changing of microphonemic unit base (very small) and the rules for the automatic generation of control instructions.

The basic idea behind concatenative speech synthesis is that speech can be generated from a limited inventory of acoustic units and that their concatenation should account for coarticulation. Therefore systems based on phonemes and words are impractical - they do not account for coarticulation and as a result unnatural synthetic speech is obtained. Moreover, an inventory consisting of all words (or syllables) occurring in a given language is too large, requires too much memory and is computationally too costly. These problems do not arise if diphones, demisyllables or microphonemes are used as acoustic units. On the one hand they account for coartilculation, and on the other their inventories are not very large.

References

1. Chibelushi, C.C., Deravi, F., Mason, J.S.D.: A review of speech-based bimodal recognition. IEEE Transactions on Multimedia 4(1), 23–37 (2002)
2. Hirschman, L., Thompson, H.S.: Overview of evaluation in speech and natural language processing. In: Cole, R., et al. (eds.) Survey of the State of the Art in Human Language Technology, Cambridge University Press, Cambridge (1997)
3. Goldsmith, J.: Dealing with prosody in a text-to-speech system. International Journal of Speech Technology 3(1), 51–63 (1999)

4. Gubrynowicz, R.: Komputerowe modelowanie artykulacji głosek języka polskiego, Prace IPPT PAN 4/2000, Warszawa (2000)
5. Jóźwik, J.: Automatyczna synteza dla języka polskiego, Master Thesis, University of Cardinal Stefan Wyszyński
6. Kiełczewski, G.: Digital Synthesis of Speech and its Prosodic Features by Means of Microphonemic Method, Sprawozdania Instytutu Informatyki UW Nr 65, Warszawa (1978)
7. Lemmetty, S.: Review of the Speech Synthesis Technology, Master's Thesis, Helsinki University of Technology
8. Tadeusiewicz, R.: Sygnał mowy. Wydawnictwa Komunikacji i Łączności, Warszawa (1988)
9. http://www.phon.ucl.ac.uk/home/sampa/polish.htm

Exploring Curvature-Based Topic Development Analysis for Detecting Event Reporting Boundaries

Jakub Piskorski

Joint Research Centre of the European Commission
Web Mining and Intelligence of IPSC
T.P. 267, Via Fermi 2749, 21027 Ispra (VA), Italy
jpiskorski@gmail.com

Abstract. In the era of proliferation of electronic news media and an ever-growing demand for prompt and concise information, natural language text processing technologies which map free texts into structured data format are becoming paramount. Recently, we have witnessed an emergence of publicly accessible news aggregation systems for facilitating navigation through news. This paper reports on some explorations of refining a real-time news event extraction system, which runs on top of the Europe Media Monitoring news aggregation system developed at the Joint Research Centre of the European Commission. Our experiments focus on the task of detecting new events in a given news story, i.e. tagging events extracted by the core event extraction system as new. Several methods ranging from simple similarity computation of event descriptions of adjacent events to more elaborate ones based on curvature-based topic development analysis which utilize global knowledge. The paper describes first the particularities of the real-time news event extraction processing chain. Next, in order to get a better insight how news stories evolve over time some statistics on event dynamics are presented. Finally, the new event detection techniques are introduced and the results of the evaluation are given.

Key words: event extraction, topic detection, security informatics, open source intelligence

1 Introduction

In the era of proliferation of electronic news media and an ever-growing demand for prompt and concise information, natural language text processing technologies which map free texts into structured data format are becoming paramount. Recently, we have witnessed an emergence of publicly accessible news aggregation systems, e.g. *Google News*[1], *Yahoo! News*[2], *Silo Breaker*[3], *NewsTin*[4],

---

[1] http://news.google.com
[2] http://news.yahoo.com
[3] http://www.silobreaker.com
[4] http://www.newstin.com

M. Marciniak and A. Mykowiecka (Eds.): Bolc Festschrift, LNCS 5070, pp. 311–331, 2009.

DayLife[5], etc., for facilitating the navigation process through a myriad of electronic news broadcasted daily worldwide. Such news aggregation systems group topically related articles into clusters and classify them according to various predefined criteria. Although such systems provide a more structured view of what is happening in the world, the amount of data to process by a human remains enormous. In the last decade, several research groups started endeavours on developing news event extraction systems, which detect key information about events from various electronic news media, and summarize this information in the form of database-like structures. In this way, even more compact event descriptions are provided that combine information from different sources.

This paper reports on some explorations of refining a real-time news event extraction system, which runs on top of the Europe Media Monitoring (EMM) [4] news aggregation engine[6] developed at the Joint Research Centre of the European Commission. This system is fine-tuned for detecting crisis-related events, e.g. violent and natural disaster events. Gathering information about crisis events is an important task for the better understanding of conflicts, and for the development of global monitoring systems aiming at automatic detection of precursors for threats in the fields of conflict and health. The work presented in this paper focuses on the task of tagging events extracted from a given news story as it evolves over time as new or old. This task is somewhat related to new topic detection in a stream of textual documents [5], but the major difference is that a given news story oscillating around one major event is significantly more topically coherent than a stream of arbitrary news articles. In this work, we investigate the usability of several computationally lightweight methods for new event detection. These methods range from simple content similarity computation of event descriptions of adjacent events in the event sequence extracted from a given news story over time to more elaborate techniques, which are based on curvature-based topic development analysis and which utilize global knowledge, e.g. the CUTS algorithm [23].

Formally, the task of event extraction is to automatically identify events in free text and to derive detailed information about them, ideally identifying *Who did what to whom, when, with what methods (instruments), where and why.* Automatically extracting events is a higher-level information extraction (IE) task which is not trivial due to the complexity of natural language and due to the fact that in news a full event description is usually scattered over several sentences and articles. In particular, event extraction relies on identifying named entities and relations between them. The research on automatic event extraction was pushed forward by the DARPA-initiated Message Understanding Conferences[7] and by the ACE (Automatic Content Extraction)[8] program.

Although, a considerable amount of work on automatic extraction of events has been reported, it still appears to be a lesser studied area in comparison to the

---

[5] http://www.daylife.com

[6] http://press.jrc.it

[7] MUC—http://www.itl.nist.gov/iaui/894.02/related/projects/muc

[8] ACE—http://projects.ldc.upenn.edu/ace

somewhat easier tasks of named-entity and relation extraction. Precision/recall figures oscillating around 60% are considered to be a good result. Two comprehensive examples of the current functionality and capabilities of event extraction technology dealing with identification of disease outbreaks and conflict incidents are given in [9] and [14] respectively. The most recent trends and developments in this area are reported in [2]. Recently, several authors reported on techniques for improving and refining event extraction through utilization of global inference. For instance, Yangarber, et al. [31,30] applied cross-document inference to correct local extraction results in the context of disease outbreak extraction. This idea was expanded to the case of more general event types in [12], where global evidence from a cluster of topically-related documents is used for refining extraction results. The event extraction system deployed in EMM uses a similar cluster-centric approach and some of the techniques for the new event detection explored in this paper also utilize global knowledge in a similar manner. Fusing locally extracted information across large document collections in order to return consensus extraction answers was also studied in other contexts, e.g. for extracting biographic facts about target individuals [15,1]. Work most similar to ours on deploying information extraction technology for merging news events from multiple news sources covering different time periods have been reported in [16,27]. New topic detection issues were also discussed in [28], which introduces an algorithm for news issue construction, through which news can be effectively and automatically constructed with real-time updates. Furthermore, [22] proposes an algorithm called time driven documents-partition (TDD) that automatically constructs an event hierarchy in a news text corpus based on a user query. Chieu et al. [6] presented a system that extracts events relevant to a query from a large collection of textual documents, which also addresses the task of identifying sentences reporting on the same event. Related work on ordering events extracted from textual corpora on a timeline and detecting relations between events was also presented in [17,3].

The rest of this paper is organized as follows. First, in section 2 the architecture of the live event extraction processing chain is described. Next, in section 3 an overview of news story dynamics in EMM and some statistics on event descriptions generated by the core event extraction engine are given. Subsequently, in section 4 the techniques for automatic detection of new violent and disaster events in the EMM news stories are presented. Some evaluation results are provided in the same section. Finally, a summary is given in section 5.

2   Real-Time Event Extraction Process

This section describes the real-time event extraction processing chain, which is depicted in Figure 1. First, before the proper event extraction process can proceed, news articles are gathered by software dedicated to electronic media monitoring, namely the EMM system [4] that receives 50000 news articles from 2000 news sources in 43 languages each day. Secondly, all articles are classified according to around 700 categories and then scanned in order to identify known

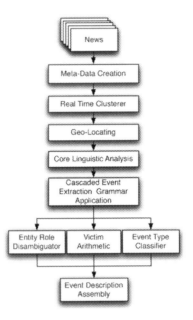

Fig. 1. Real-time event extraction processing chain.

entities (e.g. geographical references, names of known people and organizations, etc.). This information is then created as meta data for each article.

Next, the articles harvested in a 4-hour time window are grouped into news clusters according to content similarity using a standard hierarchical agglomerative clustering algorithm. The article feature vectors are simple word count vectors, and cosine measure is used for computing cluster similarity. The words used for the construction of the word-document space are selected from the information available in the time window under consideration. This process is performed every 10 minutes. Newly created clusters are compared with those calculated in previous iterations. If there is any overlap between clusters, the articles appearing in 'old' clusters which do not appear in the current cluster, are linked to the current cluster. Alternatively, clusters may split due to changes in word-document space. A cluster is maintained as long as there is some reporting in the time window. In this manner news stories can be tracked over time.

Subsequently, each cluster is geo-located using various heuristics ranging from filtering out known person names and organisations, which might also refer to place names and organisations [21] to cluster-level voting of candidate locations selected at article level [20]. Further, clusters describing security-related events are selected via the application of key-word based heuristics.

Next, each of these clusters is processed by NEXUS (News cluster Event eXtraction Using language Structures), the core event extraction engine. For each cluster it tries to detect and extract only the main event by analyzing all articles in the cluster. For each detected crisis-related event NEXUS produces a frame, whose main slots are: date, location, number of killed and injured, kid-napped people, perpetrators, and type of event. In an initial step, each article in the cluster is linguistically preprocessed in order to produce a more abstract representation of its text. This encompasses the following steps: fine-grained tokenization, sentence splitting, domain-specific dictionary look-up (e.g. recognizing numbers, quantifiers, persons' titles), labelling of key terms indicating unnamed person groups (e.g. *civilians, policemen, Shiite*), and morphological analysis. The aforementioned tasks are accomplished by CORLEONE (Core Linguistic Entity Online Extraction), our in-house core linguistic engine [19].

Once the linguistic preprocessing is complete, a cascade of finite-state extraction grammars is applied to each article within a cluster. Following this approach was mainly motivated by the fact that: (a) finite-state grammars can be efficiently processed, and (b) using cascades of smaller grammars facilitates the maintenance of underlying linguistic resources and extending the system to new domains. In particular, we use ExPRESS, a highly efficient IE-oriented extraction pattern engine [18], which allows for the digestion massive amounts of textual data in real time.

The event extraction grammar in its current version consists of two sub-grammars. The first-level subgrammar contains patterns for the recognition of named-entities, e.g. person names (*Osama bin Laden*), unnamed person groups (*at least five civilians*), and named person groups (e.g. *More than thousands of Iraqis*). This grammar has been created manually. The second-level subgrammar consists of 1/2 slot extraction patterns, similar in nature to the ones used in AutoSlog [24], for extracting partial information on events: actors, victims, number of killed, number of injured, type of event, etc. Since the event extraction system is intended to process news articles which refer to security and crisis events, the second-level grammar models only domain-specific language constructions. For creating the second-level grammar consisting of around 3000 extraction patterns a weakly supervised machine learning technique similar in spirit to the ones described in [13,29] has been used. Contrary to other approaches, the learning phase is done by exploiting clustered news, which intuitively guarantees better precision of the learned patterns. The details are described more thoroughly in [25].

The cascaded grammar is applied only to the first sentence and the title of each article from the cluster, where the main facts are summarized in a straight-forward manner, usually without using coreferences, sub-ordinated sentences and structurally complex phrases. Moreover, news clusters contain articles about the same topic from many sources, which refer to the same event description with different linguistic expressions. This redundancy additionally mitigates the effect of phenomena like anaphora, ellipsis and complex syntactic constructions. Furthermore, by processing only the top sentence and the title, the system is more likely to capture facts about the most important event in the cluster. The text, which goes beyond the first sentence is discarded for three main reasons: (a) processing larger text units usually involves handling more complex language phenomena, which might require knowledge-intensive processing, (b) the most crucial information we are seeking is included in the title or first sentence, and

(c) if some crucial information has not been captured from one article in the cluster we might extract it from other article in the same cluster.

Once the event extraction grammars have been applied locally at document level, the single pieces of information are validated and merged into fully-fledged event descriptions. This process encompasses three tasks, entity role disambiguation (as a result of extraction pattern application the same entity might be assigned different roles), victim counting and event type classification.

If one and the same entity has two roles assigned in the same cluster, a preference is given to the role assigned by the most reliable group of patterns, e.g. 2-slot patterns are considered more reliable than 1-slot patterns, whereas in the case of 1-slot constructions, patterns for detection of more generic entities are considered less reliable than others. Another ambiguity arises from the contradictory information which news sources give about the number of victims. For calculating the most probable estimate an ad-hoc technique is applied. First, at document level, a small taxonomy of person classes is used for computing a local estimate. Next, the largest group of local estimates which are close to each other is found, and subsequently the number closest to their average is selected. All articles , which report on the number of victims, which significantly differs from the estimated cluster-level victim estimate, are discarded.

The event type classification algorithm first assigns ranks to each potential event type based on the number of occurrences of type-specific keywords in the articles in a given cluster, where additional boost is given to a more specific type provided that the rank of the type which subsumes it (Types are ordered in a event type hierarchy) is non zero. Finally, the type with the highest rank is selected, unless some domain-specific event type classification rule can be applied. As an example, consider the following domain specific rule: if the event description includes named entities, which are assigned the semantic role *kidnapped*, as well as entities which are assigned the semantic role *released*, then the type of the event is *Hostage Release*, rather than *Kidnapping*.

The core event-extraction process is synchronized with the real-time news article clustering system in EMM in order to keep up-to-date with the most recent events. Initially, the system has been designed for processing English news. Recently, it has been adapted to the processing of Italian, French and Spanish. Adapting to other languages is envisaged. The results of the event extraction are accessible in two ways: (a) via *Google Earth* application which is passed event descriptions in KML format[9], and (b) via a publicly accessible web client[10] that exploits the *Google Maps* technologies.

Currently the event extraction system extracts only one main event from a given cluster of news articles. However, some types of events typically take place across a span of time and space [11], e.g. a kidnapping might include capturing a hostage, a statement and video release by the kidnappers, police action, and

---

[9] For English language: start Google Earth application with: `http://press.jrc.it/geo?type=event&format=kml&language=en`. For other languages change the value of the language attribute accordingly.

[10] `http://press.jrc.it/geo?type=event&format=html&language=en`

liberation. In such cases, the sub-events or incidents can be captured over time as the news story evolves.

A more thorough description of the event extraction processing chain and NEXUS itself, including: real-time news clustering, geo-locating clusters, creating event extraction grammars, extraction pattern engine, automatic resource acquisition, and information fusion is given in [20,26,25]. The details of the process of adapting the system to the processing of Italian are presented in [32].

3   Event Dynamics

As mentioned in the previous section the major goal of applying an event extraction engine on top of EMM news clusters is to gather and update information about crisis-related events reported in the news worldwide over time. For each news story the events extracted by NEXUS at different time points are collected and stored in a database. One of the major problems in this context is to detect duplicate events or to put it in other words to detect new events in a current story. We consider a news story to be a sequence of news article clusters, where each such cluster contains news articles on a given topic gathered within a certain time window as the story evolves. Each such cluster is also associated with an event extracted by NEXUS from this particular cluster. To be more precise, the i-th element in the sequence of clusters refers to the stories' cluster at time $t \cdot i$ (counting from the creation time of the news story), where t stands for the time interval, at which the news articles are re-clustered.

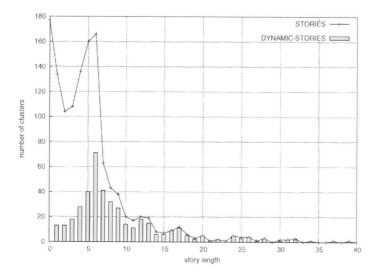

Fig. 2. The histograms for story length for crisis-related stories including dynamic stories. The vertical bars reflect the fraction of the stories, which are dynamic.

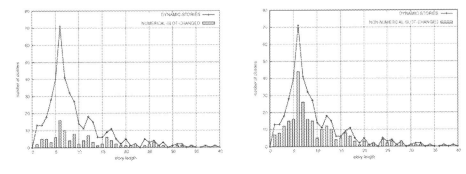

Fig. 3. The histogram for story length of news stories with changes in numerical slots (left) and non-numerical slots (right). The vertical bars reflect the fraction of the dynamic stories for which the change of the respective type occurred.

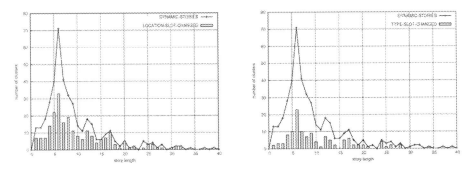

Fig. 4. The histogram for story length of news stories with changes in the location slot (left) and type slot (right). The vertical bars reflect the fraction of the dynamic stories for which the change of the respective type occurred.

In order to obtain a deeper insight into how news stories evolve over time, we have collected some data and computed event statistics. In particular, we have applied NEXUS at 30 minute intervals on the English news clusters produced by EMM between 6th and 16th October 2008. In this time period 6664 different news stories were collected, where 1296 of them were related to violent and natural disaster events. About one third of the latter (442) were dynamic, i.e. at least two distinct event descriptions were generated for such stories. Figure 2 gives the histograms for story length, where the vertical bars refer to dynamic story length distribution. As can be observed most of the stories are short, i.e. their life length does not exceed 6 time intervals. Furthermore, only a small fraction of short stories undergo changes with respect to the event descriptions generated by NEXUS. Contrary to this, NEXUS produces at least two distinct event descriptions for most of the longer stories, i.e. stories whose length is 6 or more.

In order to explore the type of changes in the generated event descriptions more thoroughly, we have analyzed the distribution of changes of: (a) event type, (b) location of the event, (c) numerical slot values, and (d) non-numerical slots except location and type. The histograms for story length of news stories that exhibit different types of changes are given in Figure 3 and 4. In the case of 125 stories out of the 442 dynamic stories, the event type in the produced event descriptions has changed at least once. As can be observed the change in location and non-numerical slots were the most prevalent ones. An empirical analysis revealed that most changes in the location slot were due to more fine-grained location descriptions, which could be extracted at the later stage of a story. Consequently, changes in location slots cannot be considered as a very reliable indicator of a new event. Analogously, changes in non-numerical slots were caused, in most cases, due to the extraction of additional or more detailed descriptions found in an ever-increasing number of articles in a given story. Although the changes in numerical slots were significantly less prevalent than changes in non-numerical slots, together with changes in the event type they appear to be a 'better' indicator for tagging a given event as a new one.

4 Detecting New Events

This section describes our experiments on using various techniques for detecting new events in a given news story.

We first explore a relatively simple method, which computes the similarity between a currently extracted event and a previously extracted one (in a previous iteration) using an event similarity metric. This metric is based on the content similarity of the corresponding slot values in the events being compared. In case the similarity is below a certain threshold the two events are considered to be distinct, i.e. the current event is tagged as new. This method is then extended by introducing additional constraints on the confidence with which the current and past events were extracted by the event extraction engine. Confidence is simply measured in terms of supporting documents, i.e. documents for which NEXUS could return an event description. Additionally, some further global constraints are introduced, e.g. comparison of the similarity between the current event and the average event similarity within a whole story.

Subsequently, we investigate whether utilization of a curvature-based approach for topical segmentation of a stream of texts can improve the performance of new event detection. The basic idea behind such approaches to topic development analysis is to compute for a given sequence of text entries $t_1, t_2, \ldots, t_k$ a sequence $s_{1,2}, s_{2,3}, \ldots, s_{k-1,k}$, where $s_{i-1,i}$ is the similarity between t_{i-1} and t_i. The latter sequence is called *similarity curve* and reflects how rapidly the content of the text stream changes between consecutive entries. In order to identify topically coherent segments one can use local minima of the similarity curve as indicators of segment boundaries, i.e. topic shift occurs at local minima. In the context of the new event detection task, our idea is to compute such a similarity curve for a given news story, and to use it for validating the decisions on tagging events as new, etc. In particular, for a given news story we compute the similarity curve for the text sequence $t_1, t_2, \ldots, t_k$, where t_i consists of the titles and first sentences of the new articles in the story cluster at time i, i.e. articles, which appeared for the first time at iteration i.[11] Subsequently, events extracted by NEXUS are tagged as new only if certain patterns in the similarity curve (indicating a topic change) can be observed.

Although the briefly sketched idea of classical curvature-based algorithms for topic development analysis is effective, we have not used it for two reasons: (a) it performs best in the case of text streams with 'clear' changes of topics, which is not necessarily the case in the context of news stories oscillating around one or two major topics, and (b) it captures solely the pairwise relationship between two adjacent entries in the text stream, whereas capturing the global relations between entries in the text stream would provide more useful and fine-grained information in topic development.[12] Therefore, we applied a different algorithm for computing the similarity curve, namely the CUTS [23] algorithm, which captures global relations between text entries in the input stream of text.

In the remaining part of this section we first describe the way that event similarity is computed in 4.1 and present the CUTS algorithm in 4.2. The details of the various techniques explored and evaluation thereof is given in 4.3.

Before we delve into the details of detecting new events in EMM news stories it is important to note that the evaluation of the core event extraction engine, i.e. NEXUS [26] revealed that different slot values can be extracted with precision ranging from circa 80 to 93%, whereas the event type can be detected correctly in 60% of the cases. This clearly impacts the performance of the new event detection task as well as the fact that news accounts of an event may vary over time.

4.1   Event Similarity

We measure the similarity of two events by calculating the overlap of the slot values in the corresponding event descriptions. First, let us denote an event as a k-tuple $e = \{(s_1, v_1), (s_2, v_2), \ldots, (s_k, v_k)\}$, where each (s_i, v_i) represents a slot-value pair in e. Each event has a type and location slot, whose names are denoted with *type* and *loc* respectively. Apart from this the remaining slots are subdivided into numerical and non-numerical slots. They are denoted as $Num(e)$ and $NonNum(e)$ respectively. Non-numerical slots may be assigned several values. Further, we denote the value of a slot x in e as $e(x)$. Next, we define the set of all instantiated numerical slots for two events e_i and e_j as follows:

$$Num^I_{(e_i, e_j)} = \{x | x \in Num(e_i) \cap Num(e_j) \wedge x \notin \{type, loc\} \wedge$$
$$(e_i(x) \neq \emptyset \vee e_j(x) \neq \emptyset)\}$$

---

[11] In case of iterations in which there are no new articles, we consider for the computation of t_i the documents which were most recently added to the story cluster.

[12] Different news media might report on the same event at different times and even the same media might re-report on the same event from time to time.

Elements of $Num^I_{(e_i,e_j)}$ contain slots, which have been assigned a value in at least one of the two event descriptions. The set of all instantiated non-numerical slots for e_i and e_j, $NonNum^I_{(e_i,e_j)}$, is defined analogously.

The method for computing the similarity of corresponding slot values in two events depends on the type of the slot. For calculating the type similarity of two events, $sim_{type}(e_i, e_j)$ a precomputed type similarity matrix is used. For instance, for two distinct types, where one of them subsumes the other, a positive score (close to 1.0) is assigned. To illustrate such a type pair consider a terrorist attack and suicide bombing, where the latter is subsumed by the first one. In case of types, which do not subsume each other, but have a common predecessor in the type hierarchy, a somewhat lower similarity score is assigned. Finally, types, which are not linked in any way are assigned a zero score.

As mentioned earlier, the location slot may be assigned several values. In such cases, different values represent different aspects of the location, e.g. country, region, province, city, etc. The location similarity of two events is computed similarly to *Jaccard* coefficient as:

$$sim_{loc}(e_i, e_j) = overlap(e_i(loc), e_j(loc))$$
$$\text{and}$$
$$overlap(A, B) = \frac{1 + |A \cap B|}{1 + |A \cup B|}$$

The similarity of non-numerical slots of two events is computed in an analogous way:

$$sim_{non-num}(e_i, e_j) = \frac{1}{|NonNum^I_{(e_i,e_j)}|} \cdot \sum_{x \in NonNum^I_{(e_i,e_j)}} overlap(e_i(x), e_j(x))$$

Next, the similarity of numerical slots for two events is defined as:

$$sim_{num}(e_i, e_j) = \frac{1}{|Num^I_{(e_i,e_j)}|} \cdot \sum_{x \in Num^I_{(e_i,e_j)}} overlap_{num}(e_i(x), e_j(x))$$

$$overlap_{num}(a, b) = 1 - \frac{|a - b|}{\max\{a, b\}}$$

Finally, the overall similarity of two events e_i and e_j is calculated as a linear combination of the four similarity scores introduced above:

$$sim_{event}(e_i, e_j) = \alpha \cdot sim_{type}(e_i, e_j) + \beta \cdot sim_{loc}(e_i, e_j)$$
$$+ \gamma \cdot sim_{num}(e_i, e_j) + \delta \cdot sim_{non-num}(e_i, e_j)$$

Based on empirical observation the weighting coefficients α, β, γ and δ have been set to 0.4, 0.2, 0.3 and 0.1 respectively. In particular, the coefficients α and γ were assigned the highest values since we observed that disagreement in type and numerical slots are strong indicators for classifying two events as distinct.

4.2   CUTS Algorithm

The CUTS algorithm for analysing topic development in text streams was originally proposed in [23]. For an ordered sequence of text entries it produces a sequence of segments, where each of them is assigned a topic development pattern. Basically, there are three topic development patterns: (a) *dominated* – topically coherent texts in the segment, (b) *drifting*—smooth transition from one topic to another one, (c) *interrupted*—sudden and significant change of the main topic for a very short period of time. The CUTS algorithm is divided into three steps.

In the first step, the input sequence of text entries is analysed, and for each entry a vector representation is computed using the TF-IDF scheme. In particular, the i-th text entry in the stream of N text entries is represented as a vector $v_i = (w_{i,1}, w_{i,2}, \ldots, w_{i,n})$, where $w_{i,j}$ is the TF-IDF weight[13] for term i in the entry j and n is the number of all terms. Subsequently, for each pair of entries i and j the similarity, $s_{i,j}$, is calculated as:

$$\sum_{k=1}^{n} w_{i,k} \cdot w_{j,k}$$

Next, a dissimilarity metric, D, is computed with $D_{i,j} = 1 - s_{i,j}$.

In the second phase of the algorithm, the sequence of entries is mapped to a curve, which reflects the topic development patterns in terms of dissimilarity between neighbouring text entries. Unlike other curvature-based topical text segmentation algorithms (e.g. [10]), which utilize solely information of pairwise (dis)similarity between adjacent entries in the input sequence, the CUTS algorithm exploits broader context, i.e. 'content' dissimilarities for all pairs of entries in the input. To be more precise, the initial sequence of entries is mapped to a 1-dimensional space so that distances between points best match the dissimilarities between the corresponding entries. The aforementioned mapping is computed through the application of a multi-dimensional scaling (MDS) technique [7] applied on the dissimilarity matrix D. Next, the time dimension is added, i.e. each point in time—index of the entry in the input sequence (x-axis) is mapped to the corresponding MDS-computed value in the 1 dimensional space (y-axis). The resulting curve is called the CUTS curve. A 'stable' topic in the CUTS curve corresponds to an almost horizontal segment (*dominated segment*). A smooth transition from one topic to another is reflected in the CUTS curve as a sloping curve (*drifting segment*), where the gradient angle measures how fast one topic fades out and new topic fades in. Finally, interruptions, i.e. sudden and temporary introduction of a new and significantly different topic, correspond to segments with a saw tooth shape. Interruptions may either occur within dominated or drifting segments. Figure 5 illustrates a drifting segment and a dominated segment with an interruption in CUTS curves computed for some of the stories in our test data described in the previous section.

---

[13] The weight $w_{i,j}$ can also incorporate some additional domain knowledge [23].

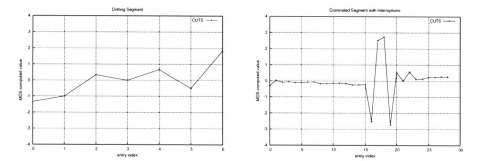

Fig. 5. An example of a drifting segment (left) and a dominated segment with an interruption (right)

The last step of the algorithm consists of analysing the CUTS curve in order to identify topic development segments. In particular, the CUTS curve is split into a series of straight line segments, where each such segment is represented as a 4-tuple $s_i = (k_i, \sigma_i, (x_{start}, y_{start})_i, (x_{end}, y_{end})_i)$, where: (a) k_i is the slope of the corresponding line segment and measures topic divergence in the segment, (b) σ_i is the average of the distances from the original points in this segment to the straight line approximating these points, which reflects the concentration of topics within the segment, i.e. the higher the value of this parameter, the more diverse set of topics is covered, (c) $(x_{start}, y_{start})_i$ and $(x_{end}, y_{end})_i$ represent the boundary points of the segment respectively. Initially, the algorithm starts with 'elementary' line segments connecting adjacent points in the CUTS curve and tries to combine consecutive segments into larger ones in an iterative manner. To be more precise, two consecutive topic segments s_i and s_{i+1} are merged into a single segment if their topic divergence and topic concentration is homogeneous. We say that two segments s_i and s_{i+1} are *homogeneous* if:

$$|k_i - k_{i+1}| < \lambda_{drifting}$$
$$|\sigma_i - \sigma_{i+1}| < (\sigma_i + \sigma_{i+1})/2$$

The parameter $\lambda_{drifting}$ is a threshold for determining whether two different topic evolution speeds can be considered as homogeneous. Its value is computed as a percentage of the overall change in the input data. Once two segments are merged into a larger one the k and σ values for the new segment are recomputed. The merging process is continued as long as there exist pairs of consecutive segments, which are homogeneous. Finally, each segment $s_i = (k_i, \sigma_i, (x_{start}, y_{start})_i, (x_{end}, y_{end})_i)$ or a subsequence of such segments in the resulting sequence of line segments is assigned a topic development pattern. In particular, if $|k_i| < \lambda_{drifting}$ the segment s_i is tagged as dominated, otherwise it is annotated as drifting. Once segments are classified either as drifting or dominated the process of identifying interrupted segments is triggered. Two consecutive segments, s_i and s_{i+1} are combined into an interrupted segment if the following holds:

$$|k_i| \geq \lambda_{drifting}$$
$$|k_{i+1}| \geq \lambda_{drifting}$$
$$k_i \cdot k_{i+1} < 0$$
$$|k_{i-1} - k_i'| + |k_{i+2} - k_i'| < \lambda_{drifting},$$

where $k_i' = |(y_{start})_i - (y_{end})_{i+1}|/|(x_{start})_i - (x_{end})_{i+1}|$. The intuition behind the above criteria is that an interrupted pattern is introduced in the case of two adjacent drifting segments, which have slopes with opposite directions and which interrupt a dominated or drifting pattern. A more in-depth description of the CUTS algorithm can be found in [23].

4.3   Experiments on Refining New Event Detection

This section describes experiments with several lightweight techniques for new event detection.

Let a story consist of a sequence of clusters $C_1, C_2, \ldots, C_k$ and a corresponding sequence of extracted events $e_1, e_2, \ldots, e_k$. Furthermore, let $Hits(C)$ denote the number of documents in C for which NEXUS returned an event description. Let us denote the CUTS value for the cluster C_i as $cuts(C_i)$, i.e. the MDS value computed by the CUTS algorithm (see section 4.2).

We have tested the following basic algorithms, which tag an event e_i as a new event if:

- **Overlap (O):** $sim_{event}(e_i, e_{i-1}) < \phi$, where ϕ is a similarity threshold

- **Average Overlap (OA):** $sim_{event}(e_i, e_{i-1}) < c \cdot \alpha$, where α is the average event overlap between two adjacent events in the current news story and c is a weighting factor

- **Overlap with Confidence (OC):** $sim_{event}(e_i, e_{i-1}) < \phi$ and $conf(e_i) > c \cdot conf(e_{i-1}) > 0.1$ and $|C_i| > 1$, where $c \in (0, 1)$, $conf(e_i) = Hits(C_i)/|C_i|$, i.e. confidence ($conf$) for event e_i is the ratio between the number of documents in cluster C_i for which en event was extracted and the total number of documents in C_i. The intuition behind OC is to filter out: (a) events whose confidence decreases (with a certain speed) compared to a previous event in the story or (b) events whose confidence is very low or (c) events extracted from clusters containing solely one document (which are considered unreliable).

- **Average Overlap with Confidence (OAC):** $sim_{event}(e_i, e_{i-1}) < c \cdot \alpha$ (as in OA) and $sim_{event}(e_i, e_{i-1}) < \phi$ and $conf(e_i) > c \cdot conf(e_{i-1}) > 0.1$ and $|C_i| > 1$ (as in OC). This is simply a combination of OC and OA.

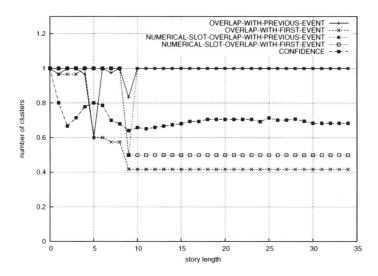

Fig. 6. An example of event statistics for a news story on a teenager accused of murdering a schoolboy. The various curves depict the similarities of adjacent events in the news story (OVERLAP-WITH-PREVIOUS-EVENT, NUMERICAL-SLOT-OVERLAP-WITH-PREVIOUS-EVENT) and similarities between events at different iterations and the event extracted at the beginning of a news story (OVERLAP-WITH-FIRST-EVENT, NUMERICAL-SLOT-OVERLAP-WITH-FIRST-EVENT).

– **CUTS:**
  - for $i \le 3$: the CUTS curve fragment at $cuts(i-t), \ldots, cuts(i)$ for a given t is predominantly interrupted or drifting and $(1/(i-t)) \cdot \sum_{i-t}^{i} cuts(i) > \phi_1$ (i.e. significant changes in the CUTS curve are observed) or if $|cuts(1) - cuts(i)| > 0.8$ (i.e. most recent documents in C_i are not similar to the documents in C_1)
  - for $i > 3$: $cuts(i-t), \ldots, cuts(i)$ is predominantly interrupted or drifting and $(1/(i-t)) \cdot \sum_{i-t}^{i} cuts(i) > \phi_2$ with $\phi_1 > \phi_2$.

The main idea behind the **CUTS** algorithm for new event detection is that events are tagged as new events only if the number of articles on a new topic or aspect of the main event of the story has reached a critical mass. Intuitively, reliable evidence for reaching such a critical mass is the occurrence of a drifting pattern or some interruptions in the CUTS curve that are directly prior to the extraction of a new event. Since many news stories are not stable in the initial iterations, the conditions, which have to be fulfilled (e.g. the threshold values) are slightly different for the beginning of the story ($i \le 3$) and subsequent iterations ($i > 3$).

Finally, we have combined the algorithms **O**, **OA**, **OC** and **OAC** with the **CUTS** algorithm. To be more precise, the **CUTS** constraints are embedded in those algorithms as additional ones. Furthermore, in case the original constraints of those algorithms do not hold, but 'almost' hold (values are within a certain distance to the thresholds, etc.) the **CUTS** constraints must hold in order to tag an event as a new one. We denote the combined algorithms as **O+CUTS**, **OA+CUTS, OC+CUTS** and **OAC+CUTS** respectively.

In order to illustrate how the combination of the various methods with the technique that utilizes **CUTS** curves work let us consider the diagram in Figure 6, which depicts several curves based on the computed similarity values between two adjacent events in the news story (i.e. e_i and e_{i+1}) and the similarity between each event and the first event (e_1) in a given news story. As can be observed the similarity of the events extracted at iteration 5 and 9 with the first event are 0,6 and 0.5 respectively. Furthermore, one can see that from the 9th iteration on the values of numerical slots are different from corresponding slots in the events in the interval [0:9]. Both facts might constitute evidence for classifying these events, i.e. e_5 and e_9, as new. However, in the corresponding CUTS curve depicted in Figure 7, the whole interval [0:23] is strongly dominated, which allows us to draw a conclusion that there are actually no new events. The changes in the event descriptions extracted at iteration 5 and 9 could be due to some error of the event extraction engine or just due to the occurrence of new and additional information contained in the new articles in the story. One can also observe that there are some interruptions in the CUTS curve in the interval [23:31]. However, in this time period the events extracted by NEXUS are identical. One could interpret the situation that there were some new and topically distinct articles, which arrived in the story cluster in the time interval [23:31], but they constituted only a small fraction of the whole news article cluster at that time and were discarded by NEXUS in the information aggregation phase. Finally, we can observe that the CUTS curve is again dominated in [31:35], but the corresponding articles in this interval are topically distinct from the ones in the initial dominated segment in [0:23], which indicates that the news story was most likely merged with another one at a later stage.

We have applied the different algorithms on the test corpus described in section 3, which consists of 125 news stories that exhibited changes in the event type in the sequence of the corresponding events extracted from these stories, and we have evaluated their precision, recall and F-measure. Let E be the set of automatically detected new events in the test corpus and let E^* be the set of events annotated as new events by a human expert. We call the latter set the ground truth. The *recall* is the ratio of correctly detected new events and all events in the ground truth, i.e. $recall = |E \cap E^*|/|E^*|$. We define *precision* as the fraction of events, which were correctly tagged by the algorithm as new events, i.e. $precision = |E \cap E^*|/|E|$. *F-measure* is defined as the mean of precision and recall, i.e. $F-measure = (2 \cdot precision \cdot recall)/(precision + recall)$. As a baseline we have applied an algorithm, which tags events as new randomly (denoted as **BASELINE**).We have also evaluated the new event detection task as a classification task and measured classification *accuracy*. For evaluation purposes all event descriptions in the test corpus, which potentially represent new events[14], were tagged by a human expert as either *new* or *old* events.

---

[14] Events, whose overlap with the first event in a story is less than a given threshold.

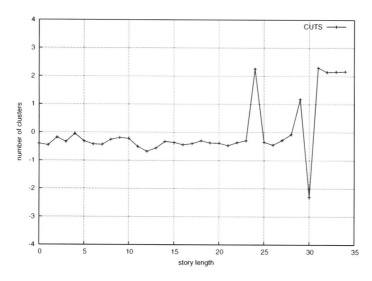

Fig. 7. CUTS curve corresponding to the story diagram in Figure 6

We have experimented with various settings for each of the algorithms and computed interpolated precision-recall curves (see Figure 8) in order to get a better picture of their performance. As can be observed, **OC** algorithm performs slightly better than other basic methods. Integration of **CUTS** curves in the basic algorithms significantly improves precision, which is penalized by deterioration in recall. Table 1 gives the ranking of the best F-measure results achieved by the algorithms accompanied by the corresponding precision, recall and classification accuracy values. Surprisingly, the F-scores for **OC** and **O** methods are the top ones, whereas best precision values and classification accuracy could be obtained with **OC+CUTS** and **OAC+CUTS** methods (with slightly smaller F-score). The base **CUTS** algorithm performed significantly worse than all other methods.

The results presented in Table 1 are not very impressive, but one has to take into account that the following factors strongly complicate the task: (a) the performance and accuracy of the clustering algorithm and event extraction engine are far from perfect, which might introduce some noise in the data, (b) media sources worldwide might not necessarily provide articles on the same topic (event) simultaneously due to time difference etc., which results in 're-appearing' events and more 'jagged' CUTS curves, (c) the beginning of some stories is frequently characterized by a small number of articles and a strongly non-dominated segments in the CUTS curve before the topic (curve) becomes more 'stable', (d) the same news clusters might report about events of the same type in the same place or region (e.g. bombings or terrorist attacks in Iraq in Baghdad or nearby areas occurring frequently on the same day), which are not distinguished by the CUTS algorithm, but they are recognized as different events by the event extraction engine. Finally, the amount of data used for the

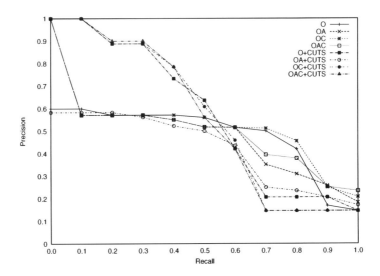

Fig. 8. Precision-Recall curves

Table 1. Top F-measure results and corresponding precision, recall and classification accuracy figures

method	precision	recall	f-measure	classification accuracy
OC	0.513	**0.769**	**0.615**	0.870
O	0.5	**0.769**	0.606	0.870
O+CUTS	0.636	0.538	0.583	0.893
OAC+CUTS	**0.75**	0.462	0.571	**0.898**
OC+CUTS	**0.75**	0.461	0.571	**0.898**
OA	0.516	0.615	0.561	0.864
OAC	0.516	0.615	0.561	0.864
OA+CUTS	0.436	0.654	0.523	0.864
CUTS	0.4	0.385	0.392	0.825
BASELINE	0.090	0.307	0.139	0.441

experiments described in this paper might not have been sufficient in order to reveal other less straightforward patterns in the CUTS curves, which might have discriminatory power for new event detection.

5 Summary

In this paper we have reported on some explorations of refining a real-time crisis-related news event extraction system developed at the Joint Research Centre of the European Commission, which runs on top of the Europe Media Monitoring news aggregation system. Our work focused on the task of detecting new events in a stream of events extracted from a given news story over time as it evolves.

This task is related to new topic detection in a stream of textual documents, but the major difference is that a given news story oscillates around one major event, and articles contained in such a story are significantly more topically coherent than articles in a stream of arbitrary news articles. In this work, we investigated the usability of several computationally lightweight methods for new event detection, ranging from simple content similarity computation for adjacent events in the event sequence extracted for a given news story, to more elaborate techniques, which are based on CUTS—a curvature-based topic development pattern analysis. Surprisingly, with respect to the F-measure, the method based on computing content similarity between event descriptions of adjacent events and using confidence scores, gave the best results. Integration of **CUTS** curves, which reflect global relationships of articles in the whole life of a news story significantly improves precision, which is penalized by deterioration in recall. We believe that performing the same experiments on a larger data set might be beneficial for fine-tuning the presented methods. In particular, the application of machine learning methods for the task of new event detection are envisaged. Since the event extraction system is continuously being extended to processing news articles in new languages the comparison of the extraction results returned by the system from articles in different languages might provide additional global knowledge for validating facts, in particular for the new event detection task. Additionally, we believe that distinguishing between streams of articles from different sources about the same topic might result in more precise **CUTS** curves, which will be explored in the future. Finally, we intend to study more throughly the usability of linguistically more sophisticated methods for refining event extraction results and for detecting relations between events [8,3,17].

Acknowledgments. The work presented in this paper was supported by the Europe Media Monitoring (EMM) Project carried out by the Web Mining and Intelligence Action in the Joint Research Centre of the European Commission. I am greatly indebted to Hristo Tanev and Jonathan Brett Crawley for fruitful discussions and to Martin Atkinson, Erik van der Goot and all other EMM colleagues without whom the presented work would not have been possible.

References

1. Alani, H., Kim, S., Millard, D., Weal, M., Hall, W., Lewis, P., Shadbolt, N.: Web based Knowledge Extraction and Consolidation for Automatic Ontology Instantiation. In: Proceedings of the Workshop on Knowledge Markup and Semantic Annotation, K-Cap'03 (2003)
2. Ashish, N., Appelt, D., Freitag, D., Zelenko, D.: Proceedings of the Workshop on Event Extraction and Synthesis, held in conjunction with the AAAI 2006 conference, Menlo Park, California, USA (2006)
3. Bejan, C., Harabagiu, S.: A Linguistic Resource for Discovering Event Structures and Resolving Event Coreference. In: ELRA, E.L.R.A. (ed.) Proceedings of the 6th International Language Resources and Evaluation (LREC'08), Marrakech, Morocco (2008)

4. Best, C., van der Goot, E., Blackler, K., Garcia, T., Horby, D.: Europe Media Monitor. Technical Report EUR 22173 EN, European Commission (2005)
5. Brants, T., Chen, F., Farahat, A.: A System for New Event Detection. In: SIGIR '03: Proceedings of the $26t^{th}$ Annual International ACM SIGIR Conference on Research and Development in Informaion Retrieval, pp. 330–337. ACM, New York (2003)
6. Chieu, H., Keok Lee, Y.: Query Based Event Extraction along a Timeline. In: Proceedings of the 27th Annual International ACM SIGIR Conference on Research and Development in Information Retrieval, pp. 425–432. ACM, New York (2004)
7. Cox, T., Cox, M.: Multidimensional Scaling, 2nd edn. Monographs on Statistics and Applied Probability. Chapman and Hall, London (2001)
8. Fillmore, C., Narayanan, S., Baker, C.: What Linguistics Can Contribute to Event Extraction. In: Proceedings of the AAAI 2006 Workshop on Event Extraction, AAAI Press, Menlo Park (2006)
9. Grishman, R., Huttunen, S., Yangarber, R.: Real-time Event Extraction for Infectious Disease Outbreaks. In: Proceedings of Human Language Technology Conference (HLT) 2002, San Diego, USA (2002)
10. Hearst, M., Plaunt, C.: Subtopic Structuring for Full-length Document Access. In: Proceedings of the 16th Annual International ACM-SIGIR Conference on Research and Development in Information Retrieval, ACM, pp. 59–68 (1993)
11. Huttunen, S., Yangarber, R., Grishman, R.: Complexity of Event Structure in IE Scenarios. In: Proceedings of the 19th International Conference on Computational Linguistics, Morristown, NJ, USA, Association for Computational Linguistics, pp. 1–7 (2002)
12. Ji, H., Grishman, R.: Refining Event Extraction through Unsupervised Cross-document Inference. In: Proceedings of 46th Annual Meeting of the Association for Computational Linguistics: Human Language Technologies, Columbus, Ohio, USA (2008)
13. Jones, R., McCallum, A., Nigam, K., Riloff, E.: Bootstrapping for Text Learning Tasks. In: Proceedings of IJCAI-99 Workshop on Text Mining: Foundations, Techniques, and Applications, Stockholm, Sweden (1999)
14. King, G., Lowe, W.: An Automated Information Extraction Tool For International Conflict Data with Performance as Good as Human Coders: A Rare Events Evaluation Design. International Organization 57, 617–642 (2003)
15. Mann, G., Yarowsky, D.: Multi-field Information Extraction and Cross-document Fusion. In: Proceedings of the 43rd Annual Meeting on Association for Computational Linguistics, Morristown, NJ, USA, Association for Computational Linguistics, pp. 483–490 (2005)
16. Naughton, M., Kushmerick, N., Carthy, J.: Event Extraction from Heterogeneous News Sources. In: AAAI 2006 Workshop on Event Extraction and Synthesis, AAAI Press, Menlo Park (2006)
17. Otterbacher, J., Radev, D.: Modeling Document Dynamics: an Evolutionary Approach. In (ELRA), E.L.R.A. (ed.) Proceedings of the 6th International Language Resources and Evaluation (LREC'08), Marrakech, Morocco (2008)
18. Piskorski, J.: ExPRESS – Extraction Pattern Recognition Engine and Specification Suite. In: Proceedings of the International Workshop Finite-State Methods and Natural language Processing 2007 (FSMNLP'2007), Potsdam, Germany (2007)
19. Piskorski, J.: CORLEONE – Core Linguistic Entity Online Extraction. In: Technical report 23393 EN, Joint Research Center of the European Commission, Ispra, Italy (2008)

20. Piskorski, J., Tanev, H., Atkinson, M., van der Goot, E.: Cluster-Centric Approach to News Event Extraction. In: Proceedings of the International Conference on Multimedia & Network Information Systems, Wroclaw, Poland, IOS Press, Amsterdam (2008)
21. Pouliquen, B., Kimler, M., Steinberger, R., Ignat, C., Oellinger, T., Blackler, K., Fuart, F., Zaghouani, W., Widiger, A., Forslund, A., Best, C.: Geocoding multilingual texts: Recognition, Disambiguation and Visualisation. In: Proceedings of LREC 2006, Genoa, Italy, pp. 24–26 (2006)
22. Pui, G., Fung, C., Yu, J., Liu, H., Yu, P.: Time-dependent Event Hierarchy Construction. In: Proceedings of the 13th ACM SIGKDD International Conference on Knowledge Discovery and Data Mining, pp. 300–309. ACM, New York (2007)
23. Qi, Y., Candan, K.S.: CUTS: Curvature-based Development Pattern Analysis and Segmentation for Blogs and Other Text Streams. In: Proceedings of Hypertext 2006, ACM Press, New York (2006)
24. Riloff, E.: Automatically Constructing a Dictionary for Information Extraction Tasks. In: Proceedings of the 11th National Conference on Artificial Intelligence (1993)
25. Tanev, H., Oezden-Wennerberg, P.: Learning to Populate an Ontology of Violent Events. In: Fogelman-Soulie, F., Perrotta, D., Piskorski, J., Steinberger, R. (eds.) Mining Massive Data Sets for Security, IOS Press, Amsterdam (2008)
26. Tanev, H., Piskorski, J., Atkinson, M.: Real-Time News Event Extraction for Global Crisis Monitoring. In: Kapetanios, E., Sugumaran, V., Spiliopoulou, M. (eds.) NLDB 2008. LNCS, vol. 5039, pp. 207–218. Springer, Heidelberg (2008)
27. Wagner, E., Liu, J., Birnbaum, L., Forbus, K., Baker, J.: Using Explicit Semantic Models to Track Situations Across News Articles. In: AAAI 2006 Workshop on Event Extraction and Synthesis, AAAI Press, Menlo Park (2006)
28. Wang, C., Zhang, M., Ma, S., Ru, L.: Automatic Online News Issue Construction in Web Environment. In: Proceedings of 17th International World Wide Web Conference, Bejing, China, pp. 457–466. ACM, New York (2008)
29. Yangarber, R.: Counter-Training in Discovery of Semantic Patterns. In: Proceedings of the 41st Annual Meeting of the ACL (2003)
30. Yangarber, R.: Verification of Facts across Document Boundaries. In: Proceedings International Workshop on Intelligent Information Access, IIIA-2006 (2006)
31. Yangarber, R., Jokipii, L.: Redundancy-based Correction of Automatically Extracted Facts. In: Proceedings of the Conference on Human Language Technology and Empirical Methods in Natural Language Processing, Morristown, NJ, USA, Association for Computational Linguistics, pp. 57–64 (2005)
32. Zavarella, V., Piskorski, J., Tanev, H.: Event Extraction for Italian using a Cascade of Finite-State Grammars. In: Proceedings of the 7th International Workshop on Finite-State Machines and Natural Language Processsing, Ispra, Italy (2008)

Domain Model for Medical Information Extraction—The LightMedOnt Ontology

Agnieszka Mykowiecka and Małgorzata Marciniak

Institute of Computer Science,Polish Academy of Sciences, Warsaw, Poland
{agn,mm}@ipipan.waw.pl

Abstract. The paper describes the creation of a domain model for an Information Extraction (IE) application in the medical domain. First, we present texts: mammography reports and diabetology patients' discharge documents, for which IE systems were created. The methodology and results of terminology extraction for both domains are described. Next, the main features and the upper part of *LightMedOnt*—medical ontology in OWL formalism are presented. In the final part of the paper we discuss the relationships between OWL ontologies and the domain model of the IE system used for our experiments.

Key words: domain model, ontology, terminology extraction, clinical data processing

1 Introduction

In the paper we describe our approach of representing domain related knowledge for an IE application in the medical domain, and a particular model which was created under these assumptions. The aim of creating the model was to establish a reliable source of knowledge for applications which operate on textual clinical data. For such programs a well defined model of the domain is crucial for the reliability of their results. One particular type of application which we have in mind is a task of Information Extraction (IE) which consists in selecting information from natural language texts. The task's formulation implies that the kind of data which is searched for has to be known in advance: the domain knowledge is necessary for defining templates which are to be filled by an IE application. For some applications a template structure is relatively easy to define (e.g. if we look for changes of prices of selected products or proper names of objects of given types), but if we want to transform complicated data into a structural form, this specialized knowledge is indispensable. Domain models can be of various types but the most popular way of representing knowledge at the moment is an ontology which includes a typology of types of objects important to the domain.

We have chosen an OWL standard to create our domain ontology as using generally accepted languages and formats allows us to exchange the results more easily. Our model was designed with different levels of generality. We represented some important subdomains very precisely, while others were only sketched. We

M. Marciniak and A. Mykowiecka (Eds.): Bolc Festschrift, LNCS 5070, pp. 333–357, 2009.

wanted to represent only information which will be used by IE applications. The created model was used in two IE applications. One was designed for analyzing mammography reports, the second one processes discharge documents of diabetic patients.

The ontology which we present here was constructed manually—all classes and properties were defined on the basis of data analysis or exchange of information with human experts. The process was incremental and ontology corrections were made in parallel to defining rules of IE applications. The manual collection of data supported by expert knowledge is thought the most reliable, but this procedure is very laborious and may lead to incomplete or erroneous descriptions. Although a completely automatic procedure in the case of complex data is still not possible, statistical methods can be of great help to human experts, as they can ease the process of model construction. In our case, we performed an automatic terms extraction procedure. The extracted terms were used to improve the completeness of the ontology, and served as comments included in the resource showing various expressions of information.

In the paper we first present data which formed the basis of our ontology. Then, in section 3, we present results of terminology extraction experiment, and in section 4 we shortly describe what we mean by an ontology and discuss our motivations and assumptions. The ontology is presented in section 5 while the next section presents the relationship between an OWL ontology and a domain model of an IE application implemented with the SProUT system [1].

2   Data Description

2.1   Mammography Reports

The first data set consists of mammography notes—comments written by physicians (radiologists) to explain mammography pictures. Our experiment was conducted using real data gathered form Warsaw health care centers. To cover the possible diversity of writing styles, data were collected from three sources. In (1) and (2)[1] we show two examples of such notes from two medical centers that differ mainly in formats of the administrative part. All documents have identification numbers within the center—documents from one hospital additionally have the date of examination and hospital internal patient's identification number.

(1)   88
       Sutki z przewagą utkania tłuszczowego, w sutku lewym pojedyncze makro-
       zwapnienia o charakterze łagodnym, nie stwierdza się zmian ogniskowych
       podejrzanych o złośliwość. Węzły pachowe niepowiększone.

       Breasts with dominant fat tissue, in left breast single microcalcifications,
       benign character, no findings suspected of being malignant were found.
       Armpit lymph nodes are not enlarged.

---

[1] In all examples, identification data are fictitious.

(2)   Badanie: MAMMOGRAFIA.
Identyfikator badania: 148514
Identyfikator pacjenta: 323251
Data badania: 2002-10-12
Rozpoznanie:
Opis:
Stan po mastektomii prawostronnej. Sutek lewy o utkaniu tłuszczowo-gruczołowym. Zmian ogniskowych i podejrzanych o złośliwość nie wyka-zano. W lewym dole pachowym widoczne niepowiększone węzły chłonne. Obraz mammograficzny nie wykazuje istotnych różnic w porównaniu z badaniem poprzednim z dnia 29.09.00r. Kontrolne badanie za rok.

Examination: MAMMOGRAPHY
Examination identifier: 148514
Patient identifier: 323251
Examination datum: 2002-10-12
Diagnosis:
Description:
State after mastectomy of right breast. Left breast with fat-glandular tissue. There are no malignant or suspicious findings. Not enlarged lymph node visible in left armpit. There are no significant changes in mammogram in comparison to the previous examination from 29.09.00r. Check-up in one year.

The notes are rather short (20–50 words long) and contain information from a very restricted domain. Texts consist mainly of breast composition descriptions and statements if there were any pathological findings observed. If so, a precise description of all their visible features and hypothetical diagnoses are written. Apart from that, notes sometimes include a comparison of the current picture with the previous one, and usually end with recommendations concerning time and type of the next examination. In some notes there is information about former breast surgery or data concerning USG examination results. No other type of information are included. In Tab. 1. we show some details about the data size and diversity.

The input data being original doctors' notes contain a lot of errors: mis-spellings, missing Polish diacritical marks, and punctuation errors. Errors very often concern domain specific names that are crucial in information extraction. Most errors are due to weak keyboard writing skills (performance errors) and writers usually do not correct them as these are not official documents. Since the original documents contained a lot of misspelled words, they were first cor-rected to increase efficiency of automatic information extraction. Initially, texts were corrected manually, then a program for automatic spelling correction was designed.[2]

Moreover, there are many non-standard abbreviations in the texts which are specific to the mammography domain. For example the abbreviation *ww* in

---

[2] A description of typical errors and the program for their correction is given in [15].

Table 1. Mammography data set characteristics

	ALL DATA		"TRAIN" SET		TEST SET	
	types	occurr.	types	occur.	types	occur.
examinations	2117	–	706	–	704	–
basic forms	940	–	705	–	440	–
inflected forms	1675	105301	1263	37992	763	30339
nonword tokens	17	45764	13	16171	11	14678

most Polish documents is interpreted as 'above mentioned' but in mammography reports its meaning is 'lymph nodes' (in plural), and *w* can be interpreted as a preposition 'in' or the abbreviation of 'lymph node' (in singular).

As the domain and its vocabulary is limited, crucial words usually have one interpretation. The word *węzeł* 'node' has different interpretations in physics, computer science, mathematics, botanics, etc., but in mammography reports always means a lymph node. Adjectives like *okrągły* 'round', *gwiazdkowaty* 'starry', *podłużny* 'elongated' always refer to the shape of a finding. There are only a few keywords having ambiguous interpretation in mammography reports: *nieregularne* 'irregular' can describe a feature of a tissue or the shape of a finding, similarly the word *plamista* 'maculated' can refer to a feature of a finding or a tissue. Of course, adverbs like *słabo* 'weakly' can be interpreted only in context of modified adjective or participle: *słabo widoczny* 'weakly visible', *słabo wyczuwalny* 'weakly palpable'.

2.2   Diabetic Data

The second set of documents consists of diabetic patients' hospital documentation, from Bródnowski Hospital in Warsaw. All documents are from a ward which specializes in treating diabetes and are written by specialty physicians. These are official documents 1.5–2.5 pages long written in MS Word. They are given to patients as the summary of his/her hospital treatment, and are also used for the reimbursement of treatments.

We have 606 documents from years 2001–2006, that we divided into training and testing sets. All 169 documents from the year 2006 are the test set and were not inspected during ontology and extraction system development.

Each document concerns one patient's visit in hospital. A particular visit is identified by two parameters: an identification number of the visit within the year and the year. The year is sometimes included in the identification number of the document, but quite often it has to be established on the basis of the patient's visit in hospital. Sometimes some results of tests are available after the patient leaves the hospital. In such cases, there are additional hospital documents referring to this visit marked as a continuation of the main document.

Each visit concerns one patient. Original documents contain patients' names and addresses. They had to be removed before making the documents accessible. We prepared a program based on the relatively strict structure of the files—personal information is given in the first lines of the documents. Information about names and addresses are substituted by symbolic identification codes. The same code is assigned to the same name and address, so as far as a patient has the same address, it is possible to link their documents.

Most information is given as free-form text but some data is written in table forms, like results of blood tests or lipid profile tests. A document starts with the identification number of the visit and the patient, the age of the patient, and date of the visit in hospital. The following information is then given in short form: past and current important diseases; diagnoses; patient's health at the beginning of the hospitalization. An example of this part of a document is given in (3).

(3)   *KARTA INFORMACYJNA (nr księgi głównej 133831)*

Pan d2005-1_ 107 lat 56
Przebywał w klinice od dnia 21.10.2005 do dnia 29.10.2005 r.

Rozpoznanie i wyniki badania klinicznego:
Cukrzyca typu 1, o chwiejnym przebiegu, powikłana retinopatią prostą,
neuropatią obwodową i autonomiczną.
Choroba zwyrodnieniowa kręgosłupa.
Choroba niedokrwienna serca pod postacią przebytego zawału w 1992 r.

INFORMATION CARD (nb 133831)

Mr. d2005-1_ 107 56 year old
He was admitted to hospital from 21.10.2005 to 29.10.2005

Clinical diagnosis:
Type 1 diabetes, uncontrolled, with simple retinopathy,
with peripheral and autonomy neuropathy.
Degenerative spine disease.
Coronary disease—myocardial infarction in 1992

After this initial data, a document contains results of examinations e.g.: basic data like height, weight, BMI (Body Mass Index), blood pressure; an ultrasound checkup of abdominal cavity; an ophthalmology examination; blood tests, lipid profile tests. Some of them, like radiology, ultrasound or ophthalmology examinations, are described in free-form text, the other results like biochemical tests are available in table form. This part of a document may also contain descriptions of attempts to select the best treatment for the patient. The most important part of the document starts from the word *Epikryza* 'Discharge abstract'. Its length is about half a page of text. It contains:

– Data about a patient's diabetes like: the type, if the illness is balanced, if
  the patient had incidences of hypoglycaemia, when diabetes was diagnosed.
– Description of diabetic complications, and other illnesses.
– Short description of examination results and surgical interventions.

- Information about education, diet observed, self monitoring, patients reactions, and other remarks.
- All recommendations including diet and treatments schema with insulin type, doses and other oral medications. This data is usually given in a template form, see example (4).

(4)   *Dieta cukrzycowa 1800 kcal. 6 posiłków/dobę spożywanych regularnie.*
'Diabetes diet 1800 kcal. 6 meals/day consumed regularly'
Leki: Insulina: 'Medications: Insulin:'
R 32 j. Mixtard 30   'Morning 32 units Mixtard 30'
P 11 j. Actrapid   'Midday 32 units Actrapid 30'
W 18 j. Mixtard 30   'Evening 32 units Mixtard 30'
Metformax 2 x 850 mg. na 30 min. przed posiłkiem
'Metformax 2 x 850 mg, 30 minutes before a meal'
Enarenal 2 x 1 tabl. a 10 mg.   'Enarenal 2x 1 tablet per 10 mg.'

Information given in free-form text can be expressed in many ways. For example the information when the diabetes was diagnosed can be expresssed: in words—*wieloletnia* 'long-standing'; as a date—*w 1990 roku* 'in the year 1990'; relatively—*20 lat temu* '20 years ago'; or *w 20 roku życia* 'in the 20th year of life'. All these expressions can be used in different contexts. For example the adjective *wieloletni* 'long-standing' can be used in the phrase *wieloletni pacjent szpitala* 'a long-standing patient of the hospital' and in phrase *wieloletnia cukrzyca* 'long-standing diabetes'. Similarly, the word *niekontrolowana* 'uncontrolled' may denote the very important feature of a diabetes *niekontrolowana cukrzyca* 'uncontrolled diabetes' as well as other uncontrolled processes like put on weight or changes of blood pressure. So, in the case of these texts, context is quite often important in the interpretation of data.

As we have already written, these texts are official documents. Therefore, they are typed more carefully, with a spelling correction, in comparison to the mammography notes. In these texts, errors are observed mainly in words that are not included in MS Word's dictionary. These are specialist medical notions: *glikemia* 'glycemia'; medication names: *Actrapid*; diseases e.g., *makroangiopatia* 'macroangiopathy'; Latin terms: *rethinopatia diabetica* 'diabetic rethinopathy'. The second group of errors are misspelled words that in result give correct other words, most of these errors are caused by the lack of Polish diacritics. For example *zawal serca* where *zawal* is an imperative of the verb 'collapse', instead of *zawał serca* 'heart attack', or ungrammatical but correct form of the same word e.g., phrase *pacjent z cukrzyca$_{nominative}$* instead of *z cukrzycą$_{instrument}$* 'patient with diabetes'. Similarly to mammography notes there are quite many punctuation errors like a datum written with commas instead of dots *25,07,2005*, or a lack of space between a number and a word like *3posiłki* '3meals'. Punctuation errors, lack of spaces and misspelled medical notions influence the process of terminology extraction and the quality of our information extraction application, especially its recall measure.

Table 2. Diabetic data set characteristics

	ALL		TRAIN SET		TEST SET		CONTROL	
	types	occurr.	types	occur.	types	occur.	types	occur.
records	773	–	442	–	169	–	162	–
basic forms	6896	–	4794	–	3759	–	3890	–
inflected forms	12847	334218	8503	182057	6377	84852	6646	67307
nonword tokens	20	261475	20	153542	20	68100	20	39833

3 Terminology Extraction

To ease the preparation of an ontology and to evaluate adopted solutions, apart from the manual domain model construction, we performed an automatic terminology extraction experiment. We analyzed a selected part of the data to extract the important nominal phrases which should be represented in a domain model. The terminology extraction procedure consisted of several phases: morphological tagging, simple phrase recognition and terms reranking using an algorithm proposed in [7]. First, the texts were morphologically analyzed and morphological tags were disambiguated. For this purpose, we used an existing tagger [18]. The tagger includes a guesser module, so its results do not contain 'unknown' part of speech tags.

The tagger results were further processed to extract simple nominal phrases. Noun phrases (groups) were defined by the means of the simple grammar given below (the initial symbol is NG, *subst* and *adj* denotes nouns and adjectives respectively, {n..m} means that the preceding symbol can occur from n to m times and the symbol '?' denotes optionality):

(5) NG1 → subst (adj(case=nom)){0..1} subst(case=nom)
     NG1 → subst(case=v1,gend=v2, nb=v3) adja interp
            adj(case=v1,gend=v2,nb=v3)
     NG2 → (adj(case=v1,gend=v2,nb=v3) interp(form=',')?) {0..3}
            NG1(case=v1,gend=v2, nb=v3)
            adj(case=v1,gend=v2,nb=v3){0..2}
     NG2 → NG1
     NG → NG2 (NG2(case=gen)){0..2}

The first rule allows for nominal modifications, the second one describes constructions like *utkanie tłuszczowo-gruczołowe* 'fat-glandular tissue' in which there is an agreement in case, gender and number between the first noun and the final adjective. The third rule allows for adjectival modifiers both before and after a noun. We allow commas between prenominal adjectival modifiers, while a comma occurring after a noun ends a group. The last rule describes the possibility of up to two nominal genitive modifiers.

Practical realization of this simple grammar consisted in applying the three elements' cascade of regular expressions responsible for recognizing word sequences deliverable from NG1, NG2 and NG symbols. As a result of applying the above rules we received a list of groups (simple phrases) with inflected words forms. On the list some groups occurred more then one time due to different case forms. To obtain only one form of each term we just form an "artificial canonical form" through word-by-word lemmatization (as lemmatization of Polish mulit-word names is complicated and requires an inflectional forms generator). To the obtained list of strings of words we applied the algorithm proposed in [8]. The experiments with two data sets differed slightly in details and will be described separately below.

3.1 Mammography Terminology

The data for the mammography terminology extraction experiment was the TRAIN data set characterized in Tab. 1. The data set was rather small, but as the subject of the texts was very limited, and they had common restricted purpose to describe results of one type of examination, they contained nearly only domain related phrases. Many word forms were repeated very frequently—50% of word forms belong to 23 types while 90% of them belong to 122 types. The number of types could have been even smaller as some types were incorrectly recognized by the tagger. There are two reasons for this fact. First, the guesser which is embedded in the tagger, for some unknown domain specific words, abbreviations, measurement units or test names, gives suggestions which are not Polish words at all. Second, the results of morphological disambiguation can be wrong and add a new type on the basis of a form which actually is a form of a different word. As this sometimes concerns words important for the domain description, and we did not have the possibility to change the tagger itself to improve its results, we decided to do some very simple postprocessing. Errors which were repaired, concerned domain related abbreviations as: *kgz* 'upper-outer quadrant', *usg, bci* 'thin-needle biopsy' which were treated as forms of 'guessed' nonexisting nouns. The second type of error concerned ambiguous wordforms of words which in texts from a chosen domain are always used in one sense, while in general the distribution of the senses is more uniform, e.g. *lewy* can be both a genitive form of the noun *lewa* ('trick') which is very unlikely to occur in mammography text and the adjective *lewy* ('left') which is quite common. As in Polish genitive modifiers are very likely to occur after the noun, the statistical tagger learned the strategy under which the sequence *sutek lewy* was tagged as it were *breast of the trick* instead of *left breast*. The next example is *sutka* which can be feminine form of the noun *sutek* ('nipple') but this form is not used by Polish physicians who prefer masculine form *sutek*. Unfortunately, a masculine plural nominative form, i.e. *sutki*, is identical to feminine singular genitive form. As the tagger preferred the feminine form, *sutki* were consequently annotated wrongly, which would have consequences in not recognizing the plural form as typical for this domain, we changed this feminine tag into the masculine in the appropriate number and case. In total, the automatic correction which preceded the term extraction procedure concerned 25 forms.

As the analyzed data practically contain texts which are only connected directly to mammography description, in this experiment we did not apply (like

[8]) "general vocabulary" filtering, i.e. removing words which are not typical for the selected type of texts. We observed that phrases with such "general" words, e.g. *prawa strona* ('the right side'), *bez zmian* ('without changes') have domain related meaning and should be taken into account.

The next phase of terminology extraction consisted in applying simple grammar rules (5) to extract nominal phrases. For the sake of improving precision, we changed the original formulation of the first rule describing nominal modification. It turned out that allowing for all nominal modifications resulted in retrieving many incorrect groups because of missing commas which would have separated subsequent noun phrases. As we observed that in our data nominal modification is practically restricted to examination types, we changed the original rule into the one below:

(6)   NG1 → subst (adj(case=nom)){0..1} subst(USG/PCI/RTG/MMG/BAC)

Extraction of noun groups (of a structure described above) resulted in 920 word-forms sequences of length from 2 to 6. After word by word lemmatization, this number was lowered to 714 (some terms were used in different forms, e.g. *badanie usg, badaniem usg, badania usg* are respectively nominative, instrumental and genitive form of the term *USG examination*).

The set was evaluated by two annotators. First results achieved are cited in Tab. 3. 87.66% of phrases were judged as describing domain related facts, e.g. *aktualny wynik badania usg sutków* ('current result of breasts USG examination'), *asymetria utkania gruczołowego* ('asymmetry of glandular tissue'), *badanie rtg* ('RTG examination'). About 7% of phrases were judged to be too general to be considered domain related terms, e.g. *brak informacji* ('no information'). 3.08% of phrases had to be rejected as they were built up from elements from more then one phrase, e.g. *rtg jakiego, rok USG* ('RTG which', 'year USG'). 1.12% of phrases were incorrect because of tagger errors (we do not count errors which were corrected before applying the term extraction algorithm). 8 phrases were built basing on incorrectly written words (although we corrected the texts, some orthographic and punctuation errors remained). The second annotator judged only 29 phrase types as too general (the inter-annotator Kappa coefficient was 0.7).

The data analyzed consisted nearly entirely of domain terms. Even such a simple identification of phrases could serve as the basis for defining a domain

Table 3. Phrase extraction results for mammography data

	types	%	occurrences	%
all extracted constructions	714	100	11704	100
domain related constructions	630	87.66	11178	95.28
too general constructions	50	7.01	499	4.26
constructions with incorrect structure	22	3.09	31	0.26
phrases incorrect due to tagging errors	8	1.12	15	0.13
phrases incorrect due to orthographic errors	8	1.12	8	0.07

vocabulary (87.66% phrases were judged to be important for the domain). To impose some order showing the degree of domain relatedness of particular phrases and to eliminate the least important ones we computed C-value proposed in [8]. After removing phrases for which the C-value was not grater than 1 from the original lists, we obtained 548 elements whose evaluation is given in Tab. 4. It can be easily observed that the improvement is not big and some of the important terms could have been eliminated. So, in this case the simplest way of selecting phrases turned out to be sufficient.

Table 4. Final terminology extraction results for mammography data

	types	%	occurrences	%
all extracted constructions	548	100	11524	100
domain related constructions	502	91.61	11098	96.23
too general constructions	23	4.2	394	3.42
constructions with incorrect structure	15	2.74	26	0.23
phrases incorrect due to tagging errors	6	1.09	13	0.11
phrases incorrect due to orthographic errors	2	0.36	2	0.02

The terms identified were used to annotate concepts in the previously built ontology, and to test their completeness. For this purpose we used term forms (830 forms in total). Case forms were contracted, but plural and singular forms were treated separately, as grammatical number can indicate important information (e.g. one quadrant or more, one calcification vs. many). In Tab. 5 we present the results of applying manually constructed IE grammar rules to the set of automatically derived terms. Newly identified concepts concern rarely occurring information (2% of phrases) such as *lack of previous mammograms, shadow of a pacemaker*, so the changes to the ontology would be minor. The results could have a bigger impact on the IE rules, as quite many previously unpredicted ways of expressing the identified concepts were found (5.6% of phrases). Nearly 84% of phrase occurrences were recognized properly.

Table 5. Comparison of manually defined phrases and automatically derived terms

	phrases	%	occurrences	%
recognized properly	561	67.6	7842	83.5
entire phrases recognized	336	40.5	6225	66.3
recognized properly on the basis of a substring	202	24.3	939	10.0
recognized only in a context	23	2.8	678	7.2
not properly recognized	171	20	704	7.5
unrecognized ways of expressing known concepts	128	15.4	512	5.6
phrases containing new concepts	43	5.18	192	2.0
too general phrases	45	5.8	783	8.39

3.2 Diabetic Terminology

As it was already described in subsection 2.2, the diabetic data differ from mammography notes in many aspects. The documents are longer and devoted to more topics. The experiment which we performed in this case had a little different aim. We did not try to extract all health relevant information included in the discharge records, but only those which are most important from the diabetes treatment point of view. We were interested in two different kinds of information. First, we needed administrative information which are typical for every patients' visit in hospital and their values are in general not connected to the diagnosis—name, gender, date of visit, or basic test values. The second group consists of information which although is in general present for all patients, but have different values in different patients groups, e.g. diagnosis, symptoms, names of medical procedures applied, names of medications used, complications, and disease specific types of treatment.

The diabetic terminology extraction experiment was performed on data from the years 2002-2005 (the TRAIN data set presented in Tab. 2). In this set there was a lot of numerical data—values of different tests or doses. As names of tests and measurement units were interpreted by the tagger as forms of non existing Polish words, we preprocessed the texts annotating all these fragments by special non-word tags. For this set, no other annotations made by the tagger were corrected. After tagging, recognition of phrases was performed using the same grammar rules as for mammography data, but in this experiment we limited nominative modifiers to cases given in (7).

(7)   NG1 → subst subst(alfa/beta/Hisa)

         NG1 → subst(lemma=lek) subst(case=nom)

Extraction of noun groups resulted in 5453 noun groups consisting of 2 to 7 words. 1835 phrases occurred more than 2 times. To eliminate phrases which do not characterize diabetic patients, we extracted word groups using the same grammar from a set of 162 documents created at the same hospital ward but for patients without diabetes (the CONTROL data set presented in Tab. 2). This experiment resulted in obtaining 3948 noun groups. Both lists were ordered using the algorithm of Frantzi and Ananiado [7]. Only terms which occurred more than 10 times and have C value of 2 or more were analyzed. The comparison of the remaining terms on the two lists was based on the frequency ratio. The "common list" contain 267 terms which are similarly distributed in both sets (frequency ratio less than 2) and 269 terms which were more frequent in the diabetic patient data. During evaluation we differentiated 5 types of terms. The distribution of these types in the two sets is shown in Tab. 6. Although the separation of concepts which are strictly related to diabetes was good, both sets include a lot of general terms. This is probably due to relatively small size of the control data set.

Table 6. Diabetic terms extraction results

		diabetic terms		common terms	
		nb	%	nb	%
1	badly formed phrases	11	4.1	5	1.9
2	too general or out of medicine domain	4	1.5	1	0.4
3	institutional patient care related terms	16	6.0	16	6.0
4	medical terms not directly connected with diabetes	169	63.3	242	91.3
5	diabetes and closely related phenomena	7	2.6	0	0.0
6	medical terms connected with diabetes	60	22.5	1	0.4
	diabetes related (5+6)	67	25.1	1	0.4
	general and non-diabetes terms (3+4)	185	69.3	258	97.4

4   Knowledge Representation by Means of Ontologies

Although a word 'ontology' has been used by the knowledge engineering community frequently for at least twenty years, it still raises many controversies as its meaning remains rather vague. But even if we omit the philosophy of being and existence, and concentrate on using the word within a knowledge representation context in which it denotes an object, the concept of an ontology remains unclear. For example, [11] lists 6 different meanings of this word and concludes the discussion on this subject with three meanings for the formal semantics domain. One of the most popular (i.e. most widely cited) definitions is this of [9] according to which an ontology is an *explicit way of conceptualization.*, where *conceptualization* is meant as *the objects, concepts, and other entities that are presumed to exist in some area of interest and the relationships that hold among them.* A new version of this definition given in [10] is more technical and states that: *An ontology defines a set of representational primitives with which to model domain of knowledge or discourse. The representational primitives are typically classes (or sets), attributes (or properties), and relationships (or relationships among class members).* A recent summary of different views on this problem can be found in [12].

An ontology is a resource built basing on the 'is-a' relation, i.e. it defines a hierarchy of classes in which every subclass represents objects/features which are specialized forms of the superclass (multi hierarchy, i.e. a hierarchy with classes having more then one superclass can also be modeled). For any class, a set of properties can be defined. Properties can have values being objects that belong to a class defined within an ontology (object properties), or have string or numeric values (data properties). One can also put constraints on the particular values (set of values) in the form of logical formulae.

Although the definition of what ontologies are, can be questioned and discussed, their role in knowledge representation is very important nowadays. Their development is tightly connected with Semantic Web which is an idea of semantically annotated WWW [2]. An important part of this idea is the establishment of shared ontologies describing knowledge concerning specialized domains. Till now, a lot of research concerning the building of various kinds of knowledge models has already been undertaken, and it might have seemed that finding one which can be adapted to the task at hand is relatively easy. Unfortunately, this is not true. Even in biomedical sciences—a domain in which studies concerning knowledge engineering were carried out very intensively—an appropriate resource may be unobtainable. First, from many existing resources only some can be named ontologies, many more are just terminological vocabularies, which contain special terms and their definitions, but do not organize them into ontological type-supertype hierarchies, e.g., SNOMED (http://www.ihtsdo.org/snomed-ct/) or UMLS (http://www.nlm.nih.gov/research/umls/).

A resource of great relevance to one of the chosen tasks is *Breast Cancer Image Ontology (BCIO)* which is the result of the MIAKT project (http://www.aktors.org/miakt). Unfortunately it is not publicly available, and the report [4] does not include many details. The huge (36,000 concepts) *NCI Thesaurus* (http://nciterms.nci.nih.gov/NCIBrowser/) does not cover mammography examination details, while *Basic Clinical Ontology for Breast Cancer* (acl.icnet.uk/~mw) although devoted to breast cancer, does not include concepts connected with mammography result descriptions. Additionally, some concepts included there, are inconsistent with Polish medical tradition. Another resource which we look into, was the publicly available ontology— *Foundational Model of Anatomy (FMA)*—built at the University of Washington (http://sig.biostr.washington.edu/projects/fm/index.html) which includes many concepts concerning human anatomy but is far too big to embed in our model.

The first version of a Polish mammography ontology, was based partially on BI-RADS and presented in [19]. The mammography part of our ontology, formulated in OWL DL standard, was inspired by that Polish ontology. The first version was described in [17].

For the second application chosen, finding an existing appropriate ontology was rather unlikely, as in this task we deal with many subdomains. In this case, from the beginning we decided to build our own dedicated resource. Another reason for this decision was the fact that even if there exists an ontology close enough to be incorporated into a given project, it would be hard to adapt and this process is not at all straightforward. [3] describes an experiment with a comparison of two ways of creating ontologies—developing a resource by a team of ontology engineers (with collaboration with domain experts when needed) and adopting an ontology which covers a similar domain. The experiment ended with the conclusion that the advantages of using already existing ontologies were not so great, and the work took even more time than would be needed for defining a new resource.

Defining an ontology requires undertaking several decisions which influence an ontology structure and its possible reusability. It is trivially true that defining a domain model one should take into account existing knowledge and resources. In practice, it quickly becomes clear that possible sources of information are numerous and frequently inconsistent. In the case of medical applications, knowledge can come from handbooks but also from specialists' experience, established procedures or can be inferred directly from existing data. These different sources can define concepts more or less formally, on different levels of details, from various points of view. Our ontology was planned to contain only information which occurs in real clinical data. In order not to overload it with unnecessary information, we built the resource basing mostly on sample data. The second source of information was expert knowledge. The granularity of description is different for different description areas—very detailed for mammography picture features but very general in anatomy part (except breasts description).

In the next section we present the basic features of our ontology defined as an OWL DL file using the *Protégé* ontology editor (http://protege.stanford. edu/). In the paper we briefly describe the highest levels of the ontology.

5   The LightMedOnt Ontology

The ontology represents concepts which are important to the task of analyzing selected types of clinical data. It covers different areas of knowledge: human anatomy, medical procedures and diagnoses, general physical concepts like shape, dimension or time, and the results of their comparisons. According to this division, the ontology comprises five main classes: *Hum(an)Anatomy*, *Medicine*, *PersonDom*, *PhysicalFeature* and *P(hysical)F(eature)ComparisonResults*. The top level of the class hierarchy is given in Fig. 1.

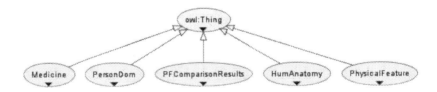

Fig. 1. Main classes of the ontology

In the following subsections we describe three of the five top classes of the ontology. The remaining two classes represent: *PersonDom*—person identification data, *P(hysical)F(eature)ComparisonResults*—classes useful for representing comparisons concerning: similarities, sizes, levels and cardinalities.

5.1   Human Anatomy

Nearly all medical ontologies contain a representation of human anatomy. Our human anatomy hierarchy is not a copy or a replacement of a full anatomy description (like e.g. FMA), but contains only those data which are important to our domains. When projections of ontologies will be possible, such small ontologies may no longer be necessary. But for now, using an ontology with thousands of concepts for a task which needs just a few of them, is not a good solution, especially when one has to learn the entire ontology to make sure which concepts to use.

Fig. 2. Hierarchy describing human tissue character

The anatomical part of the ontology comprises two classes: *HumanBodyParts* and *HumanTissue* divided into *TissueType* and *TissueCharacter*, see Fig. 2. *HumanTissue* is meant to describe all human tissue, but at the moment includes 9 *BreastTissue* subtypes. The first part of the name describes which type of tissue is dominant. *TissueCharacter* concept describes various types of irregularities which can be observed within tissue structure.

The *HumanBodyPart* class includes descriptions of body parts divided into those which occur only once, those which can be numerous, and those occurring symmetrically. In the mammography domain, we only considered the concepts which occurred in the analyzed documents.

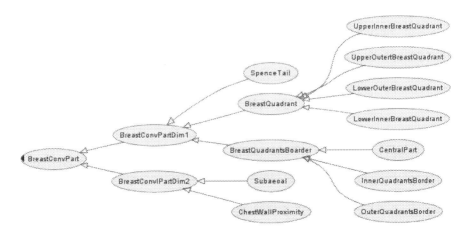

Fig. 3. Description of breast division

Although we would like to make the ontology simple, we do not want to loose any information which is usually given in mammography reports. Thus, apart from body part names, we introduce conventional names of body subparts, which are traditionally used for more precise localization of the findings (e.g. upper right quadrant), in the hierarchy rooted at *ConvHumanBodyPart*. We introduced two such conventional breast divisions: in one dimension it is a division into breast quadrants, and in the orthogonal dimension into a depth closer to the nipple or closer to the chest, see Fig. 3.

5.2   Medicine

The most important part of the ontology is located in *Medicine* hierarchy which describes various aspects of medical knowledge and procedures, see Fig. 4. The subclasses are: *AnatomicalPathology, MedicalProcedure* with *Med(ical)Examination* subclass, *DiseasesOrSymptoms, Med(ical)Judgment. Drug and Diet* class contains treatment and diet schemas and medications. The ontology covers only the relevant subsets of the chosen medicine subdomains. Although in our ontology we try to represent data which really occur in the selected type of clinical text, we have to be careful when analyzing some simplifications which often occur in less formal texts. An important example of this kind of situation is the description of pathological changes observed on the mammography images. Theoretically, radiologists should first describe what they see exactly, e.g. a darkness or an irregularity. Only after formulating a description of a picture using objective terms, they should interpret it and give their judgments. But in practice, physicians who are pretty sure about what they see, just give their interpretation. To cover both levels of description, we defined two mian classes: *AnatomicalPatInterpretation* included in *AnatomicalPathologhy* hierarchy, and *AnatomicalFinding* defined as a *MedJudgement* subclass.

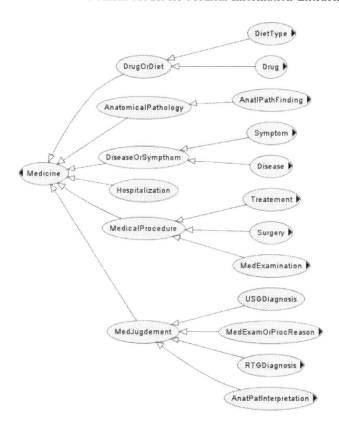

Fig. 4. Medicine hierarchy

The next two important areas of medicine which are partially covered in the ontology are medical examinations and diseases. *MedExamination* class includes two traditionally distinguished types of direct and additional examinations. *DiseasesOrSymptoms* subclass covers diseases and symptoms related to diabetes, see Fig. 5. It includes: three types of diabetes, diseases that are typical complications of diabetes, and symptoms indicating diseases.

5.3 Physical Features

Apart from the description of the strictly medical subdomains, some more general concepts have been defined within our ontology. A significant problem was to chose the way of defining concepts which are relevant not only to the medical domain. To enable further ontology development we had to foresee different usage of the same terms in different contexts. The chosen solution was to define two sets of features: the first one includes general features and the second one describes features with special meaning in the medical domain (some of them can also be in the general hierarchy, but it is not shown here).

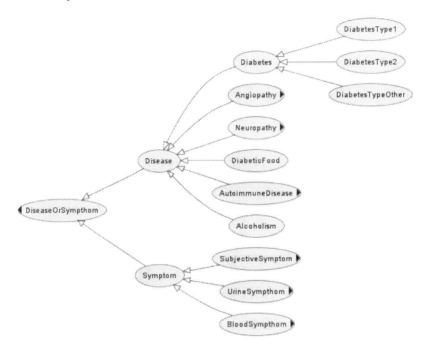

Fig. 5. Diseases and symptoms hierarchy

General features are described in two subclasses of *PhysicalFeatures*: *General-PhysicalFeature* and *MeasurementUnit*. The first subclass covers such physical features as: size, contour, aggregation, density, projection, regularity, cardinality, quantity, shape, side and time. The second one contains typical units for measuring length, frequency, time, weight, and medical units (like pills). The *Time* ontology is very simple and covers only those cases which occur in the analyzed documents. It covers: periods of time in years, months and weeks, precise and imprecise dates, and also repetitive constructions like *every year*.

The physical features hierarchy with concepts which have special meaning in the medical domain (*MedicalPhysicalFeature* class) is shown in Fig. 6.

5.4 Properties

Defining a taxonomy for a domain is the most important task in creating the domain ontology. However, it is also important to connect classes with appropriate relations. In OWL, properties serve this purpose. In our ontology pathological findings have the greatest number of properties defined for their individuals. For describing anatomy, the most commonly used property is a relation of being a part of some bigger structure (*isPartOf* and the inverse relation *hasAsPart*). For other classes, properties describing a relation of having a feature are common, e.g. *hasSex* property defined for individuals belonging to *Person* class.

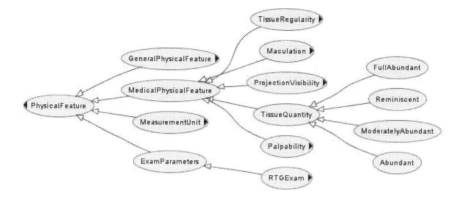

Fig. 6. Physical features hierarchy

Table 7. Properties of *AnatPathFinfing* class

AnatPathFinding		
hasRTGDiagnosis	Instance	RTGDiagnosis
hasLocaalization	Instance	MedLoc
hasAccompFinding	Instance	Calcification
hasSize	Instance	Size
hasShape	Instance	Shape
hasContour	Instance	Contour
hasInterpretation	Instance*	AnatPatInterpretation
HasMultiplicity	Instance	Quantity
hasSaturation	Instance	Saturation
hasAppendices		Boolean
hasAppendicesOfShape	Instance	Shape
HasPalpability	Instance	Palpability

An example of a class with defined properties is given in Fig. 7. It shows 12 properties which are defined for instances of *AnatPathFinding* class together with their values. Of these properties only *hasInterpretation* can have multiple values.

6   Using Ontology in IE Applications

In this section we discuss relationships between OWL DL ontologies and the domain model of SProUT extraction system [5], used for our IE experiments. We describe what elements of an OWL ontology can be easily translated into a SProUT domain model, and define the appropriate translation rules. OWL is a powerful formalism, and only a subset of facts and relations defined in an ontology can be represented by a model defined within SProUT. Apart from dissimilarities in their theoretical background, some differences are a result of the difference in the purposes they serve. The LightMedOnt ontology was created to represent all relevant domain knowledge, while the SProUT model was designed specially for the purpose of IE.

In SProUT, the domain model is represented by a multi-hierarchy of TFSs (typed feature structures, [6]). Every TFS has a name of a type assigned and a set of features (attribute names and values). A feature's value can be an atomic type, another TFS, or a list of atomic types or TFSs. Each type structure corresponds roughly to a class from the ontology, whereas attributes correspond to properties of the class.

TFS structures define templates for extraction. Let us consider an example representing a dose of insulin treatment, see (8). The structure is of *insulin_treat_str* type and consists of three attributes. The first one I_TYPE has value of the *insulin_t* type which has 36 subtypes representing insulin medication. The next two parameters DOSE_MIN and DOSE_MAX indicate the minimal and maximal units of insulin within a dose, and have values of the type *string*. A whole treatment usually consists of several doses.

$$(8) \quad \begin{bmatrix} insulin\_treat\_str \\ \text{I\_TYPE} & insulin\_t \\ \text{DOSE\_MIN} & string \\ \text{DOSE\_MAX} & string \end{bmatrix}$$

An OWL ontology consists of individuals, which represent objects from the ontology domain, classes, and properties. Classes are defined as a set of conditions describing the requirements imposed on individuals which can belong to them. OWL properties are relations between individuals. Hence, OWL ontology is designed not only for creating domain descriptions, but also as a way of storing data.

In the SProUT domain model, an important notion is "type", the unification operation is defined on types. Usually, TFSs are used to represent pieces of information, that are further processed to determine complex relationships between them. In SProUT, there are no special mechanisms for representing individuals and storing data. The extraction results are an XML file which can be stored in the original form or be further processed.

6.1   Classes

OWL classes consist of individuals that fulfill conditions imposed on class members. Classes are organized in a superclass-subclass hierarchy: subclasses are subsumed by their superclasses.

Classes defined on the highest level in OWL, are represented in SProUT hierarchy as subtypes of the special **top** type, referring in OWL to the *owl:Thing* class. An OWL superclass-subclass relation of *Class* and *SubClass*, in SProUT is represented as in (9), in the case when there are no properties defined for individuals belonging to *SubClass*. A representation of classes with properties defined for their individuals is described in section 6.2.

(9)   *SubClass*:< *Class.*

In Protégé 3.4 OWL, class names cannot be repeated in two different places of the hierarchy, so to represent the idea that a class is a subclass of two classes, equivalent classes are used. For example if *C1* and *C2* are classes with a common subclass, it is necessary to introduce two subclasses *Sub_C1* and *Sub_C2* of respective classes, and to define them as equivalent. SProUT allows for the direct definition of a multihierarchy, so a type can be a subtype of two (or more) types. The relation that *Sub_C1* is a subtype of *C1* and *C2* is represented in (10). The type *Sub_C2* is unnecessary, moreover it is not possible to represent the above relation in a different way. Therefore, while equivalent OWL classes are translated into SProUT domain model, only the name of one of them can be represented in the hierarchy. In SProUT, a type can be defined as a subtype only once, so the relation (10) cannot be split into two simpler relations as in (9).

(10)   *Sub_C1* := *C1* & *C2.*

OWL classes are assumed to be overlapping. If we want to separate classes, so individuals can belong to only one of them, we have to declare the classes as disjoint. In SProUT we do not consider whether TFSs are disjoint or overlapping but only the possibility of their unification. Two types that are not in any subtype relation (direct or indirect) might be considered as disjoint unless they have a common subtype.

6.2   Properties

Apart from a class hierarchy, in OWL it is possible to define properties that represent relationships between individuals. Properties are defined for a set of classes (property domain) and can have values taken from another set of classes (property range). Individuals belonging to the domain of a property are in relation with individuals from the range. OWL properties roughly correspond to attributes of TFSs in a SProUT type hierarchy, but not all types of properties can be represented in the SProUT domain model.

At the beginning, let us consider a property of the *functional* type. Such a property can relate an individual to one other individual at most. It is translated into an attribute of the TFS, whose type refers to the domain class of this property. Suppose *Class* with a subclass *Class_with_property*, for whose individuals *functional* property *Prop* is defined, and the range of the property is *Class_value*. In SProUT, it is represented as unification of: the predefined structure **avm** (indicating usage of attribute-value structure); the type referring to the superclass; and the attribute structure with PROP attribute whose value is of *Class_value* type, see (11). If there are more properties defined for this class, the appropriate TFS will have more attributes.

(11)   *Class_with_property* := **avm** & *Class* & [PROP *Class_value*].

There are several restrictions imposed on the type hierarchy in SProUT. For example, *Class_ value* cannot be in any supertype-subtype relation with *Class_ with_ property* type—cycles in type definitions are not allowed. Moreover, attribute names have to be unique within a type system. Let us consider an example of *functional* property *Prop* with the range of *Class_ value*, and the domain consisting of more than one class, e.g., *Dom_ C1* and *Dom_ C2*. In this case there are two solutions. The first one is to introduce a supertype— *Dom_ Class* for types referring to these two domain classes, and define the attribute property for this supertype, see (12).

(12)   *Dom_ Class* := *avm* & [PROP *Class_ value*].
        Dom_ C1 :< *Dom_ Class*.
        Dom_ C2 :< *Dom_ Class*.

Another solution is to introduce different names of attributes representing the property (e.g.: PROP1, PROP2) to types referring to classes from its domain, see (13).

(13)   *Dom_ C1* := *avm* & [PROP1 *Class_ value*].
        Dom_ C2 := *avm* & [PROP2 *Class_ value*].

When the range of a property consists of several classes, it is necessary to introduce in the SProUT type hierarchy, a supertype of types referring to all range classes.

When an OWL property has the inverse property defined, we have to chose which one we want to represent in the SProUT type system. Both properties cannot be represented, because we would get a definition of type system with cycles, which are forbidden in SProUT. For the same reason, *symmetric* and *transitive* properties cannot be represented in the SProUT type system.

So far we have considered *functional* properties, but OWL assumes that properties have more than one value unless we state otherwise. All properties that are not *functional*, in SProUT should be represented as attributes which have lists of TFSs as value. A list can be represented in SProUT with the help of the special predefined type *cons*, see (14).

(14)   *cons* := *list* & [FIRST *top*, REST *list*].

So *Class* with a subclass *Class_ with_ property*, which has a non-*functional* property *Prop* with the range *Class_ value*, can be represented in SProUT as in (15). The property is represented as the attribute PROP with the value defined as the list consisting of *Class_ value* type elements.

(15)   *Class_ with_ property* := *avm* & *Class* & [PROP *Class_ value_ list*].
        Class_ value_ list := *cons* &
                                [FIRST *Class_ value*, REST *Class_ value_ list*].

If we don't want to operate on lists, and the cardinality of the value set is known, it is possible to introduce one attribute for each value. For example, if a recommended diet consists of maximally 6 meals, it is possible to represent them not as a list but as a TFS structure with 6 attributes [MEAL1 ... MEAL6].

The expressive power of OWL properties is much stronger than the capability of SProUT's type hierarchy. Many restrictions on properties that can be expressed in OWL cannot be represented in SProUT, only those described above are straightforward to translate.

6.3 Pragmatic Decisions

For the task of information extraction, we prepared two separate systems: one for processing mammogram reports, and the other for analyzing discharge documents of diabetic patients. Whereas it is useful to design one ontology describing patients, medical examinations and diseases, it is reasonable to select only important parts of such an ontology for systems processing specific texts.

While the domain models of the extraction systems were created, many decisions where imposed by the task requirements. For example, as our SProUT grammar process separated documents, we know that all data extracted from one document concerns one visit of one patient. In the OWL ontology we assume the representation of several patients, examinations etc., so all information concerning a patient should be connected to that patient with the help of properties. As result, in the OWL ontology we have many properties that are not necessary in domain models of our extraction systems. Let us consider extraction of BMI level from discharge documents. We know that it concerns the patient whose discharge document is currently processed, and that this BMI was valid during the patient's visit in hospital. In the OWL ontology we have to create properties connecting this BMI level, both to the appropriate patient and to his/her hospitalization. These two properties are not needed in the domain model of our extraction system.

SProUT uses shallow parsing, so it extracts pieces of information. They are connected into bigger structures in the postprocessing phase. Let us consider an example from the system processing mammography reports. In mammogram reports localizations are described and can refer to tissue description or lesions. During the extraction process we do not have enough information for connecting these pieces of information, so the decision on how to connect them is undertaken in the postprocessing phases, see [13] and [14]. Therefore, TFSs representing tissue description and lesion description do not contain localizations, while in the OWL ontology both appropriate classes (representing a lesion and a tissue) have assigned localizations.

7   Summary

Automatic processing of natural language clinical data can be of great value for medical research. For example, IE applications can serve as an efficient method of identifying the values of interesting features. The quality of the results of such an application is highly dependent on the quality of a domain model used for describing the data. In the paper, we presented our approach to the task of domain model construction for two medical document types: mammography reports and hospital records of diabetic patients. The XTDL models for IE applications and the OWL ontology common to both domains, were created manually on the basis of expert knowledge and sample data. These two applications have different underlying assumptions. In the case of mammography reports, we extract and therefore describe, all information contained in texts. In the case of hospital discharge documents, we extract (and describe) only information indicated by experts.

For both domains we performed terminology extraction experiments to check if phrases typical for the domain terminology are represented in our ontology. The experiment for the mammography domain showed that our domain model was thoroughly constructed. There were only a few cases of concepts which were not predicted by our model, e.g. we did not define a pacemaker among interpretations of changes visible on mammography pictures. Moreover, in our domain there were many concepts that may appear in the data but they are very rare. In fact, for 66 attributes from the mammography domain model defined for the IE application, 14 did not appear in our test set, another 14 appeared less than 10 times. In the case of diabetic domain it is difficult to discuss its completeness because we were interested only in information indicated by the experts, so only it is described in the model. The results of terminology extraction was also less clear because of a broader scope of the documents—they describe all diseases and exams of a patient, not only those connected directly to diabetes. Because of that, extracted terminology contained many phrases unimportant in the context of the application.

The description of the IE applications based on this model and their evaluation is presented in [16]. For all attributes recognition, we obtain F-measure 99.58% for mammography domain, and 97.86% for diabetes domain. These results show that in the case of complicated structures of extracted information and rather small corpora of available texts, our approach to domain model construction and information extraction is effective.

Acknowledgments. We would like to express our appreciation to Teresa Podsiadły-Marczykowska and Roman Kuczerowski. Work on this paper would not be possible without their help with medical data interpretation.

References

1. Becker, M., Drożdżyński, W., Krieger, H., Piskorski, J., Schaefer, U., Becker, F.X.: SProUT — Shallow Processing with Typed Feature Structures and Unification. In: Proceedings of ICON 2002, Mumbai, India (2002)
2. Berners-Lee, T., Hendler, J., Lasila, O.: The Semantic Web. Scientific American (May 2001)

3. Bontas, E.P., Mochol, M., Tolksdorf, R.: Case studies on ontology reuse. In: Proc. of 5th International Conference on Knowledge Management (I-Know'05), Graz, Austria (2005)
4. Dasmahapatra, S., Dupplaw, D., Bo, H., Lewis, H., Lewis, P., Shadbolt, N.: Facilitating multi-disciplinary knowledge-based support for breast cancer screening. Int. J. of Healthcare Technology and Management 7, 403–420 (2006)
5. Drożdżyński, W., Krieger, H.-U., Piskorski, J., Schäfer, U., Xu, F.: Shallow Processing with Unification and Typed Feature Structures — Foundations and Applications. German AI Journal KI-Zeitschrift 01/04 (2004)
6. Emele, M.C.: The typed feature structure representation formalism. In: Proceedings of the International Workshop on Sharable Natural Language Resources, Ikoma, Nara, Japan (1994)
7. Frantzi, K., Ananiado, S., Mima, H.: Automatic recognition of multi-word terms: the C-value /NC-value method. International Journal on Digital Libraries, 15–130 (2000)
8. Frantzi, K., Ananiadou, S., Mima, H.: Automatic recognition of multi-word terms: the C-value/NC-value method. International Journal of Digital Libraries (2000)
9. Gruber, T.R.: A translation approach to portable ontology specifications. Knowledge Acquisition 5, 199–220 (1993)
10. Gruber, T.R.: Ontolgy. In: Ozsu, M.T., Liu, L. (eds.) Encyclopedia of Database Systems, Springer, Heidelberg (2009)
11. Guarino, N., Giaretta, P.: Ontologies and knowledge bases: Towards a terminological clarification. In: Mars, N.J.I. (ed.) Towards Very Large Knowledge Bases, IOS Press, Amsterdam (1995)
12. Guizzardi, G., TerryHalpin: Ontological foundations for conceptual modelling. Applied Ontology, 1–12 (2008)
13. Marciniak, M., Mykowiecka, A., Kupść, A., Piskorski, J.: Intelligent content extraction from Polish medical reports. In: Bolc, L., Michalewicz, Z., Nishida, T. (eds.) IMTCI 2004. LNCS (LNAI), vol. 3490, pp. 68–78. Springer, Heidelberg (2005)
14. Mykowiecka, A., Kupść, A., Marciniak, M.: Rule-based medical content extraction and classification. In: Intelligent Information Processing and Web Mining Proceedings of the International IIS: IIPWM'05, Springer, Heidelberg (2005)
15. Mykowiecka, A., Marciniak, M.: Domain-driven automatic spelling correction for mammography reports. In: Intelligent Information Processing and Web Mining Proceedings of the International IIS: IIPWM'06. Advances in Soft Computing, Springer, Heidelberg (2006)
16. Mykowiecka, A., Marciniak, M., Kupść, A.: Rule-based information extraction from patient's clinical data. J Biomed Inform (in press), doi:10.1016/j.jbi.2009. 07.007
17. Mykowiecka, A., Marciniak, M., Podsiadły-Marczynkowska, T.: A "data-driven" ontology for an information extraction system from mammography reports. In: Proceedings of 10th Intl. Protégé Conference (2007)
18. Piasecki, M., Godlewski, G.: Reductionistic, tree and rule based tagger for Polish. In: Intelligent Information Processing and Web Mining Proceedings of the International IIS: IIPWM'06, pp. 531–540. Springer, Heidelberg (2006)
19. Podsiadły-Marczykowska, T., Guzik, A.: Mammographic ontology - conceptual model of the domain. The International Journal of Artificial Organs (2004)

A Survey of Text Processing Tools for the Automatic Analysis of Molecular Sequences

Andrzej Polański[1,2], Rafał Pokrzywa[1], and Marek Kimmel[3,1]

[1] Silesian University of Technology, Gliwice, Poland
andrzej.polanski@polsl.pl, pokrzywek@poczta.onet.pl,
[2] Polish Japanese Institute of Information Technology, Bytom, Poland
[3] Rice University, Houston, USA
kimmel@rice.edu

Abstract. Automatic analysis of molecular sequences is an interdisciplinary field of science, with many analogies to the methodologies of analyses and understanding of natural languages. In both these fields the object of the study has a complex, hierarchical character, which results from natural evolution. In this paper we have presented a survey of textual processing algorithms in the aspect of their applications to molecular sequences. We have shown methods for solving problems for exact and approximate searches of patterns in texts: aligning and block aligning of molecular sequences and analyzing molecular sequences by using indexed structures and transformations. We have covered some recent developments in these fields and we have provided some examples of inferring biological knowledge by using text processing algorithms for molecular sequences.

Key words: text processing, molecular sequences, suffix trees, suffix arrays, Burrows-Wheeler transform

1 Introduction

Automatic analysis of textual data is one of the major directions in the contemporary information sciences. There is vast literature, including monographs, journal and conference papers, devoted to many aspects of the construction of text processing algorithms, some examples are [32,9,10,11,18,27] and many references cited therein.

A very interesting area related to textual data is the analysis of written texts of natural languages. The analysis of natural texts is a multidisciplinary branch of science [4,5] located between information sciences and linguistics [20,2,3]. The most important issue in developing useful systems for the analysis of natural language texts is combining the linguistics knowledge concerning structures, grammars and semantics of natural languages with appropriate mathematical formalisms and computational tools. Natural communication has undergone selective evolution which led to the complexity and structure of natural languages. Due to the complexity and hierarchical arrangement of the information passed in natural communication, areas of research into the analysis of written natural

M. Marciniak and A. Mykowiecka (Eds.): Bolc Festschrift, LNCS 5070, pp. 359–378, 2009.
© Springer-Verlag Berlin Heidelberg 2009

language have spilt into several fields, such as concording and related transformations of the textual data, frequency analyses, statistical analyses, modeling of natural languages with grammars, syntactic and semantic analyses of texts, contents analyses, retrieval of information from natural text messages. Each of the listed research fields has its specific problems and uses specialized tools to formulate and solve them. Algorithms for the automatic analysis of natural languages developed by scientists aim at capturing the true structures and complexities of languages.

Among many areas of research in text analysis, a new, interesting branch is expanding in the area of the analyses of molecular sequences. Molecular sequences are strings (words) over alphabets defined by symbols representing nucleotides in DNA, amino acids in primary structures of proteins or ribonucleotides in RNA. The problems of constructing algorithms for their analysis belong to the new interdisciplinary field of bioinformatics. The rapid advances in bioinformatics involving developing methodologies for the construction of algorithms for molecular sequence processing is motivated by the spectacular increase in the size of bioinformatic data repositories available in the Internet and by the need for introducing structure in these data and for retrieving useful information, on the basis of these data sets. There is one very strong common element between analyses of natural languages and analyses of molecular sequences, namely the interdisciplinary character of both these fields. Analogously to the processing of natural languages, developing methodologies for the analysis of molecular sequences requires combining the knowledge of several disciplines of science, genomics, genetics, proteomics, molecular biology, mathematical modeling and computational algorithms.

The goal of this paper is to survey some of the problems of molecular sequence processing and the methods developed for their solution, including recent advances in the field. Along with presenting and characterizing some of the problems which arise in the area of molecular sequences processing, and showing what text processing tools allow us to solve these problems, we also try to specify what biological knowledge we gain from these solutions. Methodologies of text processing are presented, their computational complexities are characterized and examples of their applications to the analysis of textual data files in bioinformatics and molecular biology are shown.

2   Analysis of Molecular Sequences

A substantial fraction of text processing methodologies used in the area of molecular sequence processing are standard text processing algorithms. These methodologies still belong, at least partly, to the field of bioinformatics and molecular biology since formulating text processing tasks and using the results of applied algorithms requires implementing biological knowledge. The area of processing molecular sequences also has many specific properties, which lead to the need of elaborating methodologies dedicated to special applications. Some important properties of molecular sequences analysis are: (i) The size of the data, which largely exceed typical data sizes in other areas of text processing. (ii) Special alphabets, in genomics the alphabet consists of four letters a, g, c, and t, coding for four different nitrogen bases, adenine, guanine, cytosine and thymine, present in the DNA polymer. A similar alphabet defines contents of RNA, except that thymine (t) is replaced by uracil (u). In proteomics the alphabet has 20 letters representing 20 amino acids used to build the primary structures of proteins. (iii) Special forms of text precessing problems, such as problems of searching for different types of polymorphisms and mutations in DNA (e.g., single nucleotide polymorphisms or tandem repeats polymorphisms), problems related to using DNA sequences as markers (tagging regions in genomes of organisms, taxonomy of organisms, phylogenetic analyses) and problems of performing various sequence comparisons.

Below we show a collection of text analysis algorithms, which we have organized in the order of increasing complexity and which we have presented from the point of their application in such fields as genomics, genetics, proteomics and molecular biology. We have also grouped the algorithms into groups using the criterion of similarity between problems formulations and/or tools for their solution.

2.1   Comparisons of Sequences

Comparisons of sequences have a variety of applications in the analyses of molecular sequences. Two basic tasks are comparing two (or more) sequences obtained in biological experiments and comparing a sequence or sequences to the contents of databases. Different variants of sequence comparison problems arise, depending on lengths and numbers of sequences compared, and on assumptions such as fragmented or whole, or strict or approximate comparisons. Below we overview some of problems and algorithms.

String Search. The string-search problem is to find all occurrences of a string P, called the pattern, in a larger string S, often called the text. For example, the pattern $P = act$ occurs in the text $S = gaactgacta$ twice, at positions 3 and 7. An obvious, naive algorithm for comparing pattern P to text S goes through sequential comparisons of characters of P and S and sliding P along S. The comparison loop terminates on mismatch or after the whole match of a pattern P. In either case, a pattern P is then shifted one character to the right and the comparison loop is repeated from the left end of P. The process is repeated until the right end of P reaches the right end of S. The worst-case computational time of this naive algorithm is $O(nm)$, where m and n are respectively lengths of sequences P and S. The worst-case bound, $O(nm)$, is rather not encountered in practical string searches. Practically the computational time of the naive string search is closer to $O(n)$ than to $O(nm)$.

More effective methods of string searches are algorithms discovered by Boyer and Moore, [1] and by Knuth, Morris and Pratt [28].

The Boyer-Moore algorithm performs character comparisons between a pattern and a text from right to left, both strings are aligned on the left. It has a data-dependent structure in the sense that the order of operations performed depends on the characters in the pattern string P and on the results of comparisons between the characters in P and S. The speed of execution of the Boyer-Moore fast search algorithm, on average, as the length of the string P increases. Its average execution time is cn, where c is less than 1. For this reason, this method is called a sublinear string search algorithm [18].

The Knuth-Morris-Pratt algorithm uses information gained during previous character comparisons to improve the length of the shift. To accomplish this goal, the algorithm preprocesses a table of skips for each prefix of a pattern. The preprocessing is based entirely upon a pattern, so its complexity is not a function of the length of a text. The Knuth-Morris-Pratt algorithm is also a sublinear string search algorithm [18].

Approximate String Matching. The problem of approximate string searching is a generalization of the exact string-matching problem discussed above, and involves finding substrings of a text string close, in some sense, to a given pattern string. Approximate string matching is an important paradigm in the domain of information retrieval, speech recognition and especially molecular biology. More precisely, the approximate string-matching problem involves some distance between the pattern P and the text S. There are two widely used models for measuring the amount of difference between two strings: the Levenshtein distance [33] and the Hamming distance [19]. The Levenshtein distance, commonly known as edit distance, between two strings is given by the minimum number of operations needed to transform one string into the other, where an operation is an insertion, deletion, or substitution of a single character. The Hamming distance between two strings is given by the minimum number of substitutions required to change one string into the other, or the number of errors that transformed one string into the other. Depending on the model used to measure the distance between two strings the problem of approximate string matching falls into one of two categories. The k-difference problem is to find all occurrences of P in S that have edit distance at most k from P. The k-mismatch problem is to find all occurrences of P in S that has Hamming distance at most k from P. The case when $k = 0$ corresponds to the exact string matching.

The approximate string-matching problem can be efficiently solved by using the dynamic programming method for minimization of the edit distance. In order to use the idea of dynamic programming for approximate matching of strings S_1 and S_2 we introduce the partial cumulative score function $D(i, j)$ representing the minimum number of edit operations needed to transform the first i characters of string S_1 into the first j characters of string S_2. By edit operations we mean character substitutions, insertions and deletions of characters of strings. If the string S_1 has n characters and S_2 has m characters then the edit distance of S_1 and S_2 is given by the value $D(n, m)$. The value of $D(n, m)$ is implied by values of $D(i, j)$ for all possible pairs (i, j), where i ranges from zero to n and j ranges from zero to m. Filling out the array $D(i, j)$ for all possible pairs (i, j) is realized by using the recurrence relation defined by the appropriate formulation of the Bellman equation.

For example, if $S_1 = gagtaact$ and $S_2 = cagtaca$ then the edit distance of S_1 and S_2 is $D(8,7) = 3$. To transform S_1 into S_2 a minimum of 3 edit operations are needed: substitution of the first character in $S1(g \rightarrow c)$, deletion of the fifth character in $S1(a \rightarrow -)$, and substitution of the last character in $S_1(t \rightarrow a)$.

Alignment of a Pair of Sequences. ¿From the computational point of view, the problem of alignment of a pair of sequences is similar the problem of approximate string matching discussed above. It involves searching for substrings with a sufficient level of similarity between each other. Nevertheless we devote a separate subsection to its discussion due to (i) its special importance in bioinformatics, where many research paths start or depend on aligning molecular sequences, and (ii) the existence of specialized literature and nomenclature. We also discuss the problem of alignment in more detail and show its two basic variants, global and local alignment. We highlight some of the evolutionary models, which lie behind formulations of molecular sequence alignment problems.

Using different variants of alignments of two sequences can lead to the detection of their overlap or to identify that one sequence is a part of the other or that the two sequences share a subsequence. The importance of the alignment of of molecular sequences, stems from the fact that establishing correspondences between bases or codons of DNA or RNA strings, or between amino acids forming linear sequences in proteins, can be used for a variety of research purposes. This method can be used to find asimilarity between two DNA sequences resulting from the existence of a recent common ancestor, which these two sequences originate from. Computing distances between the aligned sequences, leads to inferences about the evolutionary processes they have gone through. This inference about the evolutionary process may involve estimating the time that has passed from the common ancestor to the present, but may also involve stating hypotheses concerning their evolutionary history or reconstructing a single evolutionary event in the past, or a sequence of them.

The alignment between two sequences is commonly represented symbolically, by printing two sequences one versus another. For example, the alignment of the sequences

$$s_1 = acctggtaaa \tag{1}$$

and

$$s_2 = acatgcgtata, \tag{2}$$

can be represented as follows:

$$
\begin{aligned}
s_1 &= a\,c\,c\,t\,g\,-\,g\,t\,a\,a\,a \\
    &\quad :\ :\quad :\ :\quad :\ :\ :\quad : \\
s_2 &= a\,c\,a\,t\,g\,c\ \ g\,t\,a\,t\,a
\end{aligned}
\tag{3}
$$

where the colon symbols indicate matches between bases. The symbol "$-$" called a gap, added to the alphabet of four bases, allows us to represent insertions and deletions (indels).

When aligning molecular sequences, a heuristic graphical methodology is often applied called dot matrices. For two sequences s_1 and s_2 of lengths n and m respectively we form a rectangular $n \times m$ matrix with rows corresponding to the characters in the first string s_1 and columns corresponding to the characters in the second string s_2, such that the order of characters is to the right and down. Then we place a dot in each matrix entry, where a base from s_1 matches a base from s_2. For sequences s_1 and s_2 given in (1) and (2) the dot matrix will be as shown below.

$$
\begin{array}{l@{\,}ccccccccccc}
 & a & c & a & t & g & c & g & t & a & t & a \\
a & \cdot & & \cdot & & & & & & \cdot & & \cdot \\
c & & \cdot & & & & \cdot & & & & & \\
c & & \cdot & & & & \cdot & & & & & \\
t & & & & \cdot & & & & \cdot & & \cdot & \\
g & & & & & \cdot & & \cdot & & & & \\
g & & & & & \cdot & & \cdot & & & & \\
t & & & & \cdot & & & & \cdot & & \cdot & \\
a & \cdot & & \cdot & & & & & & \cdot & & \cdot \\
a & \cdot & & \cdot & & & & & & \cdot & & \cdot \\
a & \cdot & & \cdot & & & & & & \cdot & & \cdot \\
\end{array}
\tag{4}
$$

We aim at detecting structural similarities between sequences using dots, but in the above dot matrix many dots are related to accidental matches between letters of the two strings. We can eliminate some of these by removing dots unlikely to represent a nonrandom correspondence between characters of the strings s_1 and s_2 with the use of some intuitive criterion. If we introduce the requirement that, in order for a dot to not be removed, there must be at least k neighboring matches along the right-down diagonal direction, this will then result in some of the random accidental matches being filtered out. If k is too small, many accidental matches will remain in the dot matrix plot. On the other hand, if it is too large, some of the true correspondences between strings may be unintentionally omitted. If we take $k = 2$ we obtain the filtered dot matrix shown below.

$$
\begin{array}{l@{\,}ccccccccccc}
 & a & c & a & t & g & c & g & t & a & t & a \\
a & \cdot & & & & & & & & & & \\
c & & \cdot & & & & & & & & & \\
c & & & & & & & & & & & \\
t & & & & \cdot & & & & & & & \\
g & & & & & \cdot & & & & & & \\
g & & & & & & \cdot & & & & & \\
t & & & & & & & \cdot & & & & \\
a & & & & & & & & \cdot & & & \\
a & & & & & & & & & & & \\
a & & & & & & & & & & & \\
\end{array}
\tag{5}
$$

which quite clearly indicates a reasonable alignment between s_1 and s_2 already depicted in (3). A reasonable alignment, close to optimal, can be found by tracing the path through the dot matrix covering as many dots as possible. Let us note that by filtering dots we have unintentionally removed one true correspondence from (3), between last characters in s_1 and s_2.

When understanding the alignment between molecular sequences as resulting from the evolutionary history we encounter there is a need for incorporating the knowledge on the evolutionary process of base or amino acid substitution, into the scoring functions for comparing alignments. The mathematical model of the evolutionary process, formulated as a Markov chain [14,40], will give us probabilities of observing substitutions between bases or between amino acids. Using these probabilities one can score alignments. One possible scoring function for alignments, using the maximum likelihood method can be the following

$$l = (p_0)^{n_m} (p_s)^{n_s} (p_g)^{n_g}. \tag{6}$$

The above scoring model is defined by probabilities of different events: p_0, a base does not change (a match occurs), p_s, a base is substituted by another one (a mismatch occurs); and p_g, an indel occurs. The value of the scoring likelihood function depends on the numbers n_m of matches, n_s of mismatches, and n_g of gaps. For example in (3) we have $n_m = 8$, $n_s = 2$, and $n_g = 1$.

Now, obtaining pairwise alignments between sequences can be realized by the dynamic programming algorithm, similar to the one described in the previous subsection, where minimization of the edit distance is replaced by maximization of the likelihood function (6). The Likelihood function (6) is most often replaced by its logarithm, log likelihood,

$$L = \ln(l) = n_m \ln p_0 + n_s \ln p_s + n_g \ln p_g, \tag{7}$$

which allows interpreting the value of the scoring function as the sum over matches, mismatches and gaps. The formulation of the alignment problem will use the scoring matrix, of the size $n \times m$, the same as in dot matrices (4) and (5). However, instead of dots, the matrices elements now contain scoring coefficients for matches ($\ln p_0$), mismatches ($\ln p_s$) and gaps ($\ln p_g$). Optimal alignment is obtained by tracing the path through the scoring matrix such that total log likelihood L is maximized. The dynamic programming algorithm using scoring functions of the type (7) is called the Needleman-Wunsch algorithm [39].

The Needleman-Wunsch algorithm leads to the global alignment of two sequences. When the aligned sequences differ substantially in their lengths or when we expect that only some fragments of them are likely to exhibit similarities, then intuitively the idea of global alignment does not work. One should allow for some modifications to the algorithm allowing for the ignoring of trailing gaps from both sides. Smith and Waterman[44] modified the Needleman and Wunsch method by allowing matches and mismatches between sequences to be scored locally. To achieve this, they introduced additional two rules in the dynamic programming iterations related to traversing the score matrix: (i) If an optimal cumulative score becomes negative, it is reset to zero, and (ii) The starting point of the alignment occurs at the largest score in the optimal cumulative score matrix. As an example [40], if the alignment of two sequences

$$s_1 = ttcgga \qquad (8)$$

and

$$s_2 = acgtgagagt \qquad (9)$$

is concerned, then with scoring coefficients 3 for matches, and -1 for both mismatches and gaps, the application of the Needleman-Wunsch algorithm leads to 7 different solutions of the dynamic programming problem

$$s_1 = -\ -\ t\ t\ c\ \ g\ -\ g\ a\ -\ -$$
$$s_2 = a\ \ c\ \ g\ t\ -\ g\ a\ \ g\ a\ g\ t \quad ,$$

$$s_1 = -\ t\ -\ t\ c\ \ g\ -\ g\ a\ -\ -$$
$$s_2 = a\ \ cg\ t\ -\ g\ a\ \ g\ a\ g\ t \quad ,$$

$$s_1 = t\ -\ -\ t\ c\ \ g\ -\ g\ a\ -\ -$$
$$s_2 = a\ c\ \ g\ t\ -\ g\ a\ \ g\ a\ g\ t \quad ,$$

$$s_1 = t\ t\ \ c\ -\ -\ g\ -\ g\ a\ -\ -$$
$$s_2 = a\ -\ c\ g\ t\ \ g\ a\ \ g\ a\ g\ t \quad ,$$

$$s_1 = t\ t\ c\ g\ -\ g\ -\ -\ a\ -\ -$$
$$s_2 = a\ c\ a\ t\ g\ \ c\ g\ t\ \ a\ t\ \ a \quad ,$$

$$s_1 = t\ \ t\ c\ -\ -\ g\ -\ g\ a\ -\ -$$
$$s_2 = -\ a\ c\ g\ t\ \ g\ a\ \ g\ a\ g\ t \quad ,$$

$$s_1 = t\ \ t\ c\ -\ -\ g\ -\ -\ a\ -\ -$$
$$s_2 = -\ a\ c\ g\ t\ \ g\ a\ \ g\ a\ g\ t \quad ,$$

all of which correspond to the same value of the global score $= 5$. Concluding, due to different lengths of the sequences, assessing their similarity by using the Needleman-Wunsch alignment procedure is rather unsuccessful, as it leads to many solutions of differing quality. In contrast, by using the Smith-Waterman algorithm we obtain exactly one optimal alignment

$$s_1 = t\ \ t\ c\ g\ -\ g\ a\ -\ -\ -\ -$$
$$s_2 = -\ a\ c\ g\ t\ \ g\ a\ g\ \ a\ \ g\ t$$

which has the score $= 11$.

Sequence Block Alignment. Block alignment concerns searching for similarities between more than two molecular sequences. It usually involves processing larger data files than when alignment of pairs of sequences is performed. Another aspect of using block alignment is the wide availability of homologous variants of molecular sequences, which naturally motivates searching for similarities between the substrings of many molecular sequences simultaneously. In problems of estimation, using more data typically results in better quality of estimation. Therefore, as dependent on more data, block alignment may lead to better estimates of parameter values concerning e.g., evolutionary history of aligned sequences.

The linear increase in the data size here, may not translate to the analogous increase in the computational complexity of the block alignment procedures. The increase in the computational complexity of the block alignment problem may be greater than linear with the linear increase of the input data size. If we imagine the alignment problem for three sequences as traversing the data array in three dimensions, alignment for four sequences as the problem of traversing array of scores in four dimensions, etc., then we see that the rate of increase of the computational complexity will be of the exponential type with respect to the number of aligned sequences.

In aligning many sequences simultaneously, the underlying model must include (i) probabilities of substitutions and indels, and (ii) (additionally) the phylogeny of the sequences. The practical approaches to sequence block alignment incorporate both the data on substitution probabilities and on the ancestry tree of the sequences into the alignment algorithm. Due to the complicated structure of the problem, heuristic algorithms are applied. They are, nevertheless used very widely and considered reliable. An example of the block alignment methodology is CLUSTAL W [47], with its associated internet server. It uses the following steps. First, all pairs of sequences are "temporarily" aligned separately. On the basis of these alignments, a distance matrix is computed. Using the distance matrix obtained, a neighbor-joining tree is built, and the final alignment is obtained by progressively aligning sequences according to the branching order in the tree.

Aligning Sequences Against Databases. Sequences of characters representing nucleotides, ribonucleotides or amino acids are not only compared one to another. It is also of basic importance to search for similar sequences in existing databases of molecular sequences of DNA, RNA, or amino acid sequences in proteins. Due to the very large sizes of internet depositories of molecular sequences, performing all possible pairwise alignments, between a given molecular sequence and the sequences in the database, is not feasible. Efficient approaches to searching databases for sequences sharing similarities with a given molecular sequence, are based on the idea of hashing [13,51].The family of algorithms including fast alignment search tools, called FASTA, proceed along the following steps. First, hash tables are looked up to establish how many subsequences of given length (typically $11 - 15$ nucleotides for DNA and RNA and $2 - 3$ amino acids for proteins) a database sequence shares with a target sequence. In the next step only the database sequences with the highest scores are selected. Finally, the distances between the selected sequences and the target sequence are recomputed on the basis of the Smith-Waterman alignments.

The idea of computing co-occurrences of subsequences between a target sequence and databases was further developed by taking into account that, statistically, not all co-occurrences are of equal importance, which is especially relevant to for amino acid sequences. A statistical theory [14,24,25] for assessing the significance of words co-occurrences in molecular sequences was developed on the basis of computing probabilities that certain series of states are detected in appropriate Markov chain models. By using this theory, an appropriate scoring system was developed, leading to efficient algorithms for aligning a sequence against a database. There are several different variants of these algorithms, they are known generally as BLAST [38], the abbreviation is for basic local alignment search tool.

2.2 Browsing and Analyzing Molecular Sequences by Using Indexed Data Structures and Text Transformations

Exact or approximate searching, matching, sequence alignments, block and database alignments, discussed in the previous section, provide a variety of powerful tools extensively used in bioinformatics, for processing molecular sequences. However, when carrying out specialized studies concerning molecular sequences, one should be aware that some of the operations on sequences can be substantially sped up by using methodologies of text processing involving indexed data structures and text transformations. Also indexed data structures and text transformations enable one to perform several more advanced operations in processing molecular sequences used in many research areas, like searching and analyzing unique patterns in molecular strings, searching for exact or approximate repeats, tandem repeats or palindromic sequences. The methodologies of textual data processing involving indexed structures and text transformations are covered below along with some examples of their possible applications in bioinformatics.

Suffix Trees. A suffix tree is an indexing structure that organizes all suffixes of a given string into a tree. By using this data structure it is possible to directly access all substrings of the string. Suffix trees provide an efficient solutions to a wide range of complex problems on strings, especially in the area of bioinformatics.

It is convenient to introduce the notion of the suffix tree by first starting from its predecessor, called a suffix trie. The name trie was originated by Fredkin [16], to stand for retrieval of information. A trie is a tree data structure for storing strings over finite alphabets and makes very fast string retrieval possible. A suffix trie is a tree-like data structure storing all possible suffixes of a given string. Each node of the suffix trie is labeled by a symbol and every path from the root to a leaf forms an input string. If some suffix appears as a prefix of some other one, the path labeled by this suffix does not lead to a leaf. This problem can be avoided by terminating the string with an additional, unique special character, e.g., $. Then no suffix can also

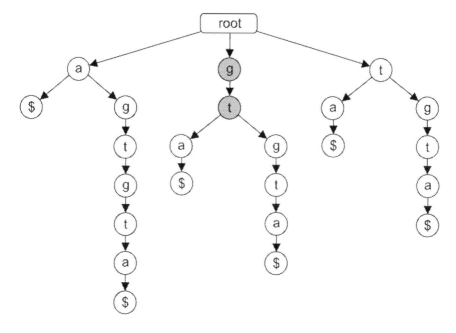

Fig. 1. The structure of the suffix trie for the sequence $S = agtgta$. The nodes g and t were shaded in order to help explaining the use of the trie structure for fast detection of pattern occurence, decribed later in the text.

be a prefix except for the entire string itself. As an example the suffix trie for a string

$$S = agtgta \tag{10}$$

is illustrated in Figure 1.

A suffix trie is an efficient data structure for carrying out various searching operations over the string S. However, as far as its memory occupancy is concerned, one encounters a problem, namely the number of nodes of the suffix trie corresponding to a string of length n is of the order of $O(n^2)$. For very large strings, likely to be encountered in bioinformatic applications, the size of the suffix trie data structure may become prohibitive for its efficient use. One method to overcome the described drawback of suffix tries, is to introduce their compact versions, suffix trees. A construction of the suffix tree corresponding to the suffix trie from Figure 1 is presented in Figure 2. In the upper part of Figure 2 the main idea of compactification is illustrated, to merge nodes if there are no branchings between them. One might have doubts concerning the true saving in memory use, since there are fewer nodes but they are occupied by longer substrings. However, in practice, the suffix tree for the string S looks like that shown in the lower panel in Figure 2. The nodes are not occupied by strings, instead they contain ranges of characters in S given by pairs of indices

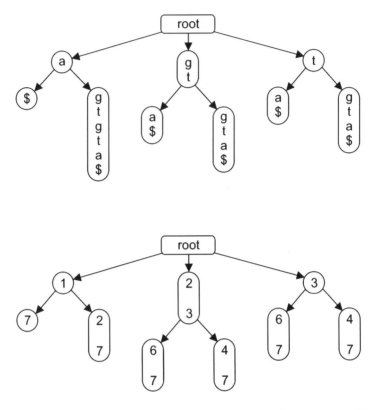

Fig. 2. Construction of the suffix tree for the sequence $S = agtgta$. The upper plot presents the main idea of compacting of the trie data structure. The lower plot illustrates the fact that suffix tree nodes contain indexes to characters in the text rather than the text itself.

of (pointers to) characters in S. Therefore, the memory capacity used by the information written in the nodes is only that necessary to hold two indices and, as a conclusion, the memory capacity of the suffix tree corresponding to a string of length n is of the order of $O(n)$.

In order that suffix trees be successfully applied to string search problems, first they must be created in the computer's memory. By creating the list of suffixes of a given sequence, and then successively growing the tree structure, one can very easily construct an algorithm for building a suffix tree working in $O(n^2)$ for a string of length n. This, as already said, may be not sufficient for many applications involving large strings. Fortunately, more time efficient algorithms for constructing suffix trees were proposed in the literature [49,37]. A well known and very widely used approach, is the algorithm developed by Ukkonen [48], which is very efficient with respect to both memory space requirements and computational time.

By using suffix tries or suffix trees many sequence analysis problems can be very efficiently solved. Here are some examples. The problem of identifying pattern occurrences can be solved by looking through the tree structure corresponding to the searched string. For example the pattern $P = gt$ occurs twice in the string S in (10). This fact can be established on the basis of reading the nodes of the suffix trie, as depicted in Figure 1, where shaded nodes correspond to the searched sequence, gt. The number of occurrences of a pattern in the string is given by the number of terminal symbols $\$$ among the descendants of the pattern P. The computational complexity of the algorithm for detecting pattern occurrence is $O(K)$, where K is the length of the pattern string. So comparing this to the results from previous subsections we see that, as expected, using indexed data structure leads to a substantial saving in computational load. By modifying the basic idea it is also easy to compute numbers of occurrences and positions of patterns in strings.

Indexing the data string by using a tree of all its suffixes makes it possible to efficiently compute many of its characteristic substrings. All substrings (patterns) that appear uniquely in the string, can be found by tracing the suffix tree from its leaves up to lowest branchings. All repeating patterns can be found by analyzing the contents of the suffix tree located above branchings. By developing appropriate algorithms one can also search for items like shortest unique patterns or longest repeating patterns. Searching for unique patterns in DNA sequences has many fundamental applications in both genomics and molecular biology. One example of an application of searching for unique patterns in DNA strings is tagging DNA sequences for the purpose of marking different regions in the genomes [22]. Another example is related to computational support for DNA amplification by using polymerase chain reactions (PCR) [12]. In this reaction short patterns (approximately 20–50 base pairs long) called primers are designed to allow for the marking of the starti for the copying contents of DNA exactly at desired positions. One more area of analysis of the uniqueness of patterns in DNA computational issues is related to the design of probe sets for DNA microarrays [17], where for the desired selectivity of DNA chips probes it is necessary that they exhibit uniqueness properties versus the rest of the genome.

Assume that two strings S, and Q are given, and form a new string, by concatenating S, and Q and adding terminating symbols, as follows:

$$T = S\$Q\#. \tag{11}$$

Two artificial terminating/separating symbols have been added, $\$$ and $\#$. Construct a suffix trie for T and search it for the longest path P such that (i) it goes from the root down to a branching into at least two children, and (ii) among the descendants of P, we find both $\$$ and $\#$. This will give us the solution to the problem.

Suffix Arrays. Suffix arrays [35] are indexed structures defined by lexicographical order of all suffixes of the text string. So, for the string S in (10), if we assign numbers $1, 2, \ldots, 10$ to its suffixes

$$1 : a$$
$$2 : ta$$
$$3 : gta$$
$$4 : tgta$$
$$5 : gtgta$$
$$6 : agtgta$$

the suffix array for S is

$$1\ 6\ 3\ 5\ 2\ 4.$$

The memory space required for a suffix array corresponding to a string of length n is again $O(n)$; however, it is lower than the memory occupancy of suffix trees in the sense that proportionality coefficients are different. If the memory requirement for suffix trees is C_1 and for suffix arrays it is $C_2 n$, then $C_2 < C_1$. Suffix array data structures can be used for on-line string searches almost as effectively as with suffix trees. Many algorithms have been published in literature, for very efficient computations of suffix arrays and for performing very efficient search tasks with the use of suffix arrays [6,26,29,36].

The Burrows-Wheeler Transform. Let us notice that the indexing structures discussed above were either pointers to positions in the analyzed string or lists of numbers of suffixes of the string. Both these cases require storing the original string being analyzed in the computer's memory as a reference.

In contrast to that, the technique of the Burrows Wheeler transform [7] allows working on just the transformed and compressed text strings. Original strings are not necessary for search tasks and the memory demand may be only a fraction of the capacity of the whole original string. The Burrows-Wheeler (BW) method was initially aimed at creating an effective, lossless compression tool for long data strings by using the idea of transforming (permuting) the initial data string to an easily compressible form [7]. However, it was later recognized that the BW transform can itself be a very fast memory-occupancy-effective search tool for substrings of any length [15,21,34].

The Burrows Wheeler transform of the string S of length n is constructed in two steps. We first build an array $Z(S)$ such that the rows of the array $Z(S)$, numbered from 0 to $n - 1$, are consecutive left-to-right cyclic rotations (shifts) of S. In the next step, we sort the rows of $Z(S)$ in lexicographic order. The last column of the sorted array $Sort[Z(S)]$ is the Burrows-Wheeler transform $BW(S)$ of S. For the string S in (10), its BW transform is

$$BW(S) = tatagg.$$

At the first sight is may seem that in order to compute $BW(S)$, for the string S of the length n we have to use a memory capacity of $O(n^2)$ to store matrix $Z(S)$. However, the problem of constructing and sorting the matrix $Z(S)$ can be simply reduced to the problem of sorting all the suffixes of the input string. The idea is to terminate the input string S with a special character, e.g., \$, that is lexicographically smaller than any other character in S. As a conclusion,

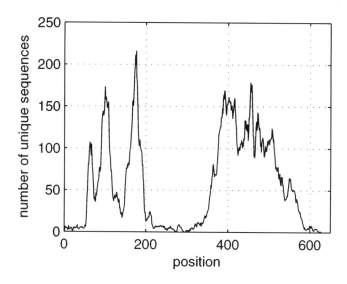

Fig. 3. Histogram showing the number of the unique sequences of up to length 10 found at each position along the 26S rDNA gene.

one can easily construct an algorithm for computing $BW(S)$, with a memory requirement of $O(n)$ and with the time complexity $O(n \log(n))$. The inverse transform, $BW^{-1}(S)$, is defined by the permutation, which transforms sorted rows of $Sort[Z(S)]$ back to $Z(S)$. One can prove that this permutation can be obtained by comparing $BW(S)$ and $Sort[S] = Sort[BW(S)]$.

The BW transform has the property, that when applied to texts coming from natural languages or to texts composed of sequences coding for some items, like molecular sequences coding for proteins, then the BW transformed sequence becomes very easily compressible. This makes the BW transform a valuable, very intensively developing, tool in bioinformatics, where reducing sizes of data is of great importance. The BW transform can be also very efficiently applied to various tasks of searching through sequences. There is very extensive research in this area, [21,34,23,41,42,43]. Below, we show some examples of very efficient use of the BW transform.

One characteristic pattern abundant in the genomes of all species is tandem repeats. A tandem repeat occurs when the number of two or more identical motifs appear in the DNA and these motifs are adjacent one to another. Detection of tandem repeats in DNA is of great importance in many areas of genomics and molecular biology. Tandem repeats serve as genomic markers for many types of studies in genomics, like studies of mechanisms of inheritance of genetic traits, for studies of genetic structures of populations, for many areas of research in population and statistical genetics. Tandem repeats are also commonly used for genetic fingerprinting of individuals and DNA forensics [8]. Databases of tandem repeat type DNA polymorphisms are under constant development in both

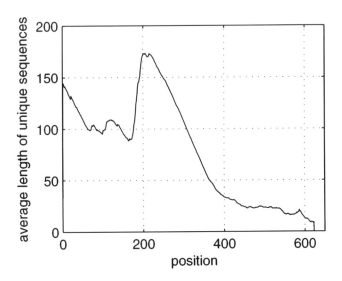

Fig. 4. Average lengths of all unique sequences detected by the algorithm at each position of 26S rDNA gene of yeast species.

size and number [46]. In the paper [43] we have shown the method of searching DNA sequences for tandem repeats, with time complexity $O(n)\log(n)$ and we have proven the superiority of this approach over some of the other existing methods.

Another example of using the BW transform for processing molecular sequences is identification of species by searching their DNA for patterns unique to each of the species. One application can be the identification of yeast species [45]. Conventionally yeast identification is done on the basis of morphological, physiological and biochemical characteristics such as the ability to utilize carbon and nitrogen compounds. However, these methods of identification are time-consuming and are unsuitable for the detection of a mixture of organisms [50]. Hence, there is a need for methods based on DNA sequence analysis. Strategies for searching for unique DNA patterns use the ribosomal DNA (rDNA) genes as a common target for the molecular identification of microorganisms. It has been shown that most of the yeast species can be identified on the basis of the sequence divergence in the 26S rDNA gene [30]. Particularly, there are two regions within the 26S rDNA gene (D1 and D2 domains), which are sufficient to deduce the relationships between species [31]. The D1/D2 variable domain sequences for almost all known yeast species are entirely sequenced. The D1 and D2 domains are approximately located within the first 650 bases in the 26S rDNA gene. These two regions have a wide application in sequence divergence among the 26S rDNA genes, and therefore can be used to identify the yeasts species.

We have applied the method based on the BW transform to the above described problem. In the first step of the algorithm, the input string S for the Burrows-Wheeler transform is created through the concatenation of all the 26S rDNA sequences separated by the character #, as already shown in (11). During this concatenation, the auxiliary array C is calculated, containing starting indices in the string S for each of n DNA sequences. This auxiliary array C is necessary to discriminate suffixes from the different DNA sequences. Because each of all the 26S rDNA sequences has approximately the same length, the origin of a given suffix can be determined with only one call to the auxiliary array C. Then, the algorithm searching for unique substrings is applied. The algorithm uses the idea of comparing successive suffixes of S. The algorithm runs in $O(n)$ time on average, where n is the total length of all the DNA sequences. In [41] and [42] we have shown that the proposed algorithm outperforms that described in [50].

Some of the outcomes of the applied algorithm are graphically presented in Figures 3 and 4. The histogram presented in Figure 3 shows the number of the unique sequences of up to length 10 found at each position along the 26S rDNA gene. This illustrates conserved and variable regions in the 26S rDNA gene. From the yeast identification point of view the most interesting are two highly variable regions. The first one is located between the positions 50 and 200, and the second one between the positions 360 and 580. Figure 4 presents the average lengths of all unique sequences detected by the algorithm at each position of 26S rDNA gene of yeast species. This histogram confirms a highly variable region located between the positions 360 and 580. Two variable regions around the position 100 and the position 170 can be observed on both histograms.

3   Conclusion

Developing text processing tools for the automatic analyses of molecular sequences is an interdisciplinary field of science, with many analogies to methodologies of analyses and understanding of natural languages. In both these fields the object of the study has a complex, hierarchical character, which resultes from natural evolution. Making progress or obtaining a new result is, very often, related to combining methods from several scientific disciplines.

The development in understanding the contents of molecular sequences proceeds along several paths. An important inspiration for progress in the field comes from the experimental side, where on one hand there is a constant increase of the throughput performance of measuring devices, and on the other there is a constant refinement in the biotechnology involving emerging techniques, possibility to study more detailed properties of biomolecules and new types of interactions between them. By using current techniques for text processing, these large data files produced in biological experiments are analyzed with the aim of extracting biological knowledge, verifying or rejecting hypotheses behind experimental plans. This can lead to a better understanding of biological models and to improvements in experimental technologies.

One important aspect of the mechanism mentioned above is that the development of experimental techniques of analyses of molecular sequences creates pressure on the computational side of textual analysis of molecular sequences, towards developing more effective and more specialized tools.

From what we have presented, it is clear that inferring information from molecular sequences is virtually impossible without specialized software tools for text processing. In this paper we have overviewed several text processing methodologies in the aspect of carrying out automatic analyses of molecular sequences. We have shown their construction, we have characterized their efficiency and computational complexity, and we have given examples of using them to infer biological knowledge.

Acknowledgment. This paper was partly supported by the European Sixth Framework Programme Project, GENEPI-lowRT, "Genetic Pathways for the Prediction of the Effects of Ionising Radiation: Low Dose Radiosensitivity and Risk to Normal Tissue after Radiotherapy", and by the University Grant BK 209/Rau-2/06

References

1. Boyer, R.S., Moore, J.S.: A fast string searching algorithm. Commun. ACM 20, 762–772 (1977)
2. Bolc, L.: Natural language generation systems. Springer, Heidelberg (1988)
3. Bolc, L. (ed.): Representation and Processing of Natural Language. Hanser-Verlag and MacMillan Press, London (1980)
4. Bolc, L., Cytowski, J.: Search Methods for Artificial Intelligence. Academic Press, London (1992)
5. Bolc, L., Borowik, P.: Many-Valued Logics: automated reasoning and practical applications. Springer, Heidelberg (1999)
6. Burkhardt, S., Kärkkäinen, J.: Fast lightweight suffix array construction and checking. In: Baeza-Yates, R., Chávez, E., Crochemore, M. (eds.) CPM 2003. LNCS, vol. 2676, pp. 55–69. Springer, Heidelberg (2003)
7. Burrows, M., Wheeler, D.J.:: A block sorting lossless data compression algorithm. Technical Report 124, Digital Equipment Corporation, Palo Alto, CA (1994)
8. Butler, J.M.: Forensic DNA Typing: Biology, Technology and Genetics of STR Markers, 2nd edn. Elsevier, Amsterdam (2005)
9. Charras, C., Lecroq, T.: Handbook of Exact String Matching Algorithms. College Publications (2004)
10. Cormen, T.H., Leiserson, C.E., Rivest, R.L., Stein, C.: Introduction to Algorithms, 2nd edn. MIT Press and McGraw-Hill, Cambridge (2002)
11. Crochemore, M., Rytter, W.: Jewels of Stringology. World Scientific Publishing Co., Singapore (2002)
12. Dieffenbach, C.W.: General Concepts for PCR Primer Design. In: PCR Methods and Applications, pp. 530-537 (1993)
13. Dumas, J.P., Ninio, J.: Efficient algorithm for folding and comparing nucleic acid sequences. Nucleic Acids Res. 10(1), 197–206 (1981)

14. Ewens, W.J., Grant, G.R.: Statistical Methods in Bioinformatics. Springer, Heidelberg (2001)
15. Ferragina, P., Manzini, G.: Opportunistic data structures with applications. In: 41st Symposium on Foundations of Computer Science, pp. 390–398 (2000)
16. Fredkin, E.: Trie Memory. Communications of the ACM 3, 490–499 (1960)
17. Gasieniec, L., Li, C.Y., Sant, P., Wong, P.W.H.: Efficient Probe Selection in Microarray Design. In: Proceedings of the IEEE Symposium on Computational Intelligence in Bioinformatics and Computational Biology, pp. 247–254 (2006)
18. Gusfield, D.: Algorithms on Strings, Trees, and Sequences. Cambridge University Press, Cambridge (1997)
19. Hamming, R.W.: Error Detecting and Error Correcting Codes. Bell System Technical Journal 26(2), 147–160 (1950)
20. Hausser, R.: A Computational Model of Natural Language Communication, Interpretation, Inference, and Production in Database Semantics. Springer, Heidelberg (2006)
21. Healy, J., Thomas, E.E., Schwartz, J.T., Wigler, M.: Annotating large genomes with exact word matches. Genome Res. 13, 2306–2315 (2003)
22. Hudson, T.J., et al.: An STS-based map of the human genome. Science 270, 1945–1954 (1995)
23. Kaplan, H., Landau, S., Verbin, E.: A Simpler Analysis of Burrows-Wheeler Based Compression. In: Lewenstein, M., Valiente, G. (eds.) CPM 2006. LNCS, vol. 4009, pp. 282–293. Springer, Heidelberg (2006)
24. Karlin, S., Altschul, S.F.: Methods for assessing the statistical significance of molecular sequence features by using general scoring schemes. Proc. Natl. Acad. Sci. USA 87, 2264–2268 (1990)
25. Karlin, S., Altschul, S.F.: Applications and statistics for multiple high scoring segments in molecular sequences. Proc. Natl. Acad. Sci. USA 90, 5873–5877 (1993)
26. Kim, D.K., Sim, J.S., Park, H., Park, K.: Linear-time construction of suffix arrays. In: Baeza-Yates, R., Chávez, E., Crochemore, M. (eds.) CPM 2003. LNCS, vol. 2676, pp. 186–199. Springer, Heidelberg (2003)
27. Knuth, D.E.: The Art of Computer Programming. Addison-Wesley, Reading (1973)
28. Knuth, D.E., Morris, J.H., Pratt, V.R.: Fast pattern matching in strings. SIAM J. Comput. 6(2), 323–350 (1977)
29. Ko, P., Aluru, S.: Space-efficient linear time construction of suffix arrays. In: Combinatorial Pattern Matching, pp. 200–210 (2003)
30. Kurtzman, C.P., Robnett, C.J.: Identification and phylogeny of ascomycetous yeasts from analysis of nuclear large subunit (26S) ribosomal DBA partial sequence. Antonie Van Leeuwenhoek Journal, 331–371 (1998)
31. Lachance, M.A., et al.: The D1/D2 domain of the large-subunit rDNA of the yeast species Clavispora lusitaniae is unusually polymorphic. FEMS Yeast Research, pp. 253–258 (2003)
32. Lebart, L., Salem, A., Barry, L.: Exploring Textual Data. Kluwer Academic Publishers, Dordrecht
33. Levenshtein, V.: Binary codes capable of correcting deletions, insertions, and reversals. Soviet Physics Doklady 10, 707–710 (1966)
34. Lippert, R.A.: Space-efficient whole genome comparisons with Burrows-Wheeler transforms. J. Comput. Biol. 12(4), 407–415 (2005)
35. Manber, U., Myers, G.: A new method for on-line searches. SIAM J. Comput. 22, 935–948 (1993)

36. Manzini, G., Ferragina, P.: Engineering a Lightweight Suffix Array Construction Algorithm. Algorithmica 40, 33–50 (2004)
37. McCreight, E.M.: A Space-Economical Suffix Tree Construction Algorithm. Journal of the ACM 23, 262–272 (1976)
38. National Center for Biotechnology Information, http://www.ncbi.nih.gov/index.html
39. Needleman, S.B., Wunsch, C.D.: A general method applicable to the search for similarities in the amino acid sequence of two proteins. J. Mol. Biol. 48, 443–453 (1970)
40. Polański, A., Kimmel, M.: Bioinformatics. Springer, Heidelberg (2007)
41. Pokrzywa, R., Polański, A.: Exact string matching with the Burrows-Wheeler Transform. In: Proceedings of the National Conference Application of Mathematics to Biology and Medicine, pp. 87–92 (2006)
42. Pokrzywa, R.: Searching for Unique DNA Sequences with the Burrows-Wheeler Transform. Biocybernetics and Biomedical Engineering 28(1), 95–104 (2008)
43. Pokrzywa, R.: Searching for tandem repeats with the Burrows-Wheeler Transform. Submitted to Journal of Computational Biology
44. Smith, T.F., Waterman, M.S.: Identification of common molecular subsequences. J. Mol. Biol. 197, 147–195 (1981)
45. Sugita, T., Nishikawa, A.: Fungal Identification Method Based on DNA Sequence Analysis: Reassessment of the Methods of the Pharmaceutical Society of Japan and the Japan Pharmacopoeia. Journal of Health Science, 531–533 (2003)
46. STRBase: Short Tandem Repeat DNA Internet Database, http://www.cstl.nist.gov/biotech/strbase/
47. Thompson, J.D., Higgins, D.G., Gibson, T.J.: CLUSTAL W: improving the sensitivity of progressive multiple sequence alignment through sequence weighting, position-specific gap penalties and weight matrix choice. Nucleic Acids Res 22, 4673–4680 (1994)
48. Ukkonen, E.: On-line construction of suffix trees. Algorithmica 14, 249–260 (1995)
49. Weiner, P.: Linear pattern matching algorithm. In: Proceedings of the 14th Annual IEEE Symposium on Switching and Automata Theory, pp. 1–11 (1973)
50. Wesselink, J., et al.: Determining a unique defining DNA sequence for yeast species using hashing techniques. Bioinformatics 18, 1004–1010 (2002)
51. Wilbur, W.J., Lipman, D.J.: Rapid similarity searches of nucleic acid and protein data banks. Proc. Natl. Acad. Sci. USA 80, 726–730 (1983)

Intelligent Decision Support: A Fuzzy Stock Ranking System

Adam Ghandar[1], Zbigniew Michalewicz[1,2,3], and Ralf Zurbruegg[4]

[1] School of Computer Science, University of Adelaide, Adelaide, SA 5005, Australia,
{adam.ghandar,zbigniew}@adelaide.edu.au,
[2] Polish-Japanese Institute of Information Technology,
ul. Koszykowa 86, 02-008 Warsaw, Poland,
[3] Institute of Computer Science, Polish Academy of Sciences,
ul. Ordona 21, 01-237 Warsaw,
[4] Business School, University of Adelaide, Adelaide, SA 5005, Australia,
ralf.zurbrugg@adelaide.edu.au

Abstract. This paper presents an intelligent decision support system for financial portfolio management. An adaptive business intelligence approach combines optimization, forecasting and adaptation with application specific financial information processing and quantitative investment paradigms.
The methodology involves constructing a ranking of stocks by strength of a buy or sell recommendation which is inferred using an adapting forecasting model that considers a range of factors. These include company balance sheet information, market price and trading volume as well as the wider economy. The system adjusts its prediction model dynamically as market conditions change. An evolving fuzzy rule base mechanism encodes a model of relationships between model factors and a recommendation to buy, sell or hold securities.

Key words: Hedge Fund Management, Fuzzy Rules, Evolutionary Computation

1 Introduction

Adaptive Business Intelligence [12] is an approach that combines elements of predictive modeling, forecasting, optimization, and adaptability. Traditionally business intelligence information systems process data to obtain information; and then use statistical data analysis techniques to infer knowledge. The resulting knowledge about the business and operations is reported to end users. There is a recent trend for intelligent business information systems to recommend courses of action or even implement decisions. An intelligent component adds another type of functionality in which knowledge is applied. Intelligent decision support requires adaptation as the operating environment changes, prediction to anticipate the future where decisions have effect, and optimization to find the best possible decisions with respect to objective outcomes. The methodology

M. Marciniak and A. Mykowiecka (Eds.): Bolc Festschrift, LNCS 5070, pp. 379–410, 2009.

has been applied successfully to a wide range of real world problems [13,1]. In this paper we apply the methodology to trading financial markets.

A financial market consists of a number of listed securities which may be bought or sold. In order to profit, it is necessary to either predict price changes of a sufficient size or to do so correctly more often not so that the reward outweighs the losses. For a number of reasons this turns out to be a very difficult task. Needless to say there is considerable literature describing theories that give insight into the nature of these difficulties. A selection of reasons includes competition between participants, the effect of information on security prices and crowd psychology. Although considerable practical and theoretical difficulties exist, a number of papers have been published that provide encouragement that it is possible to trade profitably over sustained periods using fixed strategies.

One of the simplest and most successful strategies relies on the behavioral tendency of market participants to buy stocks that have recently performed well [11]. To implement the strategy a simple tactic is used in which stocks that have had the greatest price gains (price momentum) over some recent epoch are bought and held for a period. At the end of the period these are then sold and replaced by a possibly completely new set of securities. In this way a portfolio is managed so that it always comprises of a subset of listed stocks with the highest price momentum relative to all other possible choices.

From a practical viewpoint, although it might be possible to construct rules for trading profitably from study of historic data it is still the case that there are considerable practical difficulties. Some rule inputs will have changing impact and relevance over time. Rules useful for forecasting the price of one security do not work well with all others. Another problem is the large set of input data. For example in a situation with an investment universe with just $1,000$ securities there are $10,000$ closing price and volume observations every week. Over a year there may be around $2-3$ million. For even a fairly simple forecasting model where the price prediction depends on several factors the problem complexity becomes very large indeed. For example [7] describes a prediction model with 11 inputs categorized using seven linguistic descriptions, this implies 10^{318} possible relationships that are able to be expressed. Trying to form some conclusion from the dense set of figures and tables provided in the market summary of the financial pages of a newspaper illustrates some of these problems.

Fuzzy systems seem appropriate for approximating the reasoning required to weigh possible decisions to buy or sell securities. Further, Evolutionary computation provides a method to search for rules and also to update them. Changes in market conditions are able to be observed through variation in forecasting performance and analysis of new data. The approach discussed in this section essentially involves encoding a time varying asset valuation model using fuzzy rules. This model is optimized using an evolutionary process. The model is updated periodically so that decisions are based on the most recent information available. This methodology inherently avoids specifying a fixed forecasting model. By using fuzzy logic, imprecise measurements of model factors and the extent the model is fitted by current observations are considered. The model itself is also updated adaptively by relearning with recent data, and from feedback of performance out of sample. An additional benefit of fuzzy rules is that their expression is able to be interpreted readily and understood. This contrasts with methods such as neural networks. As a result the implementation avoids having a black box giving signals without justification. This allows a fund manager to validate and tune the model parameters to incorporate additional knowledge that may not be apparent from historic data.

The remainder of this paper is organized as follows. In part 2 we examine the problem in more detail and some of the issues which are considered by our methodology. Part 3 provides a description of the design of the methodology. Finally, performance test results and concluding remarks are given in part 4.

2   Description of the Problem

Our objective may be defined precisely, although perhaps naively, as a problem involving a tradeoff between two fundamentally conflicting objectives. The first is to maximize the return from investment. The second is to minimize risk. Risk in this context may be viewed as the volatility of return.

In the following subsections we discuss these issues because of their relation with the procedures actualized in the application. These procedures include portfolio construction, financial modeling and portfolio performance measurement.

2.1   Portfolio Construction

Modern portfolio theory decomposes volatility into *systematic risk* and *unsystematic risk*. The systematic risk component reflects how the changes in market conditions affect portfolio values. The unsystematic risk component is unique for each portfolio. By enforcing constraints on portfolio contents to have minimum number of different stocks from several industry sectors, it is possible to reduce the unsystematic risk component significantly so that the main source of risk is systematic which enables general application of some basic principles for risk management. This is termed *diversification*. A well diversified portfolio should have return that compensates for the systematic risk component. In this way the return may be managed with respect to risk by using mathematical models.

Let us introduce some more precise notation. Denote $r_{m,t}$ the returns at time t of the market, and $r_{p,t}$ is portfolio return, then the systematic risk β_p of portfolio p is determined by the *Capital Asset Pricing Model* (CAPM) equation:

$$r_{p,t} - r_{f,t} = \alpha_p + \beta_p(r_{m,t} - r_{f,t}) + e. \tag{1}$$

The excess return $r_{p,t} - r_{f,t}$ of any portfolio should be fully explained by its level of systematic risk β_p and the market risk premium $r_{m,t} - r_{f,t}$. In an efficient market the alpha value for portfolio returns, α_p, should be zero because it would not be possible for traders to make a profit from past data as all relevant information for pricing a security today would be incorporated in today's price. The β (risk) term explains the difference in returns by additional risk of the portfolio above the market. A positive alpha of a portfolio (or asset) can be explained predominantly by one of two possible reasons: good stock picking ability of the portfolio manager or exposure to unaccounted risk factors beyond the scope of the CAPM model. Patterns in average returns that are not explained by the standard CAPM are termed *anomalies*.

An important stage in the development of modern financial theory was the investigation of empirical evidence of these so called anomalies in the returns of some common stocks, significantly in [6]. The possibility that a number of additional factors relating to stock prices and underlying companies explain excess returns above the market index. The occurrence of pricing anomalies is greatly reduced when only two additional factors are considered as in the three-factor Fama and French asset pricing model which extends the CAPM to include company size (total market value) and price to book value in addition to the market index [6] . The price to book value ratio is the ratio of the market value of a stock to the book value. The definition of alpha is able to be extended in terms of other factors as well.

A generalizable multi-factor alpha regression model that relates return to several risk factor premiums is able to be defined precisely. If there are k factors with each having a return f_k then:

$$r_{p,t} = \alpha_p + \beta_1 f_1 + \beta_2 f_2 + \ldots + \beta_k f_k + e. \tag{2}$$

A four factor model where price momentum is the fourth factor in addition to the three factor Fama and French model is a standard used in industry and academia. As a tool to understand portfolio dynamics all additional returns or positive alpha values can be explained in terms of unconsidered risk factors. Then factors can be added to the regression model to achieve a better fit and by assumption better explain returns.

In cases where markets are not efficient and participants actions are not always rational, other explanations for anomalies can be considered. In this case the dynamics of the group behavior of market participants would be factors. If the pricing of listed market items do not accurately reflect the risk premium because of irrational pricing tendencies of market participants, then corrections would take place leading to excess returns being observed from time to time. For example securities could become *undersold* or *oversold* by participants so that prices become unreasonably high or low. Such events could be discovered by analyzing time series of stock prices in a process termed *technical analysis*.

2.2   Financial Modeling

Financial thinking has evolved during recent decades with a shift away from absolute faith in market efficiency to the position that markets are only "almost" efficient and behavioral explanations are required to account for exceptions. In an efficient market the CAPM alpha should always be zero. The implication of the changing understanding of the market is to imply that the best strategy is not necessarily to attempt to *passively* attain returns that follow the market index. Instead *active* stock picking approaches are used to attempt to attain return on investment in excess of the market.

One of the main reasons for this shift has been an increasing body of empirical results that contradict the hypothesis that the prices of stocks and other market instruments are, for the purposes of prediction, random. As a consequence *behavioral models* are used to explain some pricing effects. Some examples of patterns found in stock prices used to obtain returns in excess of the market over long periods include the profitability of momentum strategies [11]. Other technical indicator strategies such as Bollinger bands, moving averages, relative strength index [3] also have been shown to promote excess risk adjusted returns. Another category is cyclical trends, for example the "January effect" [5] where the previous year's underperforming stocks outperform in the following January because investors and managed funds sell off of underperforming assets in January.

To some extent the growing body of empirical evidence cited from academic research above is a result of developments in information technology. In fact in recent decades, computers have had a very large impact on operations at all levels in the financial sector. Advances in computing power and availability encourages application of complex mathematical models and statistical methods that may be leveraged by easy access to large volumes of data in electronic format. The culmination of this influence on portfolio management is in the rapidly expanding field of Quantitative Investment (QI). Applied to portfolio management QI is defined flexibly as "an approach to portfolio management that takes full advantage of today's better understanding of the market and greater technological capacity for sophisticated investing" [4]. Conceptually and in practice, QI involves utilization of these ideas and techniques for three main activities: return forecasting, portfolio construction and optimization, and performance measurement of resulting portfolios (Fig. 1). There is a clear feedback loop as the performance of managed portfolios over time logically should cause the model and portfolio construction methods to be either maintained or adjusted. Although it nevertheless remains an open problem to adapt quantitative trading models as quickly as a traditional analyst because of reliance on performance analysis and historical data.

Portfolio building, implicitly or explicitly, involves a valuation based on forecasting future asset prices from a basis of current knowledge. In order to effectively understand and adjust a model it is necessary to analyze the performance of portfolios managed using the forecasting model and resulting constructed portfolios over time (Fig. 1). A (conceptual) multi-factor model relating risk and return with a time component is expressed as follows:

$$r_{i,t} = \alpha_i + \beta_{i,1}f_{1,t} + \beta_{i,2}f_{2,t} + \ldots + \beta_{i,k}f_{k,t} + e_{i,t}, \tag{3}$$

where $r_{i,t}$ is the return of a stock i at time t, $f_{1,t}, \ldots, f_{k,t}$ are k returns due to factors, $\beta_{i,1}, \ldots, \beta_{i,k}, f_{k,t}$ are multipliers for the risk of including facts and $e_{i,t}$

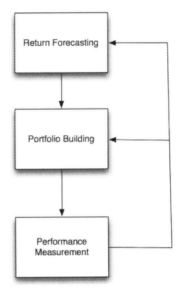

Fig. 1. The three main processes of Quantitative Equity Portfolio Management and their relation. The portfolio optimization process takes, as at least one input, information from the forecasting model. The effectiveness of the forecasting model and portfolio construction methodology can be gauged by performance measurement and includes features such as consistency of returns over time, comparison to benchmarks and so on as well as standards such as the annual rate of return.

is an error term [4]. This expression is a prototype for a prediction model that relates return to risk (by the β terms) and is also divided into model factors. It also is the case that the terms $f_{1,t}, \ldots, f_{k,t}$ can change over time to model a changing impact of factors over time. We apply computational intelligence to dynamically build a multi factor price forecasting model that anticipates and adapts the weight, possibly zero, of factors over time. The paradigm of computational intelligence is distinct from philosophies of artificial intelligence that attempt to precisely imitate human reasoning in that it involves harnessing the unique abilities of computers to produce "intelligence".

2.3 Model Factors

Now that we have presented some aspects of the forecasting model we discuss the model factors, f in equation 3. There are at least three distinct classifications of information that have been used to explain returns:

- Market and macro economic indicators
- Fundamental indicators
- Technical indicators

Macro economic indicators such as a country's gross domestic product, interest rates and also variables such as currency exchange rates, commodity prices have a significant impact on the market. Fundamental analysis is a natural approach involving consideration of the assets underlying market securities, for instance the companies whose stock is listed on the stock exchange. Using sources such as accounting data and even natural language data such as news and other reports, it is accepted that it is possible to identify assets that provide good value. Some important criteria include cash flow, total company earnings, derivative information such as the ratio of earnings to share price and others. Analysts give different importance to these factors depending on industry groups or sectors, market conditions, economic conditions and even personal experience. For example importance may be given to earnings before tax and other liabilities to give an indication of the underlying strength of a company's position, there are many reasons for variations, if for instance a firm operates in an industry that is highly regulated and subject to many taxes it may be the case that important aspects of its position relative to companies in other industries are hidden. In academia the explanatory performance of potential models is often compared with a standard such as the four factor model discussed in the previous subsection.

Technical analysis is widely used in practice. It involves constructing and applying technical analyses of price and volume movements. This approach is thought to extract information about market expectations, particularly behavioral effects. These indicators are divided into the following categories by their use in modeling different types of price movements: moving average, momentum, oscillation, and breakout indicators and also indicators based on volume, or price and volume rather than only price. Moving averages are often used to identify trends and to smooth out fluctuations due to daily or short, unsustained changes, depending on the period to calculate the average. New trends are identified when a moving average series crosses the price, or a shorter period average crosses a longer average. Oscillating indicators are used to identify cyclic patterns in price movements by compressing observations into a range, possibly giving more weight to recent points, and then generating buy or sell signals appropriately when extremes in the range are reached. Breakout indicators, as suggested by their name, are designed to catch significant changes in price direction at an early stage, for example a movement well outside the standard deviation of the mean historic returns is an indication that an unusual trend is emerging as opposed to a cyclic occurrence. Volume data is an important input component and an indicator of market sentiment with links to behavioral aspects of market activity. In general a market is considered strong by technical analysts if price and volume are both increasing.

2.4   Performance Measurement

Performance measurement is an important consideration. In this subsection we briefly discuss some of the main issues and point to some references that provide further information. A key concept is the relationship between risk and returns.

Fig. 2 shows the meaning of the relationship between risk and return illustrating sample return distributions and possible corresponding returns over time, more risky portfolios where return is justified however by the risk would have long tails in the negative area but a higher probability of positive return. Ideally by adjusting the forecasting and portfolio optimization, returns can be shaped to accurately target a specific risk profile. Although clearly the frequency of returns is not usually ideally normal, commonly, a general understanding of return properties can be obtained by a normal approximation. portfolio with a positive skewness is more likely to exhibit large returns, kurtosis is an indication of the likelihood of deviation from the mean, a lower kurtosis implies a less regular and also smaller swing away from the mean return value. Fig. 3 shows an example of the return frequency distribution obtained by testing the system described in this section.

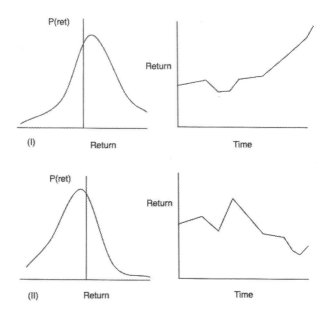

Fig. 2. Two return distributions and corresponding performance over time. The top panel is positively skewed showing returns are on average greater than they are negative, the bottom panel shows the reverse.

The *sharpe ratio* [14] measures how much excess returns (portfolio return r_p above the risk free rate r_f) investors are awarded for each unit of volatility, i.e.

$$sharpe = \frac{r_p - r_f}{\sigma_p},$$

where σ_p is the standard deviation of returns.

Portfolio Returns

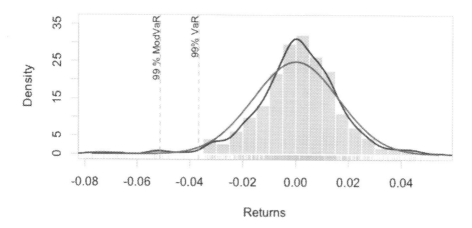

Fig. 3. Daily portfolio return frequency distribution. The actual return histogram is superimposed by a normal approximation and density. A measure of the extreme loss probability with confidence (the lowest 1 percentile approximation of expected return) is marked VaR, Value at Risk. The modified VaR is a prediction estimation which accounts for the positive skew and the kurtosis of the observed returns and the plain VaR is an estimation that assumes a normal distribution.

Another important concept is relating performance to a *benchmark*. Two useful measures are the market index and the risk free rate of return. Comparison to the market as discussed earlier in this section is an important yardstick for theoretical reasons and also techniques exist to attempt to passively follow the index without necessarily holding all stocks. If a strategy is not able to outperform the risk free rate it has no practical benefit. Examples of portfolio performance measurement are provided in [7,10]. For detailed information about this subject we refer readers to [9], of particular relevance are information content and net selectivity measures.

3   Approach for Decision Support System

In this section we describe the approach we used in constructing an adaptive stock ranking model that provides recommendations to buy, sell and hold stocks. Fig. 4 summarizes the Adaptive Business Intelligence (ABI) methodology. This approach may be divided into three main components: optimization, prediction and adaptation.

   The system generates a multi factor model for stock selection. A potential for prediction exists in the universe of model factors according to academic financial research and practices in industry. This latent possibility is actualized by opti-

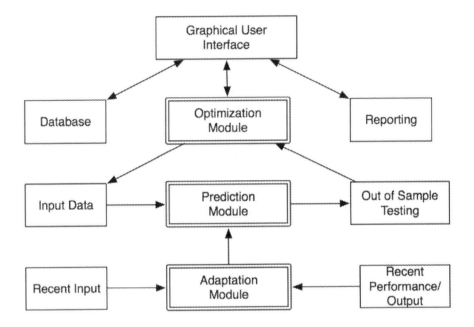

Fig. 4. Adaptive Business Intelligence (ABI). The components for optimization, prediction and adaptation are the main logical divisions in design, each requires its sub components (and each other). The optimization loop controls the recommendation presented to the user, it is predominantly a function of the input data and the result of predictions which are updated when new data is loaded and in response to feedback from recent management portfolio performance.

mization to produce a usable model using a heuristic search process. The model is represented using a fuzzy rule base structure. Optimization and prediction are coupled more closely in this financial systems than many other applications of ABI. For example in logistics or scheduling, operational requirements have a greater impact on solutions. Therefore we emphasize a distinction in terminology. Optimization refers here to the search process to optimize the forecasting model using historical data. Prediction refers to the application of this forecasting model in the future. Finally specific techniques are introduced for adapting the optimization for prediction as time passes.

3.1   Model Factors

The model inputs are the kernel of the methodology from a financial view point because the relationships between these factors and returns are from a high level the model definition. We assume and implement the methodology for equity markets and in the discussion we use the term stock. However the principles can be applied to other listed market items such as options, warrants, CFD's and other instruments.

From a top down view point the inputs are divided into macro economic factors which operate at a global level such as GDP, the value of the whole market and also information related to sectors of the economy in which individual companies conduct their business such as the resources sector, consumer discretionary sector etc. Fundamental company information also exists outside the market but is specific to each stock, this information includes information about cash flows, earnings an so on (see section 2). A separate category is data from trading activity within the market and includes things such as price and volume series and market capitalization. The raw inputs and their meaning is given in table 1. The processing of these raw inputs is summarized in Fig. 6.

The model factors are constructed from these elements. In this approach using processed data, a structure for the universe of models is imposed in a way that influences the model and the optimization process to use some predefined values. Techniques such as genetic programming [10] could be used to optimize the factors and find equations directly from raw data. In practice there is a balance between on the one hand providing no pre processing and on the other defining the factors restrictively so as to prevent exploration outside a specific type of model. Fig. 6 shows a matrix of the type of data and an analysis method for this structure to summarize how technical, fundamental and macro economic data is processed to examine attributes of the economy, stocks and investor behavior. The number of processed model factors is termed the breadth of the model.

The fundamental value factors are derived from 10 basic data sources (see Fig. 6). For each fundamental element with respect to a stock s and at time t, $X_{s,t}$ the factors X-industry and X-industry-momentum are calculated as follows

$$X\text{-industry}_{s,t} = \frac{X_{s,t}}{X_{1,t} + \ldots + X_{k,t}},$$

$$X\text{-industry-momentum}_{s,t} = \frac{X\text{-industry}_{s,t}}{X\text{-industry}_{s,t-period}},$$

where $X_{1,t}, \ldots, X_{k,t}$ are the values of the X element for all stocks in the industry sector of s and $period$ is a number of time units. X-industry normalizes X to stocks industry sector, X-industry-momentum tracks the rate of change in X. This normalization with respect to industry is because companies in different sectors have common attributes shared with other businesses in the same sector, the growth or decline with respect to the sector and in general is tracked by the momentum factors. In addition a variable industry sector identifies a stocks sector and also the relative placement in the sector with respect to market capitalization are included here.

The market factors include market capitalization and a number of technical indicators found using price and volume. Market capitalization from the MV input file is directly imported as a factor and enables the model to differentiate between small and large capitalization stocks. The money flow index (MFI)

Table 1. Raw input types. The model factors are derived from these basic data types.

Name	Description
Macro Data	
INDEX	the market index, a weighted average by market cap of the value of listed equities.
RF RATE	interest rates for short term government bonds (3 months).
GOLD	the spot price of gold in USD.
OIL	the price of a barrel of crude oil in USD.
Fundamental Data	
DY	dividend yield for the company. A percentage value of the dividend income earned over the stock price.
PTBV	price to book value for a company. Literally calculated as stock value over accounting book value.
PE	price earnings ratio for a company. Calculated as the price of a stock divided by earnings per share.
PE2	a forecast of price earnings ratio for the next year by financial analysts.
MV	the market capitalization of a company. Calculated as the company stock price multiplied by the number of shares.
EPS	earnings per share.
TDE	total debt to equity ratio.
LDE	long term debt to equity ratio (> 1 year) .
EBITDA	earnings before interest and tax.
ROA	return on assets.
ROE	return on equity.
Technical Market Data	
PRICE	daily close prices.
VOLUME	daily trading volume.

attempts to track the rate of capital flow into and out of stocks by relating price and volume. The calculation is with respect to a period, p, at time t:

1. if $price_{t-p} > price_{t-p-1}$ then $MF_t^+ = MF_{t-1}^+ - (price \times volume)$,
2. if $price_{t-p} < price_{t-p-1}$ then $MF_t^- = MF_{t-1}^- - (price \times volume)$,
3. if $price_t > price_{t-1}$ then $MF_t^+ = MF_{t-1}^+ + (price \times volume)$,
4. if $price_t t < price_{t-1}$ then $MF_t^- = MF_{t-1}^- + (price \times volume)$,
5. $MFI_t = MF_t^+ / MF_t^-$.

The technical factors listed in Fig. 6 analyze various characteristics of price and volume time series. PPO (percentage price oscillator) and PVO (percentage volume oscillator) emphasize cyclical patterns. The percentage oscillator is calculated by taking the ratio between a longer and shorter moving average. Std. Dev is a running standard deviation over the previous 3 months. A longer and shorter period price momentum tracks the rate of change in price during a previous period. Bollinger bands are lines a standard deviation from the mean around a stocks price — when the price moves outside, a signal is generated. In addition a variable for each stocks CAPM alpha and beta that provide an indication of if a stock is over or under priced with respect to the capital asset pricing model are included (see section 3). The macro indicators (Fig. 6) relate the previous analysis to changing economic conditions over the longer term by measuring the growth of the index, change in interest rates, gold and oil prices. The MA variables contain a ratio of the current value of the indicator to a long term moving average. Momentum again tracks the rate of change in either positive or negative directions. This enables the model to consider wider economic trends and cycles depending on the length of the averages.

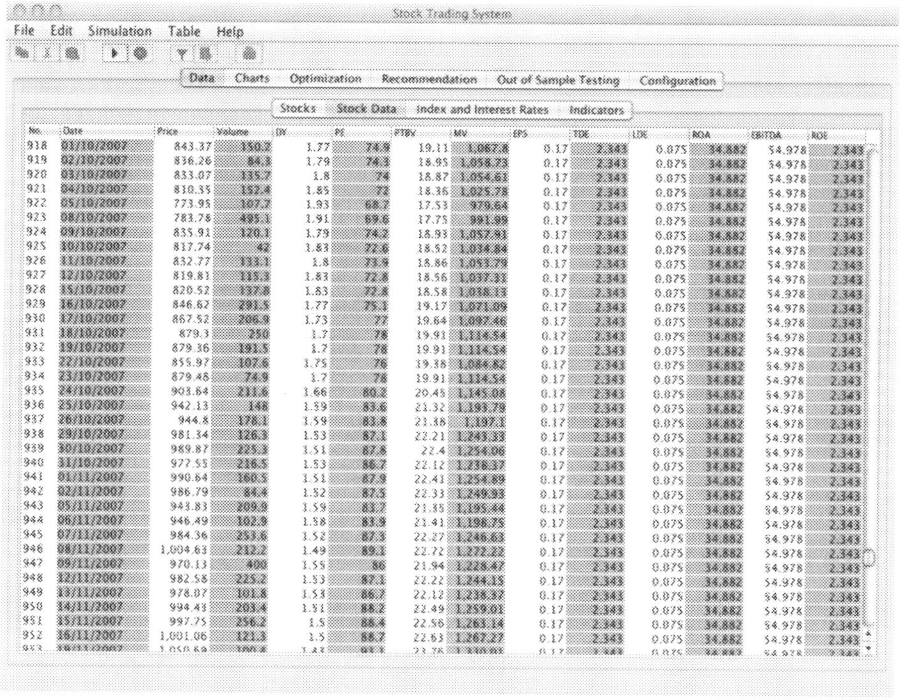

Fig. 5. Screenshot from the application showing raw data view.

3.2 Model Representation

A fuzzy rule is a causal statement that has an *if-then* format. The *if* part is a series of conjunctions describing properties of some linguistic variables using

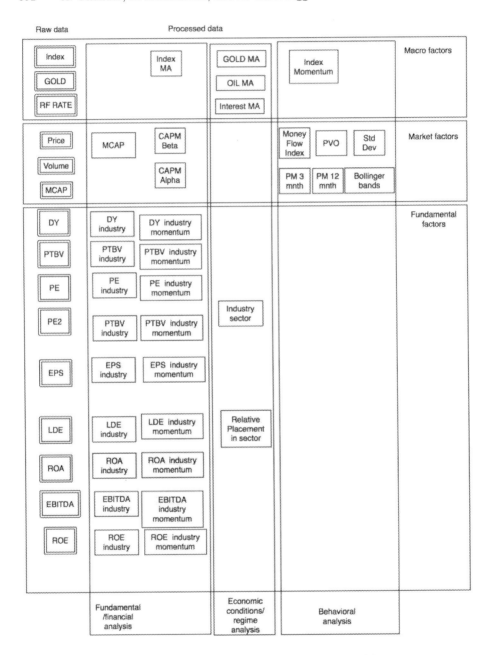

Fig. 6. Multi factor stock market valuation model.

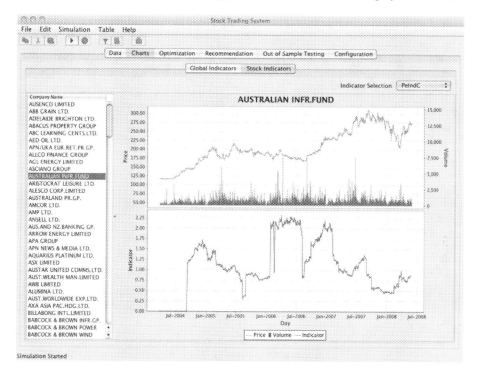

Fig. 7. Screenshot from the application showing momentum indicator for price to earnings ratio.

fuzzy sets that, if observed, give rise to the *then* part. The *then* part is a value that reflects the consequence given the case that the *if* part occurs in full. A rule base consists of several such rules. For some input observations of the linguistic variables a rule base is evaluated using fuzzy operators to obtain a classification given the (possibly partial) fulfillment of each rule by that input.

A rule base model is able to be represented as follows:

- If *PE Industry Momentum* is *Extremely Low* then *rating* = 0.9
- If *Price Momentum* is *High* and *DY Industry* is *Very High* then *rating* = 0.4

The *if* part specifies a relationship between factors (linguistic variables) and output of the rule is a weighting for the rule when the set combined. The output or rule weighting is a discrete value between 0 and 1, let us define d as the number of discrete output ratings. The rule base specifies a pattern in the data. For a data vector containing observations of crisp values the evaluation of the rule base is 1 if the observations are exactly the pattern the rule specifies. It is less than 1 but greater than 0 for partial matching of the pattern. And 0 implies no match at all.

For a rule base prediction model with breadth L there is one linguistic variable for each factor $\{f^1, ..., f^L\}$. Each of these variables, f^i, is described by a set of m_i linguistic descriptions of that factor using a set of corresponding triangular membership functions μ_j^i, $j = 1, ..., m_i$. In the example above DY Industry and Price Momentum are model factors and High, Extremely Low etc are linguistic descriptions. The membership functions map an observation of the corresponding variable to a degree of membership in fuzzy sets Low, High etc. They have a standard form where each is characterized by a triple $(min, center, max)$. min and max are the extreme values that comprise the membership set, $center$ is a value that maps to full membership of the set:

$$\mu_j^i(x) = \begin{cases} 0 & : x < min \text{ or } x > max \\ (x - min)/(center - min) & : min \leq x < center \\ (x - max)/(center - max) & : center \leq x \leq max \end{cases}$$

Given a vector x of L observations, one for each factor, a fuzzy rule r_1 containing linguistic descriptions $D_{i_1} ... D_{i_K}$, $K < L$, of factors $f^1, ..., f^L$ is written as follows:

$$r_1 = \text{ If } f_r^{i_1} \text{ is } D_{i_1} \cdots \text{ and } f_r^{i_K} \text{ is } D_{i_K} \text{ then } o_1$$

is evaluated by:

$$rule_r(x) = o_r \prod_{i=1}^{K} \mu_j^i(x_i).$$

In general a rule base B may contain several such rules $r_1, ..., r_R$. Rules in a rule base are aggregated accounting for the weight of each rule, $o_1, ... o_R$, to obtain a real value as follows:

$$\rho_B(x) = \frac{\sum_{m=1}^{R} rule_m(x)}{\sum_{m=1}^{R} o_m}. \tag{4}$$

The number of possible rule bases is quite large. Each rule may specify up to L factors, each of these factors f^i is described by one of M_i membership sets that relate to it. There are also d possible output ratings. So for a single rule the number of possible variations is

$$v_r = d \times \sum_{k=1}^{L} M_i^k \times \binom{L}{k}$$

and the number of possible rule bases with R rules is therefore of the order v_r^R.

Figure 7 gives a typical example of a chart showing a stock price and volume time series along with an indicator. Graphs similar to this are found in any trading or stock price report and a trading system would be incomplete without such a presentation. Many books have been written discussing the intricacies of interpreting such graphs. Although visualization is an improvement over data tables as a way to make trading decisions it is clearly not possible for an analyst to examine any more than a few assets in this way.

3.3   Comparing Models

Using equation 4 a rating for a stock by a rule base on a day t is obtained using observations of the values of the model factors for the stock on the day t. Let us use a subscript to denote this rating such that $rating_{s,t}$, means the rating for a particular stock s on day t. A ranking is defined here as a set of stocks ordered by some value associated with each stock. in this case let us assign this value to be the output of a rule base given data for the stock on a particular day. For a set of stocks $M = \{s_1, s_2, \ldots, s_m\}$ a rule base can be applied to order a set of stocks to obtain a ranking of all stocks in the set on the day t as follows:

$$R_t^\rho(M) = [(s_{1,t}, \rho(\boldsymbol{x}_{s_1,t})), \ldots, (s_{m,t}, \rho(\boldsymbol{x}_{s_m,t})], \tag{5}$$

where $\rho(\boldsymbol{x}_{s_i,t}) \geq \rho(\boldsymbol{x}_{s_{i+1},t})$. Each element of a ranking is a pair comprising the ranked stock and its rating, $(s_{i,t}, \rho(\boldsymbol{x}_{s_i,t}))$, has rank $i \in \mathbb{Z}, i \geq 1$.

The fitness of a rule base is defined to be a measure of its ability to rank stocks by return ordering over a specific period of time in the future termed a forecast horizon of length H days. An ideal return ordering $R_{ideal,t}$ for a day t is constructed by looking forward H days into the future within the available training data. Then we find the average price for a stock, $p_{s,H}$ during a P day period starting after the H'th day in the training data window. As this operation takes place in training data we may assume that for every stock the return during the period is simply:

$$r_{s,t,H} = p_{s,H} - p_{s,t},$$

where $p_{s,H}$ is the average price over a period starting H days after t. The reason the average price is used is to avoid the fitness being overly sensitive to fluctuations in stock price on particular dates. An ideal ranking for days in the training window for comparing the rule base ranking with is able to be defined as follows:

$$R_t^{ideal}(M) = [(s_{1,t}, r_{s_1,t,H}), \ldots, (s_{m,t}, r_{s_m,t,H})]. \tag{6}$$

Evaluation functions are defined to compare the similarity of rankings, $R_{\rho,t}$, from rule bases to optimal rankings with hindsight ,$R_{ideal,t}$. In the application rule bases are tested using input data from several days so that rule base fitness is not overly dependent on patterns in a single day of training data.

As a step to constructing an evaluation function let us initially define a comparison operator for comparing rankings. An obvious method for comparing the ordering of A and B is to count the number of times the same stock has an identical rank in both. However it is preferable that the method should be more lax for a number of reasons, including for accuracy. Two rankings would be defined as very different if the rankings were out of sync by even a single element. Trying to find rules to predict a very specific ranking property would likely lead to over fitting and loss of generalization. In addition the ordering of stocks within the top percentile is not relevant since all stocks in this group are, relative to the others, recommended to buy. For these reasons and to make the optimization task easier we use a flexible approach for comparison designed to be sensitive to very small changes in similarity due to any change in the ranking order.

Given two rankings A and B that order stocks in a set $M = \{s_1, s_2, \ldots, s_m\}$ we define two corresponding sub rankings:

$$a = [(s_{a_1}, r_{s_{a_1}}), \ldots, (s_{a_{u_1}}, r_{s_{a_{u_1}}})],$$

and

$$b = [(s_{b_1}, r_{s_{b_1}}), \ldots, (s_{b_{u_2}}, r_{s_{b_{u_2}}})],$$

with sizes $u_1, u_2 < m$ containing the highest u_1 and u_2 rated stocks in A and B respectively. We construct two sets of stocks which are subsets of M

$$a_s = \left\{ s_{a_1}, \ldots, s_{a_{u_1}} \right\}$$

and

$$b_s = \left\{ s_{b_1}, \ldots, s_{b_{u_2}} \right\}.$$

A real value measure of similarity of two rankings A_M, B_M defined over the set of m stocks M is then found by the operator:

$$\text{similarity} : R_M \times R_M \mapsto \mathbb{R}$$

$$\text{similarity}_{u_1, u_2}(A_M, B_M) = \frac{\mid a_s \cap b_s \mid}{\max(|a_s|, |b_s|)}. \tag{7}$$

where $u_1, u_2 \in \{u \in \mathbb{Z} | 0 \leq u \leq m\}$. The meaning is interpreted as the number of stocks from the top u_1 of A_M that are also in the top u_2 of B_M.

Now we define the evaluation function that uses a training window of length $horizon + 2 \times period$ where H is the forecast horizon and P is a fixed period of sequential days that is both the number of days used to test the rule base and also the number of days used to calculate average values for the ideal ranking. Let the first day in the training window be denoted day T then:

$$eval_{buy, H}(\rho_B) = \sum_{t=0}^{P} \frac{\text{similarity}_{l,q}(R_{T+t}^{\rho_B}, R_{T+P}^{ideal})}{P}, \tag{8}$$

where, $R_t^{\rho_B}$ is a ranking from a rule base ρ_B with respect to a set of listed stocks and R_{T+P}^{ideal} is an ideal ranking of stocks at a day taken at the end of the possible training testing days. The parameters l and q may be tuned by experimentation or adaptively. In optimization it is easier to try to find any stocks in the top of the ideal ranking rather than a specific ordering. Another fitness function is also used in the system to measure the ability of rule bases to rank stocks by likelihood of decreasing value. This function is defined in a similar way, the only difference is that the order of the ideal ranking is reversed:

$$eval_{sell, H}(\rho_B) = \sum_{t=0}^{P} \frac{\text{similarity}_{l,q}(R_{T+t}^{\rho_B}, \text{reverse}(R_{T+P}^{ideal}))}{P}. \tag{9}$$

3.4   Optimization

This component optimizes a rule base to perform well in ranking stocks relative to their increase in price over a prediction horizon. Input data consisting of a historic data window is used as training data, this window is updated each time the system is used to rank stocks for trading decisions. In the subsection 3.3 we discuss the fitness function for ranking stocks. There are a very large number of possible rule bases. To handle this a combination of local search algorithms and an evolutionary algorithm are used in sequence:

1. Restricted initial search,
2. Evolutionary algorithm,
3. Hill climber search.

The processes are also initially guided by past searches, a new optimization with a fixed length sliding data window takes place each time the system is used for trading and a repair operator fixes the rules to be focused around the area neighborhood of the previous optimal solution. In addition a rule structure is fixed over the whole population periodically during the evolutionary search in a way inspired from natural human reasoning to enable the system to follow a "line" of reasoning through the solution space. Figure 8 shows a screenshot of a panel to view the progress of optimization in the graphical user interface.

A measure of the generalization ability of rule bases is also used in the optimization process to try to find rules that work better in general rather than only in the training window. Two generalization criteria are used, the first counts the average number of ranked stocks over a period before and in the end of the training window, both are separate from the days used to generate rankings to test.

Representation and Alteration of the Genotype. First of all, the rules are encoded using array structures. The maximum number of possible rules in a prototype solution is fixed but the actual number can vary in the range 0 to this maximum setting $rules_{max}$. A boolean vector UR of length $rules_{max}$ indicates if a rule is used. A $rules_{max} \times L$ matrix I indicates by integer values from 1 to j corresponding to membership functions μ_j^i for each column f_i that is specified in each rule. An additional boolean $rules_{max} \times L$ matrix UI indicates if a factor f^i is used in a rule. A vector of floating point values O with length $rules_{max}$ indicates the output weighting for each rule. Fig. 3.2 clarifies this definition and superimposes these four arrays, UR, I, UI and O, into a single genotype visualization.

There are a number of variation operators that are defined over the genotype representation to search for new solutions by mutation (variation of an existing rule base to get a new one) and cross over (combination of two or more existing rule bases to get a new one). Table 2 lists these operators.

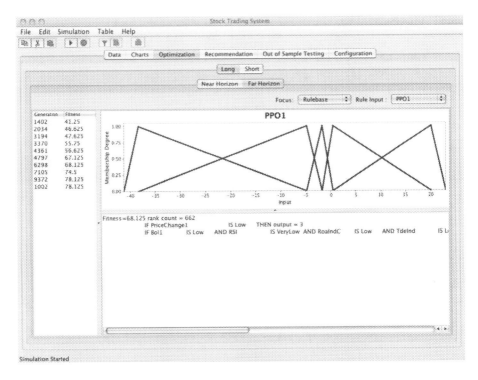

Fig. 8. The optimization process in the application GUI.

If PPO is EL	then rating = 0.8
If DYIND is L and DYINDM is VH	then rating = 1.0
If MFI is M and PEIND is VH	then rating = 0.2

(a) Example phenotype rule base representation

use	f_1	f_2	f_3	f_4	f_5	f_6	f_7	$\cdots$	f_L	o
B	B I	B I	B I	B I	B I	B I	B I	$\cdots$	B I	F
B	B I	B I	B I	B I	B I	B I	B I	$\cdots$	B I	F
B	B I	B I	B I	B I	B I	B I	B I	$\cdots$	B I	F
B	B I	B I	B I	B I	B I	B I	B I	$\cdots$	B I	F
B	B I	B I	B I	B I	B I	B I	B I	$\cdots$	B I	F

(b) Example genotype rule base representation.

Fig. 9. The phenotype rule base representation is shown in (a) and the geno-type or internal encoding is depicted in (b). The genotype shows a rule base rep-resentation for a rule base with 5 rules and L factors). B indicates a boolean value: $B \in \{T, F\}$; I an integer: $I \in \{1, 2, 3, 4, 5, 6, 7\}$; and F a float: $F \in \{0.1, 0.2, 0.3, 0.4, 0.5, 0.6, 0.7, 0.8, 0.9, 1.0\}$.

Table 2. Operators, all act on rule bases ρ.

	Operators	
Type	Description	Output (new ρ)
Mutation	μ-SMALL $\rho \mapsto \rho$	selects an element with equal probability from UR, I, UI or O and increments or decrements with equal probability by 1 appropriate unit (e.g. if boolean goes to false, if integer add or subtract 1).
Mutation	μ-LARGE $\rho \mapsto \rho$	selects an element with equal probability from UR, I, UI or O and sets to a new legal value chosen with equal probability.
Mutation	μ-SMART $\rho \mapsto \rho$	selects an element with equal probability from I or O such that UR, UI are true and increments or decrements with equal probability by 1
Crossover	γ-UX $\rho \times \rho \mapsto \rho$	uniformly select elements from UR, I, UI or O with equal probability to construct a new rule base from two parents
Crossover	γ-RULE $\rho \times \rho \mapsto \rho$	swap rows from UR, I, UI or O to construct a new rule base with whole rules from two parents
Repair	r-SAME $\rho \times \rho \times [0,1] \mapsto \rho$	genes in from ρ_1 are changed to be the same as ρ_2 until p, $0 < p < 1$ percent genes in ρ_1 are the same as ρ_2
Repair	r-LEGAL $\rho \mapsto \rho$	genes in from ρ are changed until there are noillegal values (e.g integers in I less than 0 or greater than the number of membership functions for the corresponding variable)
Repair	r-FIXED $\rho \times \rho \mapsto \rho$	genes from I and O in ρ_1 where UR and UI are true in ρ_2 are overwritten by corresponding values from ρ_2

Heuristics. Initially, we search by enumeration the C_2^L single rules that are a possible combinations of 2 factors. A percentage of the best of these are inserted into the initial population before the evolutionary search. The best rule from the initial search is fixed across the population by applying the operator r-FIXED (see table 2). The evolutionary algorithm is as follows:

1. Initialize population $P = \langle \rho_1, \rho_2, \ldots, \rho_n \rangle$ of n rule base individuals where $g\%$ are from the enumeration search and the remainder are random
2. Initialize variables from parameter file: *cBEST, BEST, SWI, generation, FIXED, rules$_{max}$*
3. Apply r-SAME to the whole population using a previous search's best rule base and double parameter $p \in [0,1]$ if available
4. Evaluate each solution: calculate $eval(\rho_v)$ for $v = 1, \ldots, n$
5. Identify the best solution, *cBEST* in P

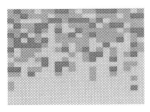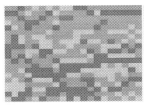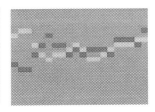

Fig. 10. Visualization of the phenotype illustrating the increase in fitness during the evolutionary process. Each pane represents the top percentile of a series of stock rankings for the same days in training data found from different fuzzy rule bases. The grey segments represent stock selections which, in training, were from the top risers over the forecast horizon (3 months), dark segments represent selections which were in the worst half of the possible choices and light grey means there was no recommendation or the selection did not rise or fall significantly. The objective function was to maximize the selection of the best choices. The first panel shows the best solution at 1402 generations, the middle panel after 3370 and the right was after 49,149 generations.

6. If $eval(cBEST) > eval(BEST)$ then $BEST = cBEST$ and $SWI = 0$
7. Alter the population by applying mutation and crossover operators (tournament selection of size 2 is used)
8. Apply r-LEGAL operator to each offspring with respect to the best solution $best_{previous}$ from the previous generation (elitism is not used)
9. If parameter to use fixed rules is used, apply r-FIXED operator to each offspring and a single global fixed rule base
10. If parameter is set, apply r-SAME operator to each offspring using the current $best$
11. $generation = generation + 1$
12. $SWI = SWI + 1$
13. Repeat steps 3–12 successively until SWI is $maxSWI$ and no improvement is recorded
14. If $generalization(BEST) > generalization(FIXED)$ set $FIXED = BEST$, $generation = 0$, $rules_{max} = rules_{max} + 1$ and return to step 3; else return P

where $generalization$ is a function to return a value that measures the generality of solutions, for example by counting the ranked stocks or testing with different historical data. Fig. 5 shows a visualization of the change in the ranking of stocks implied by rule bases that occurs during the evolutionary process.

After the evolutionary algorithm has completed a hill climber algorithm successively improves the the top $T\%$ of the individuals from the evolutionary algorithm using the mutation operators. Its is as continues from the EA as follows:

1. $HCP = $ top $T\%$ of P
2. evaluate each solution ρ in HCP
3. apply mutation to each to get the next generation of the HC heuristic
4. repeat steps 2–3 hc-step times
5. return the final best individual from HCP

3.5 Prediction

Partly to enhance predictive capacity, a meta repair operator is used across subsequent search processes which links real out of sample performance to each optimization over a historic window. This operator, r-SAME, [7] focuses the search within the neighborhood of previous solutions that worked well out of the sample. In addition several other methods were found to be useful for this purpose of increasing the generality of solutions including lengthening the data window beyond the forecast horizon so that the model is "fitted" on several different periods, a greater weight is then given to recent data to balance this requirement with an adaptive capacity.

3.6 Adapting the Model

The system is applied using a sliding data window framework so that as new data is observed the optimization procedures are rerun to obtain a model that matches the latest data. In addition the model is adjusted depending on the performance of a real portfolio managed by the system which is of course out of sample.

The repair operator r-SAME is used across fitness procedures to alter the genotype in such a way that it is no more than p percent different from a second genotype. When the optimization process is initialized the second genotype is the solution from the previous window. The parameter p is adjusted depending on the performance of a real portfolio in relation to the index which serves as a benchmark. It is reduced when performance is worse than the benchmark and increased if the real portfolio is out performing the benchmark. The rationale is to focus the search close to solutions while they give good performance and to broaden the search when this performance decays.

To calculate p such that it varies depending on portfolio performance is as follows:

$$p = \begin{cases} \frac{sharpe(r_{p,t})(1+k)}{2} & \text{if } r_{p,t} < r_{p,t-l} \\ \frac{sharpe(r_{p,t})(1-k)}{2} & \text{if } r_{p,t} > r_{p,t-l} \\ 0 & \text{otherwise,} \end{cases}$$

where $r_{p,t}$ is the return recorded for a portfolio managed by the system over one month prior to day t, $k, 0 < k < 1$ is a constant for the sensitivity to changes, l is a parameter of the distance to look behind period to measure performance. l was set equal to the interval in days between portfolio rebalancing events. Further $sharpe(r_{p,t}) := (r_{p,t} - r_{f,t})/(\sigma(\mathbf{r}_{p,t}))$ for risk free rate return $r_{f,t}$ and a vector of daily portfolio returns for the month before t, $\mathbf{r}_{p,t}$.

The length of a sliding window is set according to observed recent performance volatility. The base window length, max length, in months is reduced according to the formula:

$$\text{window length} = e^{kln(sharpe(r_{p,t}) - sharpe(r_{m,t}))} \times \text{max length,}$$

where k is constant set to control the sensitivity of the window change, $sharpe(r_{p,t})$ and $sharpe(r_{m,t})$ are the sharpe ratios of the managed portfolio and the benchmark index retuns, and $maxLength$ is the maximum window length. The new window date is never set earlier than the previous window start date. In the case that the window length is calculated as earlier than this, the previous start date is used.

3.7   Constructing Recommendations

Four separate optimization procedures are run using variations of the fitness evaluation functions to obtain rankings of stocks that are predicted to fall over a short and long horizon and that are forecast to rise also over a short and long horizon. This results in four rule base solution models with fitness x_1, x_2, x_3 and x_4 respectively

1. $\rho_{buy,H_1}^{x_1}$
2. $\rho_{buy,H_2}^{x_2}$
3. $\rho_{sell,H_1}^{x_3}$
4. $\rho_{sell,H_2}^{x_4}$

From these rule bases we can find four corresponding stock rankings on a day T after the end of the training data when we wish to apply the rules to construct a recommendation. Buy rankings $R_T^{\rho_{buy,H_1}^{x_1}}$ and $R_T^{\rho_{buy,H_2}^{x_2}}$ are generated using an evaluation function of type $eval_{H,buy}$ (equation 7) that is designed to pick stocks that increase in value over the forecast horizon. And the other two rankings highly rank stocks that are forecasted to decline in price. $R_T^{\rho_{sell,H_1}^{x_3}}$ and $R_T^{\rho_{sell,H_2}^{x_4}}$ are produced using $eval_{H,sell}$ (equation 9). Two forecast horizons, $H_1 > H_2$, are used to obtain predictions to a longer horizon (generally one year) and a shorter horizon (generally 3 months). The four rankings are combined to produce two rankings for taking long and short positions and for buying and selling. We combine all four together to obtain recommendation rankings by using some operations to amalgamate rankings. Two operations on rankings are used by the decoder.

One operator, $\odot$, takes more than one ranking as input and returns a single output ranking in which the rating for each element is an average of the ratings for the element in each of the input rankings. Symbolically two rankings, $A^{\rho_{x_1}}$ and $B^{\rho_{x_2}}$ are generated from rule bases with fitness x_1 and x_2 respectively. Both imply orderings of a set of listed stocks $M = \{s_1, s_2, \ldots, s_m\}$. We have the following definition:

$$A^{\rho_{x_1}} = \left\{ \left(s_{a_1}, a_{s_{a_1}}\right), \left(s_{a_2}, a_{s_{a_2}}\right), \ldots, \left(s_{a_m}, a_{s_{a_m}}\right) \right\},$$

where s_{a_i} is a stock with index a_i in M, i.e. $0 \le i \le m$; also $a_{s_{a_i}}$ is the rating of s_{a_i}. Note that by the ordering definition it is the case that $a_{s_{a_i}} \ge a_{s_{a_{i+1}}}$. Similarly there is another ranking B:

$$B^{\rho_{x_2}} = \left\{ \left(s_{b_1}, b_{s_{b_1}}\right), \left(s_{b_2}, b_{s_{b_2}}\right), \ldots, \left(s_{b_m}, a_{s_{b_m}}\right) \right\}.$$

An operator to combine rankings by weighted average of the fitness of the underlying rule bases is then defined with reference to A and B as follows:

$$A^{\rho_{x1}} \odot B^{\rho_{x2}} = \left\{ \left(s_{c_1}, c_{s_{c_1}}\right), \left(s_{c_2}, c_{s_{c_2}}\right), \ldots, \left(s_{c_m}, c_{s_{c_m}}\right) \right\}, \tag{10}$$

where the rating of each stock $s_{c_i} \in M$ is

$$c_{s_{c_i}} = \frac{x_1 a_{s_{c_i}} + x_2 b_{s_{c_i}}}{x_1 + x_2}.$$

The second operation, $\ominus$ takes a ranking and processes it so as to assign a rating of zero for any elements that have a non zero rating in the other. With reference to A and B above it has the following action

$$A^{\rho_{x1}} \ominus B^{\rho_{x2}} = \left\{ \left(s_{c_1}, c_{s_{c_1}}\right), \left(s_{c_2}, c_{s_{c_2}}\right), \ldots, \left(s_{c_m}, c_{s_{c_m}}\right) \right\}, \tag{11}$$

where

$$c_{s_{c_i}} = \begin{cases} 0 & \text{if } b_{s_{c_i}} \geq 0 \\ a_{s_{c_i}} & \text{otherwise.} \end{cases}$$

To combine rankings a decoder looks at the ratings given to stocks by each of the four rule bases found using the optimization and prediction components. A buy recommendation ranking is produced relevant to an application day T which is the first trading day after the training data window. In this recommendation each stock is assigned a rating that is an average of its rating in each of the rule buy rankings. In addition any stock that has a non zero rating from the sell rule bases is given a zero rating:

$$R_T^{buy} = \left(\left(R_T^{x1} \rho_{buy,H_1} \odot R_T^{x2} \rho_{buy,H_2} \right) \ominus R_T^{x3} \rho_{sell,H_1} \right) \ominus R_T^{x4} \rho_{sell,H_2}.$$

The average is weighted by the training fitness of rule bases used to construct the original rankings. The sell recommendation is found using the same process to interpret the two sell rankings. In this case we combine the sell rankings for longer and shorter horizons, and remove any stock with a buy recommendation.

$$R_T^{sell} = \left(\left(R_T^{x3} \rho_{sell,H_1} \odot R_T^{x4} \rho_{sell,H_2} \right) \ominus R_T^{x1} \rho_{buy,H_1} \right) \ominus R_T^{x2} \rho_{buy,H_2}.$$

Finally, to recommend transactions the recommendation takes into account the existing portfolio and the available budget. An equal weighting from available cash is allocated to each stocks with a buy recommendation with the constraints that the resulting portfolio contains stocks from at least 3 industry sectors and a minimum portfolio size. All stocks in the existing portfolio that have a sell recommendation are sold as are stocks for which the original buy recommendation horizon has expired. A recommendation to hold stocks already in the portfolio occurs if no sell signal is given and the holding horizon has not elapsed. The holding horizon is the average of $horizon_1$ and $horizon_2$. Fig. 12 summarizes the process to obtain recommendations. The output of the decoder is a series of recommendations to hold stock over time, the figure 13 shows a screenshot of a recommendation from the application. By following the recommendations an allocation of capital occurs over time. Fig. 11 shows the allocation by sector during a test run.

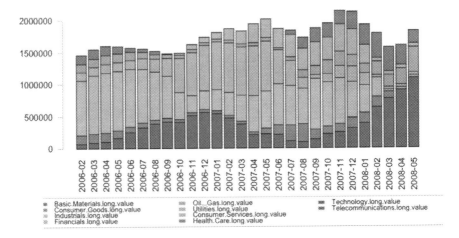

Fig. 11. Sector value allocated to long positions each month during a test run. The period included a term where financial stocks were highly recommended when this sector was performing well, subsequently these stocks became less attractive after the global liquidity crisis.

4   Performance and Concluding Remarks

We consider two different test scenarios and compare the performance with alternative approaches. Tests between 1991 and 2005 using stocks that comprise the MSCI Europe index are reported in [7]. Further results between 2001 to 2006 for another test data set from the Australian S&P ASX 200 are given in [8]. The testing process involved trading with (out of sample) historical data. Transaction costs and limitations according to reported trading volume were incorporated, initial parameter settings for the evolutionary algorithm and fuzzy representation was done using analysis of earlier data periods.

Performance of a portfolio managed using the system was compared with a passive index tracking portfolio benchmark and also several portfolios managed using alternate methods. Fig. 14 and Fig. 15 show comparisons of overall return performance for adaptive management and alternative strategies for the MSCI and S&P ASX200 test cases respectively. Table 3 shows return distribution characteristics for a portfolio managed using the methodology (EA) compared with a benchmark and also alternative strategies. And Fig. 14 shows the cumulative returns over the test period. Portfolios managed using buy and hold (B&H), alpha and price momentum (PM) strategies are considered as is a hill climber optimization heuristic (HC). The buy and hold portfolio is constructed by holding the same stocks at the beginning of the simulation until the end. The price momentum strategy [11] involves buying the top 10% performance stocks every 20 days and selling any stocks not in this set. A Jenson's alpha strategy (Alpha) is constructed using the single factor CAPM (see subsection 2.1), the portfolio is managed such that stocks with the highest alpha values

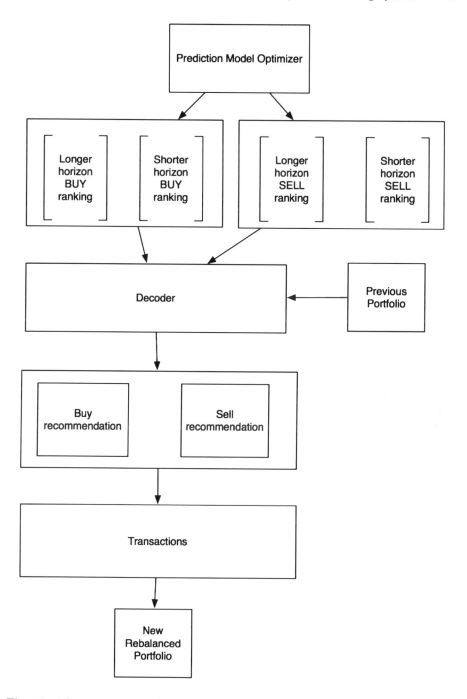

Fig. 12. The process to make a recommendation uses input from the optimization process to predict future events and the existing portfolio to consider diversification constraints. The result is recommended transactions that lead to a new portfolio.

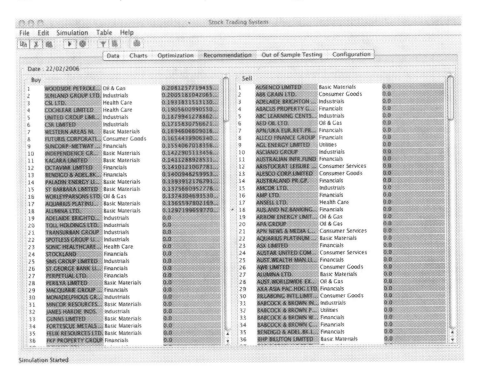

Fig. 13. A recommendation view from the application.

are always held at each rebalancing operation every 20 days. Finally the benefit of the algorithm is considered by comparison to a hill climbing optimizer (see subsection 3.4).

Let us discuss the MSCI and ASX test cases in turn. It is clear from Fig. 14 that over the holding period the application was able to outperform the other strategies. Over the 14 years of data the system would have increased an initial capital by 782.98 percent value compared with 187.25 for the index, 282.72 for the alpha strategy, buy and hold 224.38, 175.94 for the price momentum and only 78.2 using a hill climber to optimize the rules. To calculate annual returns a standard method in financial reporting is to multiply the mean daily log returns, $\ln(val_t/val_{t-1})$, by 260 trading days per year. On this basis the annual return performance was 19% for the system compared with $8.61, 10.78, 10.25, 8.46$, and 4.70 percent for the other index, alpha, buy and hold, price momentum and hill climber. A common measure of risk is annual volatility which is defined as the standard deviation of daily log returns times the square root of 260 (approximation of the number of trading days in a year). The annual volatility of a portfolio managed by the system was 18.96% compared with $18.07, 25.81, 19.69, 25.63$ and 16.98 for the others in the above order. From these figures we can conclude that the overall return from using the system was greater than all the other methods.

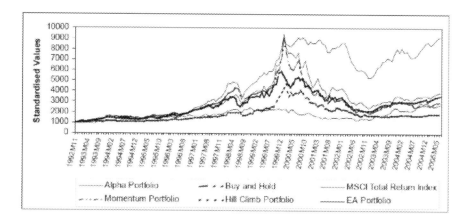

Fig. 14. Return performance for using the MSCI Europe data set. The methodology (EA) is compared with an CAPM alpha strategy, buy and hold, an index, price momentum and a hill climber heuristic.

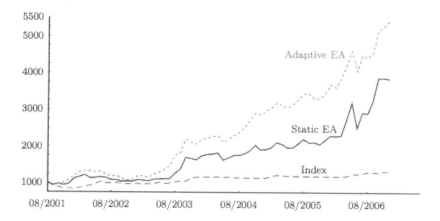

Fig. 15. Return performance in the Australian stock exchange. The adaptive management methodology is compared to static and index approaches.

The volatility observed in the index and hill climber were both better than that observed using the system.

To ascertain whether the additional returns are justified by the level of risk it is necessary to relate the two. A standard measure of this type is the sharpe ratio [14], to calculate this value the annual excess return above the risk free interest rate is divided by the annual volatility. The sharpe ratio for the portfolio managed by the system was found to be 1.0603. The index and hill climber were only 0.5307 and 0.2825 respectively. This indicates that the system managed its

Table 3. Standard return characteristics of a portfolio the system managed using the MSCI data set. The methodology (EA) is compared with an CAPM alpha strategy, buy and hold, an index, price momentum and a hill climber heuristic. Monthly returns are calculated on a discrete basis as a percentage change from one day to the next. The Jarque-Bera statistic is a chi-square distributed test for normality within the series. [a] signifies rejection of the null hypothesis of a normal distribution at the 1% significance level. The abbreviations of the headers indicate portfolios managed by the following methods: HC, hill climber heuristic; PM, price momentum strategy; B&H, buy and hold strategy; EA, evolutionary algorithm managed adaptively by the system; MSCI, an index tracking portfolio. The row description abbreviations refer to the following performance measures: MR, average monthly return; MMR, median monthly return; LPR, largest negative monthly return; LNR, largest negative monthly return; AMV, average monthly volatility; PLoss probability of a loss greater than 10% in any given month; PGain, probability of a gain greater than 10% in any given month; NM, number of months before a negative monthly return.

	MSCI	EA	Alpha	B&H	PM	HC
MR	0.8280%	1.6629%	1.2095%	1.0152%	1.0272%	0.5196%
MMR	1.1648%	2.0782%	1.7063%	1.3757%	1.5490%	0.7591%
LPR	14.0260%	17.0171%	28.6675%	22.3981%	29.8286%	18.0645%
LNR	-12.8476	-17.0083	-19.5658	-16.1300	-24.3551	-25.24%
AMV	5.4607%	5.1996%	7.4242%	5.6528%	7.3720%	4.9016%
Skewness	-0.0247	-0.1863	0.2633	-0.0322	0.0552	-0.667352
Kurtosis	3.0383	4.3629	5.3485	4.2690	5.8977	8.907055
Jarque-Bera	0.02491	12.7255[a]	36.9289[a]	10.2929[a]	53.6072[a]	226.1609[a]
PLoss	4.58%	2.61%	11.76%	5.88%	16.99%	4.70%
PGain	3.92%	4.58%	7.84%	0.65%	1.31%	4.00%
NM	2.4	3.1	2.7	2.5	2.7	2.3

portfolio in such a way as to outperform the others in return justified by the level of risk exposure.

Table 3 summarizes characteristics of the return distribution. The Jarque-Berra test for normality [2] indicated none of the portfolios tested are standard Gaussian distributions except for the index. The skew of the return distribution for all the portfolios except the price momentum was negative. On consideration of the returns and the probability of loss and gain greater than 10% given in the table we can see it is the case that large negative outliers are the reason for this in the case of the EA system portfolio. Kurtosis is able to be interpreted as an indication of tendencies towards larger and more regular swings away from the mean. The last line in table 3 shows the average number of months before a negative return occurs. Overall, despite the presence of some negative outliers, the system obtained a distribution with a "fat" positive tail with a balance of probability in positive returns to a greater extent than the others. A return distribution function that is desirable for a rational risk-averse investor.

Another set of test results are reported in [8]. Fig. 15 emphasizes the benefit of performance of the system using adaptive mechanisms (see section 3.6) including relearning the model each time the system is used for trading and using a varying length sliding window and different parameters based on feedback of system performance. A portfolio managed using the adaptive methodology is compared with an alternate "static" portfolio managed using an unchanging set of rules optimized at the beginning of the historic data set. It is observed in Fig 15 that the performance remains similar for the first 2 years after which the adapting portfolio begins to clearly outperform the static. Over the holding period the adaptive methodology obtained returns of 468% versus 304% for the static. Annualized return performance was 28% for the managed portfolio compared to 23% for the static. Annualized volatility was 20% for the adaptive method compared with 25%, indicating the adaptive mechanisms also had advantages in managing risk.

Performance analysis in the two scenarios suggest the potential of the approach to find general models with good predictive characteristics. This method can be used for guiding investment decisions by portfolio managers or to manage asset allocation without external input. The application recommends decisions consistently according to information available to it using a multi factor stock forecasting model it learns and adapts to present conditions. Evolutionary computation augmented by local search heuristics is used to optimize a forecasting model. Market instruments are ranked according to the extent they fit a prototype fuzzy specification of the model. In further steps the ranking is decoded and used to recommend buy and sell transactions for a portfolio of stocks over time. An adaptive component enables adjustments depending on observed conditions in the market and also management performance.

Acknowledgments. This work was supported by grant N516 384734 from the Polish Ministry of Science and Higher Education (MNiSW) and by the ARC Discovery Grant DP0985723.

References

1. Bäck, T.: Adaptive business intelligence based on evolution strategies: some application examples of self-adaptive software. Inf. Sci. Appl. 148(1-4), 113–121 (2002)
2. Bera, A., Jarque, C.: Efficient tests for normality, heteroscedasticity, and serial independence of regression residuals. Economics Letters 6, 255–259 (1980)
3. Brock, W., Lakonishok, J., LeBaron, B.: Simple technical trading rules and the stochastic properties of stock returns. Journal of Finance 47, 1731–1764 (1992)
4. Chincarini, L.B., Kim, D.: Quantitative Equity Portfolio Management. McGraw-Hill, New York (2006)
5. D'Mello, R., Ferris, S.P., Hwang, C.Y.: The tax-loss selling hypothesis, market liquidity, and price pressure around the turn-of-the-year. Journal of Financial Markets 6(1), 73–98 (2003)

6. Fama, E.F., French, K.R.: Multifactor explanations of asset pricing anomalies. Journal of Finance 51(1), 55–84 (1996)
7. Ghandar, A., Michalewicz, Z., Schmidt, M., To, T.D., Zurbruegg, R.: Computational intelligence for evolving trading rules. To appear in IEEE Transactions On Evolutionary Computation, Special Issue on Evolutionary Computation for Finance and Economics (2008)
8. Ghandar, A., Michalewicz, Z., Schmidt, M., To, T.D., Zurbruegg, R.: The performance of an adaptive portfolio management system. In: CEC 2008: Proceedings of the 2008 IEEE Congress on Evolutionary Computation, IEEE Computer Society Press, Los Alamitos (2008)
9. Grinold, R.C., Kahn, R.N.: Active Portfolio Management, 2nd edn. McGraw-Hill, New York (2000)
10. Iba, H., Nickolaev, N.: Genetic programming and polynomial models of financial time series. In: CEC 2000: Proceedings of the 2000 IEEE Congress on Evolutionary Computation, pp. 1459–1466. IEEE Computer Society Press, Los Alamitos (2000)
11. Jegadeesh, N., Titman, S.: Profitability of Momentum Strategies: An Evaluation of Alternative Explanations. Journal of Finance 56(2), 699–720 (2001)
12. Michalewicz, Z., Schmidt, M., Michalewicz, M., Chiriac, C.: Adaptive Business Intelligence, Secaucus, NJ, USA. Springer, New York (2006)
13. Michalewicz, Z., Schmidt, M., Michalewicz, M.T., Chiriac, C.: Adaptive business intelligence: Three case studies (2007)
14. Sharpe, W.F.: Mutual fund performance. Journal of Business 39(1), 119–138 (1966)

COLLANE: An Experiment in Computer-Mediated Tacit Collaboration

Tomek Strzalkowski[1,2], Sarah Taylor[3], Samira Shaikh[1], Ben-Ami Lipetz[1],
Hilda Hardy[1], Nick Webb[1], Tony Cresswell[4], Ting Liu[1], Min Wu[1], Yu Zhan[1],
and Song Chen[1]

[1] ILS Institute, SUNY Albany, Albany, NY USA
[2] Institute of Computer Science, Polish Academy of Sciences, Poland
[3] Lockheed Martin Corporation, Arlington, VA USA
[4] Center for Technology in Government, SUNY Albany, NY USA

Abstract. We introduce COLLANE, an experimental collaborative analytic environment that allows a group of professional analysts to work together effectively on complex, multifaceted information problems. COLLANE has been developed to investigate innovative ways of harnessing the power of collaboration so that to maximize the quality of the analytical product while at the same time controlling for its hidden costs: bias, groupthink, compromise, suppression of dissent and individual initiative. The key innovation that we are advancing in this project is the concept of *ubiquitous tacit collaboration* enabled through computer-mediated information sharing between the participants. By design, tacit collaboration requires no extraneous effort from the users since the information exchange is both automatic and targeted to what each analyst is currently doing. It also requires no specific "engagement" with subject matter experts since their continuous virtual presence assures ubiquity of collaborative opportunities. In this paper we describe an initial prototype of COLLANE, explaining its basic functions and components.

Key words: collaborative analysis, information sharing

1 Introduction

Collaborative work can be both highly efficient and tremendously constraining. A dedicated team can often quickly solve a difficult problem that would stump an individual analyst for a long time. One key advantage of teamwork is its efficiency: a complex task can be subdivided among the participants into manageable subtasks that may be accomplished in parallel, matching individuals' strengths and capabilities. Another important advantage of a team is its diversity of ideas and viewpoints: in an optimal situation, the strongest, most plausible solution is created that reflects the contributions of all group members. However, those apparent strengths of collaboration are also sources of significant problems. For a team to deliver efficiency, the task must be divided into discrete, coherent pieces that align well with the capabilities of individual analysts, and achieving this requires skillful leadership. At the same time,

M. Marciniak and A. Mykowiecka (Eds.): Bolc Festschrift, LNCS 5070, pp. 411–448, 2009.

too rigid a management structure may easily drive the group to an early consensus by promoting groupthink and suppressing alternative ideas or less likely hypotheses. This is clearly an undesirable side effect, which is often considered crippling in investigative analytic tasks where plausible conclusions need to be drawn from fragmentary evidence.

It appears then, that collaborative work may be a mixed blessing unless new ways are found to take an advantage of it while avoiding the pitfalls. In looking for a suitable model it is instructive to observe how analysts, in the government as well as in business, law, and other investigative professions, organize their work. Until recently, these organizations have been traditionally depending upon the work of individual analysts who have deep knowledge and expertise in specific areas (countries, organizations, technologies, etc.) and who tend to work independently producing reports and analyses as tasked by their agencies. Of course analysts do not work alone; in fact, they interact constantly with one another seeking advice, bouncing off ideas, or looking for leads. Specifically, they often consult experts in areas where they may have less experience. From a traditional viewpoint, none of this normally counts as collaboration, since the analysts may have independent tasks for which they are individually responsible (and also receive individual credits). Nonetheless, this informal networking is a vital part of the information gathering process; it also has some hallmarks of "good" though indirect collaboration: pulling in multiple perspectives, including alternative views, and adding critical feedback, while also keeping the overall case management coherent and motivated. Can this model be replicated and expanded into a true collaboration? Can a new generation of analytic tools be designed through which tacit collaboration be harnessed and managed in a way that improves the quality of intelligence overall?

The COLLANE project has been established to address these problems and to develop a computer-assisted analytic environment that can support effective collaboration while avoiding the drawbacks associated with more traditional forms of teamwork. The model that we advance in COLLANE is *ubiquitous tacit collaboration* where we attempt to capture some of the benefits of informal networking mentioned above but without the disruption of having to stop one's work and call another person for advice. In our model tacit collaboration is more than networking though; it is true collaboration, focused on the task at hand to which all participants contribute, albeit indirectly. Tacit collaboration does not require the participants to subdivide their work or to actively coordinate their activities; instead, they are assumed to pursue their individual lines of analysis on the same or related problems. Collaboration occurs, tacitly, because the system: (a) captures the associative knowledge generated by each participant when they query data sources and retrieve and retain information; (b) keeps track of what each participant is doing at any given time; and (c) shares relevant information and knowledge among the participants based on its relevance, timeliness, and usefulness. This continuous targeted information sharing has an effect similar to having several colleagues walk into your office as if on a cue and offer the information and advice that you require at this precise moment; however, this effect is achieved without the distraction normally associated with such activities. In other words, the relevant information is exchanged but no extraneous effort is needed to obtain it. As a result, the participants are aware of others' relevant activities and progress, past and present, which in turn informs and influences their own activities. Our hypothesis is that tacit collaboration, broadly defined, is more efficient and produces better quality analytic results than what can be achieved through individual work or through work in traditional open collaboration teams.

The COLLANE system has been developed to instantiate the above concept and to provide an experimental vehicle for exploring this and other forms of computer-assisted collaboration. The current, preliminary prototype can support up to 4 analysts working simultaneously on the same topic, and it incorporates basic information sharing capabilities sufficient for conducting meaningful evaluation experiments. To put things in perspective, the fully developed COLLANE system will eventually support a community of users and user groups working asynchronously on related topics. Furthermore, it will enable the automatic exchange of complex episodic and associative knowledge that is created by the participants' research activities. The current prototype was designed to support both tacit and open collaboration, as well as individual work by single analysts; this was essential for comparison between different work modes and also for deciding which of the system features need to be retained or expanded, and which new capabilities may be needed.

We have also designed an initial set of metrics for comparing both the efficiency and the effectiveness of each work mode, as well as for quantifying the user experience in each case. Some of the metrics were adopted from earlier evaluations conducted with single-user interactive information systems ([6]; [21]; [13]) to the extent that these metrics could be applicable in the collaborative setting. Nonetheless, developing a meaningful evaluation strategy for a collaborative system turned out to be a significant challenge. The focus of this paper is therefore as much on a description of COLLANE and the analytic experiments we conducted with it, as it is on the design of evaluation metrics that can effectively measure system performance in future experiments. Furthermore, due to a relatively small scale of the evaluation conducted to date, the results reported here can only be regarded as indicative of certain phenomena occurring in collaborative work. These will serve as a basis for developing more formal evaluations in the future.

In order to design a realistic exercise we turned to professional analysts representing various government agencies; we also asked these agencies to develop realistic analytic tasks. The analysts were presented with brief descriptions of problems, and asked to prepare comprehensive reports within a preset time limit. Analysts were using the COLLANE prototype through which they could search a fixed subset of web-mined data and collaborate.

The preliminary results from this study suggest that COLLANE-supported tacit collaboration has the potential to produce significantly higher quality analytic reports than would be possible when working alone or in open collaboration groups. This assessment is not easy to quantify using the existing methods for measuring the quality of an information product, as we will elaborate further in this paper arguing for new quality metrics. A better information product does not simply mean finding the most relevant evidence (although it matters, of course), or even arriving at the most likely explanation of this evidence (while this definitely counts too). It also means alternative interpretations of what is relevant and how the different pieces interconnect, and moreover how these different interpretations stack and rank against one another. This latter effect is almost impossible to obtain in a single analyst environment, and it is very difficult to see in a traditional, open collaboration because it is normally driven towards a consensus.

The following is a summary of key observations from the analytic workshop with COLLANE. We elaborate each point in the rest of this paper.

- Information sharing improves the analytic process. Analysts working in collaborative teams (open and tacit) are exposed to more topic-relevant information than analysts working alone. Therefore, we may infer that the collaborating analysts are using more evidence to arrive at more informed conclusions.
- The quality of the reported intelligence improves. Qualitative assessment of the reports prepared by the collaborating teams suggests that better and more conclusions are drawn by collaborating analysts than by analysts working alone.
- Tacit collaboration improves the analytic process by introducing constructive competition. Analysts working in tacit collaboration tend to pursue alternative or complementary interpretations of evidence. Unlike in the traditional teamwork, they are not being driven into consensus or compromise.
- Tacit collaboration exposes multiple interpretations of the available evidence: the outcome is a diverse portfolio of reports that facilitates the survival of all sound hypotheses.
- Tacit collaboration improves analytic productivity by inducing analysts to do more useful work per time unit. Tacit collaboration allows the less experienced analysts to benefit from the more experienced ones without necessarily slowing them.

The study also revealed that current methods for measuring quality of information products, based primarily on content precision and coverage, need to be revised:

- Current evaluation methodology is inadequate because it is geared towards a single output of the analytic process (report quality), does not support the evaluation of multiple outcomes of tacit collaboration, and penalizes minority dissenting views.
- New evaluation metrics are required to assess the relative value of a multi-faceted portfolio of reports and hypotheses covering a complex information problem. Ways are needed to rank the hypotheses and to quantify the information value of the set.

– Revised evaluation design is required in order to control for confounding factors such as level of experience, subject matter expertise, and analytic skills of the participants. In addition, we need ways of measuring the effects of differently skilled participants on team performance.

> *No one really works alone. [Analysts] always work in a collab-*
> *orative mode. [I need] to be able to ask collaborators, midway,*
> *take a look at my work and see if I am going in the right di-*
> *rection.* (Analyst D, COLLANE Experiment, 2007)

2   The COLLANE System

2.1   Overview

COLLANE is a collaborative analytical environment designed to enable ubiquitous tacit collaboration among a group of analysts working on the same or related information problems. The current prototype also provides a platform for evaluating analytic effectiveness and for experimenting with various collaborative settings. The two key capabilities of COLLANE that enable effective collaborative work are: interactivity and information sharing. Interactive features include question answering, question refinement, answer negotiation, and data navigation capabilities, which can be accomplished through natural language dialogue as well as through a visual interface. Information sharing includes the creation and maintenance of a combined answer space, targeted and time sensitive delivery of relevant data items, as well as distillation and exchange of exploratory knowledge accumulated throughout the analytic session. This exploratory knowledge arises from analysts' information access, retrieval, and assessment activities that interlink queries, data items, and any relevance or utility tags assigned by the analysts to the data items they view. Subsequent *automatic* interchange of knowledge thus captured allows analysts to continuously take advantage of each other's *relevant* actions and insights. One must note in this context that relevant information and knowledge are not limited to what may be considered supporting evidence for a particular query but includes complementary, tangential, even contradictory items that may be part of alternative hypotheses advanced by other analysts.

Analysts using COLLANE may do so remotely and in an asynchronous manner. This extends the notion of tacit collaboration to situations where some of the participants may be offline or otherwise unavailable; nonetheless, the collected information and associative knowledge they leave behind remains accessible and sharable under the same rules as before. Furthermore, we can include in our design "collaboration" with legacy analyses completed in the past by people who are no longer around, as well as with an analyst's own prior work or alternative approaches. In order to manage the totality of information and knowledge created by such a complex collaborative effort, COLLANE maintains the Combined Answer Space (CAS), an efficient data storage, which is continuously updated and always accessible. Any analytical action initiated by one participant, such as information search or relevance assessment, is automatically checked against CAS for the presence of any relevant items and their usage by other participants. This way the cognitive power of each analyst is maximized without creating an undue distraction or information overload.

2.2   COLLANE Design

COLLANE expands the interactive question answering technology in HITIQA ([15]; [12]) into a multi-channel, mixed-initiative interactivity that covers the entire analytical history of an information task. Unlike the more standard one-question/one-answer mode typical for the internet search engines such as Google, COLLANE can accept a series of interlocking questions keeping track of what the users have seen thus far. In addition to answering direct questions from the analysts, COLLANE acts as a coordinator and a facilitator, using various modes of interaction (both verbal and visual) to communicate similarities and differences among the information requested, collected, retained, and discarded by different analysts. Depending upon the session progress and the state of the emerging solution, the system may use different techniques, ranging from subtle to strong, to alert analysts of relevant activities by other participants in their group.

Any actions taken by an analyst while logged into COLLANE will effect changes in the Combined Answer Space. For example, new data items retrieved in response to the analyst's current questions are placed into CAS and their relevance to any previously logged questions, whether they came from this or another analyst are automatically assessed. The multi-modal dialogue manager (MDM) then decides how to notify each analyst about the changes that affect their individual workspaces. Some changes may be reflected in a dynamically evolving visualization, which can be immediately seen by all collaborating analysts, usually as minor background alterations. For other, more consequential changes (e.g., new or contradictory evidence) verbal alerts are used, i.e., a dialogue act is generated by the system in the form of a textual message, usually as a question or an offer that necessitates a response. COLLANE maintains a virtual individual working space for each analyst. In essence, an individual working space is a view of CAS in which the data items of interest to one analyst are made salient while all other data items are in the background or hidden from direct view. This allows MDM to conduct focused and meaningful interaction with each analyst rather than simply addressing them as a group. This interaction extends to the visual panels: each visual panel on a COLLANE client interface is a private view into the public space, reflecting a single user perspective. It allows an analyst to concentrate on his or her own work while also taking advantage of relevant aspects of other analysts' work. In particular, analysts may view and assess relevance of data items that belong to another analyst's primary view, thus altering their salience and indirectly affecting the other's workspace. It is the key function of MDM to make sure that such tacit collaboration has a positive effect on the performance of each analyst.

The analyst's interaction with the system occurs through a multi-modal dialogue that combines verbal (textual) exchanges and direct manipulations of the visual panel. As they work on a case, analysts ask questions, view system responses, save some items while ignoring others, inserting comments and annotations. By asking specific questions and by making certain decisions regarding the viewed data items, the analysts add their own knowledge, preferences, and biases to the system, and this knowledge, preferences, and biases are then shared with other analysts connected to the system. Such multi-way interaction has two effects: (1) it causes the individual workspaces to be populated with information related to their owners' queries, and (2) it alters the content of the Combined Answer Space. Unlike the localized changes in the individual work-spaces, any changes affecting the Combined Answer Space will be propagated to all participating analysts based on their relevance, thus further affecting their workspaces. Moreover, any explicit changes to one's individual view will immediately affect other views, although these effects may be only marginal, for example, other analysts may only notice changes that affect items currently visible to them. Depending upon the significance of these changes, the Dialogue Manager may utilize amplification messages to induce a desired reaction. The objective here is to keep all analysts current to the present state of interactions by facilitating but not forcing them to see each other's work and the effect of that work on the data.

The key advantages of the collaborative model outlined above are *completeness* and *efficiency*. Completeness is achieved through inclusion of multiple individual approaches and perspectives of a complex problem. Each analyst is now able to quickly identify alternative hypotheses to the present problem, by looking at the views over the same data created by other analysts' interactions. This helps, in turn, to identify where an analyst has perhaps missed certain evidence, or even where evidence directly contradicts the current working hypothesis. In this case, supporting documents can be identified, by looking into another analyst's folder. The Combined Answer Space plays a critical function in showing all emerging approaches in relation to one another. For example, the visual map of CAS may induce analysts to ask new questions, investigate new lines of inquiry or drop existing lines as no longer promising.

Efficiency is gained by accelerated collection and vetting of evidence. As all participating analysts work toward the same goal and share their insights and partial results, we gain the effect of parallel processing so that much of the duplication of work common in individual work settings can be avoided. Furthermore, since COLLANE does not preselect any specific roles for the analysts, the team can maintain dynamic flexibility by allowing each participant to take the initiative and explore avenues they find most promising. By asking questions to follow up competing or complementary hypotheses, the analysts effect changes on the combined answer space, thus *rapidly* (but indirectly) communicating the effects of their actions to others.

This model also significantly increases the *quality* of analytical products (which is subject to experimental verification, discussed in a later section).

One reason for this is the expanded evidential basis for the team report where each piece of evidence has been thoroughly vetted as potentially competing hypotheses are considered. This can be contrasted with one-analyst/one-system setting where the analyst advances a hypothesis, and the system then retrieves supporting or refuting evidence but it cannot on its own come up with an alternative. Another reason for increased quality may be the combined expertise brought into the task by all participating analysts. Experienced senior analysts are expected to exert considerable guiding influence on junior team members thus improving the quality of their work without overt instruction. A further reason for increased quality is the competitive element that promotes diversity of approaches in tacit collaboration: analysts are not driven towards a consensus or a compromise; instead, a portfolio of alternative and complementary analyses is generated.

2.3   Key Functionalities

The overall architecture of COLLANE is illustrated in Fig. 1. CAS is the central shared working space on any given information task; however, it is actively managed by the Dialogue Manager (MDM), which maintains individual views to best support each analyst. For clarity, we omit some components that have not been integrated yet, but will be discussed briefly later.

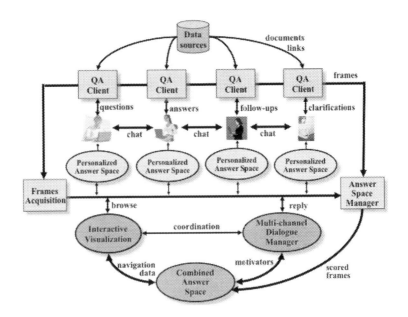

Fig. 1. COLLANE-1 system architecture

Combined Answer Space. The Combined Answer Space contains all information units under consideration by a group of analysts working on an information task (a case). Each time an information request is initiated by any of the analysts, all data items matching this request will be pulled out of available data sources. These data items, currently text snippets but in the future also multimedia files, are normally expected to have different degrees of relevance to the analytical problem at hand, as well as to the questions that analysts pose through their client interfaces. The initial assessment of relevance is computed by the system in response to specific questions, but it may be subsequently modified through interaction and other analyst actions such as saving an item in a "shoebox" or a draft report. Due to differences of opinions and approaches between the analysts, this scoring system is inherently multi-valued.

CAS is built out of all the retrieved information units, not just those that may be used by analysts in their final reports. In order to facilitate uniform handling of all information types by the system, each information unit (or a group of like units) is assigned one or more *event frames*. Event frames are template-like structures that represent the content of the underlying information unit: usually an event (e.g., an agreement, a transfer, an attack, etc.). Frames are classified into a dozen or so types, which are automatically defined for each subject domain (e.g., trade, politics, terrorism, emerging technologies, etc).

Each frame provides an "access handle" to the original information unit through which the system can compare and manipulate information content *regardless of their origins*. Our experience has thus far been primarily with text-extracted frames; however, similar structures can be extracted from e.g., video clips, after which they can be handled transparently by COLLANE.

A frame represents only a portion of the information contained in the original unit: the predicate, key attributes and selected modal operators. For example, $ATTACK(X,Y,Z)$ represents an event where X attacks Y using (weapon) Z. Additional attributes specifying time, location, and modality (e.g., past, future, alleged, denied, etc.) are extracted as well; *for instance, In northern Baghdad, the owner of an ice cream shop was shot dead outside his store on Sunday morning...* We have developed an ontology of basic event frames that cover a number of analytic domains. For example, the weapons proliferation domain includes several basic events that characterize this domain from a national security viewpoint: *TRANSFER, DEVELOP, AGREE*, and *ATTACK*. These basic event frames instantiate to specific events reported in the information sources; for example, Iraq importing uranium from France would be an instance of *TRANSFER* frame, as shown in Fig. 2. Other domains will use similar sets of basic relations.[5]

CAS is built and managed by COLLANE as the case analysis progresses, and over time, it can become quite complex. Below is a summary of the key elements:

---

[5] For more details about the system of event frames and how they are acquired, the reader is referred to ([5]).

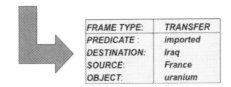

Fig. 2. A text snippet is assigned a TRANSFER frame

1. *Frames representing the retrieved data*: (text passages, XML snippets, other media types) with one or more frames per information unit. In addition, frames representing the same exact event are merged with their attributes combined, in which case a single frame may represent a group of information units.

2. *Relevance scores for each item*: Since information units may be retrieved in response to different questions, multiple scores are assigned to each along with the corresponding pedigree (which question, whose question, any amendments). Relevance is then assessed to all questions posed, including those answered previously. In addition, any direct actions by any analyst with respect to a particular item (mark as relevant, mark as non-relevant) are captured. This allows COLLANE to calculate and display a combined score of each item with respect to the overall analytic case.

3. *"Ownership" information*: Items retrieved in response to questions posed by individual analysts are assigned to them so that an individual answer space can be identified. Clearly, such individual spaces may overlap in various ways, but it is important for an analyst to know where their work is located vs. other analysts. Furthermore, information sharing is expected to be more effective when it is passed along with such essential context as "who's got it?", "who's seen it?", "what they did with it?", etc.

4. *Cross frame links*: While this feature is not currently implemented, frames will be linked by shared attributes forming various chains: temporal, geospatial, person/organization. Additional linkages may be inserted by link analysis components external to COLLANE (e.g., social, communication, etc.)

A schematic (and highly simplified) illustration of CAS is shown in Fig. 3. We should note that the overall information model does not have to be consistent—inconsistencies are expected to arise from analysts pursuing different approaches and forming incompatible hypotheses. In particular, relevance assessments for each data item will likely vary between analysts for a variety of reasons, including differences of opinion but also utility of a particular item to each analyst's workspace.

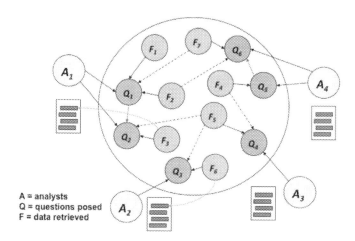

Fig. 3. A schematic illustration of CAS showing questions posed and data items collected by 4 analysts. Links between question nodes (Qn) indicate order in which they were asked; links between frames (Fn) representing data items and query nodes indicate relevance.

Multiple Views of CAS. The Combined Answer Space holds the entire evidential history of an analytic case. While all information is available and persistent, it is not necessarily viewable all at once. Instead, *multiple views* of it are created as required to support each analyst. In a typical case, each analyst's primary workspace is in focus while the remaining parts of the Combined Answer Space form a background. This is necessary to communicate its content effectively, whether by graphical means (visualization techniques) or by verbal dialogue. Here is how COLLANE uses the Combined Answer Space to interact and to facilitate teamwork:

1. *Supporting Dialogue Manager*: CAS is the primary data structure supporting the Multi-channel Dialogue Manager. Much of the dialogue generated by the system arises in order to resolve any perceived inconsistencies in the model as well as to negotiate the scope of the answer space.
2. *Rendering into interactive visual display*: For each analyst, her/his workspace is displayed and organized around the questions they ask. The display includes only these data elements that are necessary to answer the questions and to support effective interaction.
3. *Rotating views*: An analyst may "rotate" available views to examine other analysts' work-spaces. This feature is only partially implemented at this time. The plan is to support switching from one analyst's scenario-level view to another's. While in another workspace, the analyst's own data items can be viewed in relation to the items collected by the other analyst.
4. *Browsing*: Analysts can browse the visual display, access original data items, change/add their relevance assessment, and copy them into their reports (or "shoeboxes"). Changing relevance assessment of a data item provides a direct

feedback from a user, and this information is propagated to other users depending upon their individual circumstances and using an appropriate communication channel (as discussed further below).

5. *Analyzing information from retrieved evidence, chat messages and user copies*: In addition to matching retrieved text passages (or other media types) with user questions, CAS keeps track of questions and statements transmitted through COLLANE's chat interface, which allows analysts to communicate directly if they so choose. If an analyst's chat statement, for example, is found to match a question some other analyst has posed to the system, CAS and the MDM will share that piece of information with the other analyst. Likewise, CAS will search among retrieved items and present to the user any that are relevant to questions he or she has posed through chat. If an analyst finds and copies a passage of text from a full-document link (that is, a passage not identified by the system as highly relevant), CAS will analyze the new passage and look for matches among questions already asked, thereby enriching the answer space for other analysts.

Hypothesis Footprints. A natural consequence of the collaborative design outlined in the previous sections is the multi-dimensionality of the analytic process in COLLANE. Each analyst on the team may pursue a different strategy and consider alternative, even contradictory hypotheses. These hypotheses are not necessarily apparent to an observer, as the analysts may be exploring various options that appear promising at one time or another. The totality of the analyst's actions while pursuing a hypothesis: evidence collected, questions asked to collect it, assessment of this evidence for relevance, responses to system suggestions, and reactions to other analysts' progress—all these elements form an information "footprint" left by this analyst while pursuing the hypothesis. In COLLANE we define a concept of *hypothesis footprint* to be the set of all actions performed by an individual analyst while considering a specific hypothesis.

The main advantage of computing hypothesis footprints is the ability to recognize that analysts may be pursuing different approaches to a problem. We should note that this is only meaningful in a collaborative environment, where analysts can be made immediately aware of such alternative approaches. The system is unlikely to guess, based solely on the information in the footprint, which hypothesis is being considered, but radically different footprints left by two analysts may signal that they are pursuing different approaches. The system may now attempt to reconcile these differences by making the analysts aware of each other's progress, thus further accelerating the analysis.

In the course of their work on a case, each analyst may pursue several hypotheses, thus an additional technical challenge is to determine where one hypothesis ends and another begins. Moreover, while a particular footprint cannot be used to prove or disprove a hypothesis, we might be able to discern from the analyst's actions (saves vs. discards, line of questions, etc.) whether he or she succeeded or abandoned an approach. This aspect of COLLANE is not yet fully implemented. In its current form a hypothesis footprint is represented as an undirected acyclic graph (Fig. 4) with analyst's actions as nodes and the data items associated with those actions as leaves. This structure allows for swift and straightforward comparisons between two distinct footprints, as well as for finding characteristic episodes within a single footprint that may indicate approach boundaries.

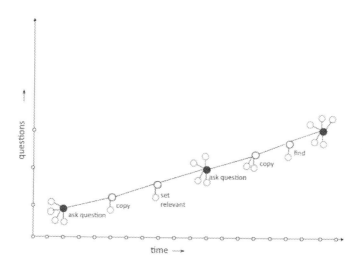

Fig. 4. Graphical representation of an analyst's Hypothesis Footprint

Multi-Channel Interactivity. A key aspect of COLLANE is its interactivity, which allows for efficient information exchange between the analyst and the system. Analysts may negotiate the exact scope of each question, receive suggestions on how a question may be reformulated or expanded, and be alerted about any related or contradictory information found. Interactivity facilitated by the system also encompasses direct and indirect communication between the participating analysts. Indirect communication occurs when one analyst's actions affect the workspace of another analyst, which may in turn cause the second analyst to alter his or her working hypothesis. The interaction may proceed verbally or visually with the Dialogue Manager selecting the optimal means depending upon the urgency of a communication, and other contextual parameters.

Given a complex information problem, an analyst would normally pose a series of questions, probing for specific details that could support an unstated hypothesis or else open avenues for further exploration. The choice of a particular line of questions may reflect the analyst's background, prior knowledge of the subject matter, or other biases. Each question may be compared to a narrow spotlight shined into massive data, which means that an answer, even if correct, produces only a "dot" of information at a time. Therefore what questions are asked and how they are asked can make a huge difference in the final outcome.

One way to improve the odds of finding and connecting the right "dots" of information is by adding a broader context "halo" around each answer returned and this can be accomplished by engaging the analyst in a *dialogue* to evaluate additional information items that appear highly related to the direct answer. This has the effect of providing a broader evidence context to the answer reported and increasing the analyst's success rate. Even if the contextual information is not relevant to the problem at hand, its presence may strengthen the answer selected; i.e., it may indicate that further exploration in a particular direction is unlikely to be useful. We have accumulated experimental evidence based on our work with active duty analysts that this contextual dialogue increases both the speed and the quality of the analysis and frequently leads to additional information nuggets that analysts utilize in their reports ([15]; [20]; [21]; [9]; [6]).

This human-machine interactive analysis is significantly accelerated in a *multi-channel, multi-thread dialogue* situation. When multiple analysts approach the same problem, they are likely to do so from different perspectives, thus cutting different evidence paths though the massive data. Each of these evidence paths may reflect the pursuit of an alternative hypothesis, thus becoming a hypothesis footprint, as discussed above. By creating a particular hypothesis footprint, an analyst communicates to COLLANE, and indirectly to other participating analysts, that they are pursuing a particular approach. When these footprints are combined and compared, a significantly larger, multi-dimensional evidence base is obtained. This provides a much wider context for each data item that the system may now provide to each analyst, creating even more opportunities to continue the search. Still more importantly, from the perspective of any one analyst on the team, the system is now significantly more responsive and *forthcoming*: it provides active feedback to all direct questions asked <u>and</u> also explicates alternative explanations for evidence, based on what other analysts are doing.

In COLLANE's collaborative environment, the quantity of information relevant to an individual analyst rapidly increases, and we require an efficient interaction mechanism without overwhelming the users with streams of communication. The key function of the Multi-Channel Dialogue Manager (MDM), in addition to "regular" human-computer interaction support, is to alert each user to new information as well as new, promising lines of investigation relevant to their enquiry, as they are being uncovered by other collaborating analysts.

Several interaction decisions are taken by MDM as to how, and when, to alert the user about developments outside their individual workspaces. For one thing, COLLANE will only engage in a dialogue when the nature of the information is such that intrusion into analysts' current work process seems warranted. For example, Analyst A has asked a question some time ago. Analyst B now asks a new question, which results in new data items, some of which match the original question of Analyst A. Our decision on how to handle this new, matching information depends on what Analyst A saw in response to the original question. If the original answer appeared satisfactory (i.e., Analyst A saved new information into their "shoebox" or a report) then the update to A's working space will be silent and unobtrusive. For example, relevant items are silently dropped into appropriately labeled folders in the analyst's workspace, and a visual "flag" is raised over the folder, mailbox style, to indicate a new item arrival. On the other hand, if the original answer was not satisfactory, as evidenced by lack of copied material and possibly several fruitless followup questions, then a more visible "highlighting" of relevant folders in the visual workspace is used to alert the analyst that the data item he was unable to find has been located and may be viewed now. In an extreme case when a newly discovered data item appears•to contradict some earlier findings, or is an entirely new data item, where no previous data was seen, the system may engage the user in a direct verbal dialogue.

---

(1)Analyst A: *Is there any evidence that man-made artificial reefs are beneficial?*

(2)COLLANE: Displays matching results to Analyst A

(3)Analyst B: *Where has sea life increased due to artificial reefs being constructed?*

(4)COLLANE: Displays matching results to Analyst B, including some results deemed relevant from the previous question of Analyst A

(5)COLLANE: Displays to Analyst A any new results retrieved by Analyst B

Fig. 5. Interaction between COLLANE and 2 analysts

In the example in Fig. 5, Analyst B asked a question which included more specific details (possibly based on prior and tacit knowledge) than the initial question of Analyst A. By using this specific hypothesis, sea life increase, COLLANE is able to retrieve new information for *both* analysts. How this new information is displayed to Analyst A depends on their prior actions. If no relevant data was seen in response to the initial question (1), COLLANE will initiate verbal dialogue at step (5), directly informing Analyst A of new, directly relevant information. If the new information complements a partial answer, COLLANE may choose a less disruptive notification through visual display, e.g., raising a "new arrivals" flag on a folder.

More generally, given an emerging solution to an analytic problem, the system employs a series of dialogue moves, which may be either verbal or visual, in order to draw the analyst's attention to a particular detail, or an issue, or some changes that may be occurring. A dialogue move is a particular manner of communicating information, which also aims at eliciting a desired reaction from the user. Some dialogue moves are more direct than others, e.g., a direct question usually compels the other party to respond ("*Would you be interested in information on new marine habitats?*"), while an open-ended offer may be ignored or put aside ("*Please check these when you have a moment*"). The selection of which dialogue move to employ and when to employ it is all-important because the dialogue should never become a distraction or nuisance to the analysts. A combination of verbal and visual communications gives the system significantly more options that could also be deployed simultaneously. For example, a continuous complex change in data may be more readily visualized than described.

Timing is also important for such actions to achieve the desired effect. The analyst may have aborted an earlier line of questioning altogether, so we must be careful as to how we inform them, not to presume that the new information is still vital to the completion of the task. This is where the tracking of hypothesis footprints becomes critical—a feature that allows COLLANE to tell which questions remain open and which are no longer active. The Combined Answer Space provides the structure necessary for efficient dialogue with all users.

To summarize, the role of the MDM module is to accept direct queries from each user, decide what additional information is needed, when it is needed, and how to get it: by asking the user, by observing the user, or by inducing some action from the user. The system continually measures disconnection between the user's interpretation of the analytical problem (a hypothesis) and the content of the answer space obtained (the evidence). This can manifest itself as a mismatch between the questions posed, the relevant information found, and the information retained by the analyst. The objective is to make the user an active and effective participant in the information-seeking process, but to do so in a manner that is unobtrusive and naturally fits with the task flow.

The approach described above should be contrasted with more standard information system approaches to interaction. For example, most current "interactive" information systems implement only very basic forms of interactivity, typically variants of passive feedback. Document retrieval engines, including Google, are good examples of this: the user must decide if and how to revise the query to get better results. In dialogue research, early forms of interactive systems used fixed menus to guide users through a maze of options often unrelated to the user's information need. While theoretical research on dialogue modeling has made good progress ([1]; [18]; [8]; [22]; [23]), the practical implementation still lags behind. A significant development was the AMITIES project ([4]), which delivered a practical implementation of the data-driven dialogue approach, and was subsequently adapted to the unstructured data in HITIQA ([17]; [11]). Another related area is research on multi-party dialogue (e.g., [19]; [7]; [3]); however, this work concentrates primarily on the structural and functional aspects of the interaction, rather than on the information exchange, which is critical for COLLANE.

Direct Communication Among Analysts. To facilitate inter-analyst communication, COLLANE provides a chat mechanism by which analysts (or teams of analysts) can communicate directly. Based on the experiments we conducted, analysts use chat primarily to bounce off ideas and to ask questions of each other related to the current scenario, e.g., to inquire if others had a better luck with a particular topic or exchange prior knowledge about a topic, etc. Other uses we noted include various forms of work coordination among analysts. While COL-

LANE does not require analysts to collaborate openly, this direct communication channel allows them to subdivide a complex task and to exchange suggestions and advice on the progress thus far. The chat channel also complements tacit collaboration particularly when information sharing appears slow ("*Can't find anything on opposition to cabinet restructuring, did you?*").

COLLANE considers chat exchanges between analysts as another source of information that may reveal analysts' prior knowledge of a topic as well as other assumptions they make. Using discourse analysis tools, such as a general domain Dialogue Act tagging mechanism ([22]), COLLANE spots key excerpts in this information interchange; specifically we identify classes of questions and statements relating to known types of named entities, e.g., people, locations, organizations, etc. In a statement, we are looking for novel data items that may represent tacit knowledge exchanged between collaborating analysts: such knowledge is captured into the CAS, although it is also clearly identified as having originated in inter-analyst chat. For questions posed through chat, COLLANE will search available data sources for candidate answers, as well as the CAS for similar questions already answered by other analysts. In this way, COLLANE augments analyst collection ability by complementing the external data sources with the shared knowledge built by the collaborating team, currently and in the past.

Intuitive Conceptual Visualization. The role of visualization in COLLANE is twofold:

- The primary role is to create a representation of the Combined Answer Space that would allow the global view of all collected information and analyst individual views to coexist on an interactive display;
- The secondary role (but no less important) is to extend the capabilities of human-machine dialogue by allowing a greater variety of means of communication: non-linear dialogue acts, low-disturbance messages, and subtler alerts.

Our main objective in COLLANE has thus far been to develop a conceptual design for the visualization interface. In the future, we plan to develop a more advanced graphical rendering. Based on a series of user-centric experiments, we have identified the following requirements for the effective interactive visualization required to support COLLANE:

1. Effective visualization must clearly communicate the current content of the Combined Answer Space and the progress of the analysis.
2. The visualization must let analysts alternate between wide (more context) and narrow (individual workspace; specific aspect subspace) views of the answer space.
3. The visualization must allow for easy viewpoint "rotations" so that different analyst's views can be switched to as needed.
4. The visualizations must complement the verbal (text window) dialogue. Ideally, both means of communication should mix seamlessly and naturally.

In the current version of COLLANE, the default view of the combined answer space is associated with the most recent question that an analyst posed to the system. This view shows all information units considered at least partly relevant to the question including these just retrieved and others that may have been found previously by other analysts. It supports the analyst's current focus and also allows for detailed content negotiations to occur via dialogue. Figure 6 illustrates this; the reader should note that the icons (representing individual information units) are organized into groups (e.g., by event types) but not necessarily by relative semantic compatibility or "distance", which are difficult to determine objectively. The focus is thus on clarity and ease of perception.

In Figure 6, we note that a user question "*When was Teflon invented?*" returned a number of information units that fall into 6 groups: two groups of development events (DEV label), one group of transfer events (TRF label), a single attack event (ATT label) and two groups of other general events (GEN label). Color-coding represents the degree of relevance computed by the system (discussed further below), while the icon shape identifies the source (direct retrieval or shared from another analyst). The answer found by COLLANE's integrated QA system appears on the right panel. The analyst may change the view to any previous question by bringing it into focus from the answer folders panel (pictured in the screenshot on the right side of Figure 6).

From the question-level view (left side of Figure 6) the analyst may zoom in on each of the event groups thus entering an event (or frame) view (right side of Fig. 6). At this level additional details about each event frame are displayed, including key attributes, such as Agent, Object, Time, Location, etc. The analyst may also consult the underlying data sources and toggle the relevance assessment of each item.

As noted above, icons on the visual panel represent information items (text passages) that are delivered to the analyst in response to the most recent question. These items may come from any number of external sources that COLLANE may search as well as from the data collected by other collaborating analysts in the Combined Answer Space. We briefly explain the significance of colors and shapes of the icons. All icons represent salient events and relationships found in the source text passages, with key attributes displayed on the periphery. The color indicates the degree of relevance to the user question as estimated by COLLANE algorithms: the dark blue are most relevant, the red are least relevant. The dark blue items are automatically saved into answer folders, while the red items are considered not relevant although potentially useful for contrast or as context. The intermediate colors (light blue, green, yellow, orange, etc.) indicate items with increasing numbers of conflicts with the question, which nonetheless may prove relevant upon closer examination. As already discussed above, COLLANE may engage the analyst in an explicit dialogue in order to clarify the relevance of some of these items; however, the color system itself is a form of silent (visual) dialogue about items which are left for the analyst to examine.

In addition to colors, COLLANE uses icon shapes to represent the origins of the information. Circular icons (shown in the pictures in Fig. 6) indicate original

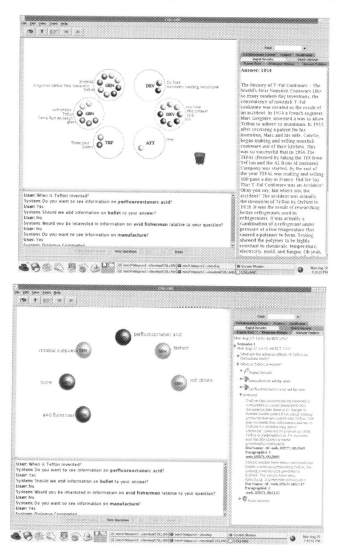

Fig. 6. Individual workspace views with frames grouped by event type and key attribute (top display) and individual data icons inside one of the groups (bottom display). Color-coding indicates the degree of relevance to the question (dark blue—seen as the darkest shaded icons—are most relevant)

new items directly retrieved from external data sources. Triangular icons (not shown) represent related items found by other analysts either in the past, or concurrently, or possibly in the future (this latter possibility arises when the analyst reexamines the answer to a prior question). Currently, only the most relevant items from other analysts' workspaces are displayed, i.e., the items that would receive dark-blue coloring on the visual panel.

From the question level the analysts may also zoom out to the Scenario-level view where they can see all their questions posed in connection with the current task. This view provides a wider perspective on the work done up to this point and helps the analyst to assess the state of task completion. At this level it is also possible to meaningfully compare the progress with other analysts on the same team. This can be accomplished simply by switching to the scenario views of other analysts and noting the questions they posed, the answers they obtained, as well as their assessment of any common data items. The scenario view is currently under development and has not been included in the version of COL-LANE we tested. It is likely to display multiple question groups in a reduced resolution that can be zoomed into by passing the mouse cursor over them.

3   Experimental Evaluation

The initial prototype of COLLANE described in the preceding sections was built during the first year of the CASE Program to support up to 4 analysts working simultaneously. While the research and implementation process has only begun, the system's development reached the point where direct feedback from potential users was required to assess the progress made thus far and to prioritize the challenges lying ahead. To do so, we have organized, in close collaboration with the U.S. National Institute of Standards and Technology and other government organizations, a collaborative analytical exercise to evaluate COLLANE performance on realistic information analytic tasks. The primary purpose of this exercise was to assess whether our testing methodology can support a meaningful evaluation of collaborative systems in general and COLLANE in particular. The secondary objective was to obtain a preliminary measurement of effectiveness of information sharing in COLLANE.

In September 2007, the team has conducted a five-day on-site analytical workshop with a group of 8 professional analysts representing various information services of the U.S. Government. During the workshop the analysts were presented with a series of realistic information problems of strategic nature and asked to prepare draft reports on each problem within a preset time limit. Analysts were divided into several groups and each group used COLLANE to collect and organize information, to prepare one or more draft reports, and, when appropriate, to collaborate. Each group worked under different conditions: open collaboration, tacit collaboration, or individually, as will be explained in more detail below. The searchable dataset consisted of approximately 2 GB of text documents premined from the Internet. We were interested in comparing the quality of reports produced by each group, as well as the effort expended and the user satisfaction; specifically:

1. Evaluating efficacy of tacit collaboration technology, specifically automated, targeted information sharing, in solving complex information tasks under limited time and resource conditions.
2. Comparing several collaborative and individual work settings and how they affect the use of the technology and the outcome of the analysis.

3. Determining if the current evaluation design is feasible and sufficient to measure the impact of various forms of collaboration on the analytic process and on the quality of the results.
4. Gaining insight into how the current COLLANE technology needs to be advanced to obtain a more effective tool.

3.1   Overall Evaluation Principles

Our objective was to design a methodology for evaluating the effectiveness of collaborative systems, such as COLLANE, for solving complex information problems by teams of analysts. This takes into account the following key dimensions:

1. *Quality of Solution*: an objective measure of how well the problem has been solved. This includes importance (criticality), coverage (completeness), precision (non-redundancy), and organization of the final report. This quality can be assessed by a panel of experts.
2. *Quality of the Process*: an objective measure of how the analytic process is affected by the technology. This includes accuracy of intermediate steps, rate and timeliness of information sharing, effectiveness of dialogue, etc. This also includes more complex measures such as the number and quality of hypotheses considered, the depth of the information search, and the rigor of the attempts to prove or disprove hypotheses, etc. This quality is measured by a combination of standard accuracy metrics (recall, precision, MRR) and the analysis of the structure of the interaction logs left by each analyst.
3. *Effort Expended*: an objective measure of the user effort expended to obtain the solution. This includes elapsed time, the number and types of steps required, number of sources consulted, etc. Effort is estimated from system logs that capture all significant task events and time stamps.
4. *User Satisfaction*: A subjective perception of difficulty of the process and confidence in the resulting solution. User satisfaction can be measured along many dimensions primarily through specially designed questionnaires.

We note that the above dimensions are partially orthogonal and which of them is more important depends upon the nature of the task. In most tasks the report quality will likely dominate other criteria; however, the process quality may be a necessary pre-requisite, i.e., it is hard to draw good quality conclusions from poorly collected evidence.

The evaluation process that we envisioned consists of two major stages. We first establish benchmarks for comparing effectiveness of multiple tools and work modes by conducting a series of end-to-end evaluations in a controlled environment. These evaluations must involve real analysts and realistic tasks and data sources in order to produce reliable outcomes, i.e., under which conditions we can expect a particular level of analytic performance. Once the benchmarks are in place, we can attempt a hands-off automated and predictive evaluation (i.e., how is a tool "doing"). In order to accomplish this, we need to isolate intermediate *performance indicators*: automatically measurable variables of which values can be aligned with specific end outcomes. Such indicators may include: the number of data items shared, the number of data items retained per unit of time, the number of messages exchanged between analysts, the time spent searching vs. reviewing, etc.

The workshop reported here constituted only a "dry-run" of the first step in the above 2-stage process and its main purpose was to test the mechanics of the first phase evaluation before a longer-term study is attempted. The key objective was to see if the existing evaluation design, as well as the instruments and metrics are in fact adequate for measuring the effect of collaboration on the quality of analysis, and if not, what other or additional instruments and experiments may be needed. As it turns out, the workshop raised more questions than it answered; but it also exposed that the current evaluation design and metrics may not be sufficient.

The Analytical Tasks. Realistic analytical tasks were prepared with assistance from the sponsoring agencies. The tasks were selected and formulated to allow analysts to complete a report within the time limit set by the workshop organizers (2.5 hours per topic including time for report editing). The topics were selected to concern recent events of potential general interest, but with which the analysts were not likely to be very familiar. This last provision was included to minimize analysts' reliance on prior knowledge and thus to place more stress upon the system. We therefore selected 7 topics for evaluation that did not assume specialized knowledge on the part of the analysts but nonetheless displayed sufficient structural complexity (also reflected in the richness of the data available) to require both an analytic strategy and discipline in order to write meaningful reports. Here are titles of the selected topics:

(0) L-3 Communications Holdings, Inc
(1) Effect of Focused Vibrations on The Human Brain
(2) Tainted Chinese Food
(3) Risk of Cancer from Teflon-coated Products
(4) Artificial Reefs
(5) Honeybee Disappearance
(6) Chinese/Hong Kong IP Counterfeiting Operations

Each task was described using a brief narrative; below is an example task formulation:

Risk of Cancer from Teflon-Coated Products
Please gather evidence and report on whether or not the use of Teflon-coated products (i.e., pans, pots, cookware) causes cancer in humans. List any reactions Teflon may cause when introduced by any means into the human body. Describe the current state of research and evaluation on Teflon. List the organizations, government or otherwise, that are responsible for Teflon product safety, and evaluate the degree of bias in studies on this product. Add to your report any other relevant information.

Table 1. Schematic Workshop Schedule

	Day 1	Days 2, 3 and 4			Day 5
9-10 AM	Orientation	Analytic Topic 1	Analytic Topic 3	Analytic Topic 5	Final Debrief
10-11 AM	Training				
11-12 AM		Peer Evaluations & Questionnaires	Peer Evaluations & Questionnaires	Peer Evaluations & Questionnaires	
12-1 PM	Lunch break				
1-3 PM	Warm-up Task	Analytic Topic 2	Analytic Topic 4	Analytic Topic 6	
3-4 PM	Peer Evaluations & Questionnaires	Peer Evaluations & Questionnaires	Peer Evaluations & Questionnaires	Peer Evaluations & Questionnaires	
4-5 PM	Group Discussion	Group Discussion	Group Discussion	Group Discussion	

Workshop Schedule. Table 1 shows the schematic training, work, and feedback schedule for the exercise. The first day was devoted to training and warm-up tasks to assess analysts' proficiency level with the system. The last day consisted of debriefs and focus groups with the participants.

Work Modes. During the workshop analysts worked in groups of different sizes. Each group operated under one of the following conditions:

1. *Individual Work, No collaboration* (INC). Analysts were expected to prepare individual, independent best reports. No contact between analysts was allowed. This was the baseline against which other groups would be measured. At least 2 analysts were in this group.
2. *Individual Work, Tacit Collaboration* (ITC). Analysts were expected to prepare individual best reports, just like in INC; however, they were allowed to tacitly collaborate, with the system facilitating information sharing on as-needed basis. This work arrangement required no top-down coordination or management; analysts were free to share information and to communicate but each was expected to pursue an independent work strategy, and produce individual (though not necessarily independent) reports.
3. *Joint Work, Tacit and Open Collaboration* (JTC). Analysts were expected to collaborate by any means available. They were also required to prepare a single combined report, thus an up front division of work was normally assumed, with a leader/assembler elected at the outset (usually a senior group member). This work arrangement required time and asset management and coordination in order to create a successful product.

Group Size. We attempted to test effectiveness of collaboration in different size groups. Given the limited scale of our experiment we compared two group sizes across different collaborative settings: "small" groups of size 2 and "large" groups of size 4.[6] With 8 analysts (A–H) and 6 tasks (1–6), a rotating assignment schedule was established, as shown in Table 2. In this arrangement, most groups

---

[6] Some aspects of collaborative group size are discussed in, among others, ([2]; [10])

Table 2. The work modes rotation schedule

	JTC	ITC	INC
Task 1	ABCD	EF	GH
Task 2	EF	GHAB	CD
Task 3	GHAB	CD	EF
Task 4	CD	EFGH	AB
Task 5	EFGH	AB	CD
Task 6	GH	ABCD	EF
#reports	1	2 or 4	2

rotate through both collaboration modes (JTC, ITC); however, due to time limitations it wasn't possible to do for every group.

Training. We have developed training materials to introduce analysts to COL-LANE. The training session was performed at the beginning of the workshop and included a tutorial, a hands-on tryout, and a warm-up task. The warm-up task was similar to the evaluation tasks but less complex. Its purpose was to allow the analysts to gain a degree of confidence in using the system, explain some "obvious" misunderstandings, and also to minimize the impact of unfamiliar technology on the first evaluation task. At the end of the warm-up task the analysts used the evaluation instruments, again to familiarize themselves with these aspects of the exercise. The entire training session took approximately 6 hours.

3.2 Preliminary Evaluation Results

COLLANE end-to-end performance was measured using the following metrics: (a) scoring of the final analytical reports for coverage, significance, organization, and other quality factors, and (b) questionnaires related to user satisfaction with the system and an assessment of their own performance using the technology, as well as other subjective factors such as workload perception. The report scoring was performed using the peer evaluation method (*cross-evaluation*) developed by the Albany team during the AQUAINT Project: in this approach all analysts act as a panel of (independent) judges producing multiple scores for each report, including their own. The questionnaires were developed to capture subjective assessment of particular aspects of the process that were not easily captured in the system logs. One of the questionnaires used was NASA TLX, which measures *individual perception of effort* put into the task. In addition, focus group interviews were conducted that solicited free form comments about participants' experience with the system and the exercise as a whole.

A number of objective performance indicators were also computed from the system logs; these included: the number and type of questions asked, the quantity of text snippets retrieved and retained into the report, relevance and utility of the returned answer elements, time needed to assemble various portions of the report, among others. These performance indicators, i.e., their content, order, and structure, are currently being correlated with end-to-end evaluation metrics for each of the task/user/mode settings in the exercise in order to isolate the performance characteristics that lead to a successful outcome (e.g., high-quality report), as well as those that may signal troubles. Once such performance indicators are isolated, we believe that they could be used to automatically monitor system performance outside of the controlled evaluation environment. This work is still ongoing and will be reported in a future publication.

Peer Evaluations. The cross-evaluation forms (Table 3) were used to facilitate scoring of the analytical reports based on their content and organization. Each participating analyst became an independent judge on the panel that reviewed and scored all reports generated during the session just completed. Since each session was devoted to a single analytical topic, the panel was, in effect, ranking the reports produced under different working conditions. An additional advantage of this method was that the judges are also the participants, and their scoring tends to reflect the experimental conditions: task difficulty, data availability, and time limits. Moreover, scoring assigned to own report (or own group report) provides additional cues on relative importance of certain information units.

The responses collected from cross-evaluation are tabulated and averaged over all judges, and then displayed as a bar chart. Figure 7 shows the cross evaluation results from one of the sessions (Artificial Reefs topic). In the chart, each bar represents an average cross-evaluation score (one for each of the 6 categories) assigned to each of the report coming out of this session. The tacit collaboration group (ITC) delivered 4 reports (4 leftmost bars in each bundle); the open collaboration group (JTC) delivered a joint report (5th bar), and the two single analysts (INC) delivered their own reports (rightmost 2 bars). We note that for this topic, the tacit collaboration group outperforms other groups in nearly all categories except for "coverage" (cat. 2). As we explain further below, this is not really the case since the reports in ITC group are often complementary, and thus should be judged as a "portfolio" rather than singly.

These results can be further averaged over all topics; however, in order to obtain a meaningful statistic we need to control for topic difficulty, analyst experience and skill, as well as for the judge bias. To do so would require a significantly larger data sample than the current experiment provided. For this reason the results reported here should only be treated as indicative of some possible trends.

Structured Questionnaires. Questionnaires were used to collect subjective opinions from the participants regarding their experience with COLLANE as

Table 3. Report Cross-Evaluation Form

Please evaluate each report using the following criteria Use 5 point scale (1=awful, 5=great) and justify
1. Includes crucially important information Score: ①②③④⑤ Justification:
2. Has sufficient coverage Score: ①②③④⑤ Justification:
3. Avoids the irrelevant materials Score: ①②③④⑤ Justification:
4. Avoids redundant information Score: ①②③④⑤ Justification:
5. Is well organized Score: ①②③④⑤ Justification:
6. Overall rating of this report Score: ①②③④⑤ Justification:

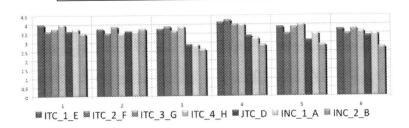

Fig. 7. Cross-evaluation scores for Artificial Reefs topic across all 6 criteria. The bars represent reports obtained in different work modes (ITC—tacit collaboration; JTC—open collaboration; INC—single mode).

well as with the various collaboration modes. We also sought analysts' opinions about the evaluation process itself and whether they felt they had sufficient exposure to the new technology to make a judgment. All questionnaires were carefully designed to control for bias and to detect inconsistencies judgment (e.g., some questions were restated in a different form, etc.). Answers were recorded on a numerical scale allowing for computing scores that then could be averaged over all participants and all topics.

Post-session Questionnaire

Post-session questionnaire consisted of 29 structured questions related to the task just completed by the analyst plus several free-form feedback questions. The questionnaire was administered immediately following each task. For collaborative tasks, each participant completed a separate questionnaire, thus multiple opinions were collected from collaborating teams. The questions sought analysts' assessment of the task itself (difficulty, appropriateness), specific features of the system (e.g., interface, speed, ease of use), and the work itself (collaboration, confidence, effort). All questions were structured so that they required answers on a 5-point Likert scale, which is normally presented as a set of "radio buttons" along with an intuitive scale, e.g., "not at all", "some", "a lot" (e.g., *How often did you use the visual interface?*) or "strongly disagree", "strongly agree" (e.g., *Did collaboration make analysis more efficient?*). The final free-form questions asked for comments on how to improve the existing system and how to make the work more efficient. The content of the questionnaire was adapted to each of the three work modes, i.e., questions concerning collaboration experience did not apply to analysts working singly. Below are a few sample questions from the beginning of post-session questionnaire:

> **1. How did this scenario compare to the tasks you perform at work?**
> (1: much less difficult 3: same 5: much more difficult)
> **2. How difficult was it to formulate questions for this task?**
> (1: much less difficult 3: same 5: much more difficult)
> **3. How confident were you about preparing a report for this task using COLLANE?**
> (1: not at all confident 3: confident 5: very confident)
> **4. How often did COLLANE respond to your questions with useful information?**
> (1: never 3: frequently 5: always)
> **5. How often did you find Rapid Results helpful?**
> (1: never 3: frequently 5: always)

Exit Questionnaire

smallskipamount An exit questionnaire consisting of 32 questions was administered at the end of the workshop after all work sessions were completed. This questionnaire reprised some of the questions from the post-session questionnaires, now in a more general form and included additional questions about overall assessment of the system, the tasks, and the collaborative arrangements. We used the same general format of structured questions with responses collected on a 5-point Likert scale. Below are a few sample questions:

> **1. The COLLANE system allowed me to easily change my line of questioning.**
> (0: strongly disagree 5: strongly agree)
> **2. It was difficult to get the COLLANE system to do what I wanted it do?**
> (0: strongly disagree 5: strongly agree)
> **3. I easily understood the relationship between the question that I asked and the answer that the COLLANE system provided.**
> (0: strongly disagree 5: strongly agree)
> **4. The COLLANE system seriously slows down my process of finding information.**
> (0: strongly disagree 5: strongly agree)
> **5. The COLLANE system helps me find important information.**
> (0: strongly disagree 5: strongly agree)
> **6. The COLLANE system helped me think of new ways to search for information.**
> (0: strongly disagree 5: strongly agree)

We tallied the responses from post session and post workshop questionnaires in Figures 8 and 9, respectively. The charts in Figures 8 show that the analysts have generally found the COLLANE system satisfactory and the exercise realistic with task difficulty at the level typical for their professional experience. The scores within the 2.5 and 3.5 range represent the middle point where the analysts' expectations are being met, e.g., the task difficulty is compatible with their experience, the system response is in line with the technology they use at work, etc. We note that analysts were quite positive about their reports and expressed a fair amount of confidence in their results. We also note that current COLLANE information sharing and collaboration support capabilities are acceptable but clearly need more work to be truly satisfying. This suggests that COLLANE development is on the right track even though much work remains to be done.

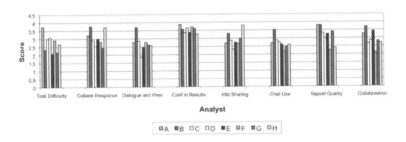

Fig. 8. Score averages from post-session questionnaires grouped into categories on 1-5 scale

Figure 9 shows average scores collected from the final questionnaire administered on the last day after all working sessions were completed. Again, we grouped the responses into several more intuitive categories. The assessment is very encouraging—analysts found COLLANE a promising technology while it is still a very preliminary prototype.

NASA TLX Questionnaire. NASA TLX instrument was used to assess subjective perception of workload during the task. Originally designed to test stress level for NASA astronauts, it has been adapted to analytic tasks with help of researchers from the National Institute of Standards and Technology. The revised questionnaire includes 6 categories of questions to rate analysts' experience with the task. Each question required a response on a 7-point scale,

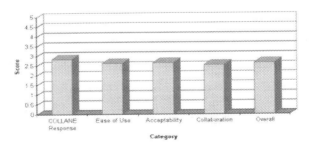

Fig. 9. Average scores from final questionnaire grouped into 5 categories

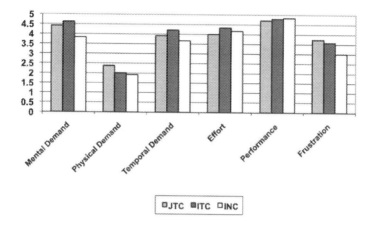

Fig. 10. TLX average scores for all sessions. The lower score is better except for the Performance category. The scale is 1 through 7.

with 1 meaning 'little' and 7 standing for 'much'. The 6 categories were defined as follows:

1. *Mental demand*: to what degree does the task affect a user's attention, brain and focus
2. *Physical demand*: to what degree does the task affect a user's health, makes a user tired, etc.
3. *Temporal demand*: to what degree does the task take time that a user can't afford
4. *Performance*: to what degree is the task heavy or light in terms of workload
5. *Frustration*: to what degree does the task make a user unhappy or frustrated
6. *Effort*: how much effort did the user spend on the task

For all of the TLX categories a higher number assigned by the analyst corresponds to subjective perception of a higher cognitive workload. Figure 10 shows the averages from the TLX-1 questionnaires obtained from all analysts across all tasks. It is interesting to note how modes of collaboration make different demands upon the analyst: tacit collaboration requires more mental effort and time pressure, while open collaboration adds primarily physical effort. None of these differences are statistically significant given the small data sample.

System Logging. During the workshop, all analysts' activities were automatically logged into several data streams. These included all analytic actions performed on COLLANE user interface (questions asked by the user, responses and questions from the system, all browsing activities on the panels, access to source documents, etc.) as well as all copy and paste events from COLLANE answer space to the analyst report (assembled in a separate text document). In addition, for the collaborating analysts, their exchanges over the chat interface were recorded, including messages sent as well as the data items exchanged. All of these data streams were time stamped allowing for easy alignment and verification of each event.

The following is a partial list of key logged events. We should note that these capture "basic" analytic events that can be combined together to obtain more meaningful "analytic episodes" (e.g., exploring, drilling down, verifying, etc.) which may lead to derivative utility-based metrics, e.g., time needed to assemble most of the information eventually included in the report, etc.

Examples of tracked user or system events:
Opening and closing an answer folder
Changing the relevancy of text passage
Displaying text through visual panel
Selecting an attribute to display on the visual panel
All dialogue between the users and the system
Bringing up a full document source
Text copied, and where it was copied from
Passages display and view
Browsing of the visual panel

One of the effect we were interested to note from the system logs was the degree and effectiveness of information sharing among the collaborating analysts, particularly in the tacit groups where the bulk of information sharing was automatically directed by the system. We wanted to see if apparently relevant but non-redundant information is correctly forwarded from one analyst to another, if it is being noticed by the recipient, and above all if it is being utilized in any way. We took copy events (into report draft) as evidence that a piece of information is being used; we also counted events where apparently viewed information (as evidenced by a passage display event) is ignored (i.e., not copied) or worse, it is labeled as non-relevant (e.g., by icon color change event). Figure 11 shows the effect of information sharing based on report usage across all three work settings. We note that the tacit collaboration group manages jointly to cover all key source citations.

We have also collected other quantitative information from the system logs. Some more interesting of these are reported below. For example, the graph in Figure 12 shows that on average, the analysts working in a collaborative setting required *less time to complete their tasks* than the analysts working alone. Specifically, analysts in the ITC mode with tacit information sharing completed their tasks faster than when working alone.

We need to note that while the open collaboration teams (JTC) completed their work faster than other analysts, the comparison is not straightforward. On the one hand, the analysts in open collaboration divide the task among themselves, which means that each analyst has a smaller problem to work with than the analysts in other groups; on the other hand, the JTC team needs to combine their partial reports, which requires extra time for assembly. It may be worth noting that time reduction also varies by the size of the JTC group: for topics 1, 3, and 5 (Vibrations, Teflon, and Honeybees), the JTC group had 4

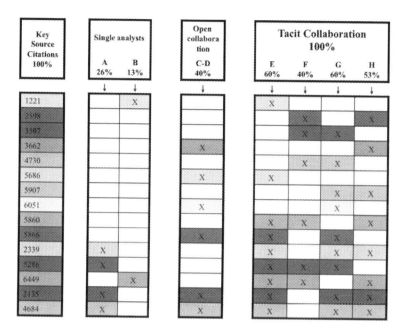

Fig. 11. Non-unique source citations among the analysts in Artificial Reefs task

analysts, thus the time reduction is more pronounced when compared to other work methods; for the other topics the JTC group had only 2 analysts and, as expected, the effect is lesser. In order to estimate the true time load in open collaboration, one would need to sum the times spent by each analyst. This would make the cost of open collaboration (measured per time unit) significantly higher than the cost of tacit collaboration. [7] Figure 13 shows that analysts in tacit collaboration (ITC) ask the system more questions, on average, than analysts working alone (INC). This is possibly a result of ITC analysts being exposed to more evidence through tacit information sharing, and thus following more leads and researching their topics more thoroughly—a clearly desirable effect. We also note that tacitly collaborating analysts, while asking more questions, spent less time on their tasks than when working alone (cf. Fig. 12 vs. Fig. 13). This seems to indicate that analysts in tacit collaboration work faster and do more than in other work modes—another highly desirable effect. As before, we can't directly compare analysts' performance in open collaboration (JTC); while they asked fewer questions, this is most likely an effect of task subdivision, which does not occur in either ITC or INC. Nonetheless, it may indicate that JTC is a less productive form of collaboration than ITC.[8]

---

[7] There may, however, be an additional cost in tacit collaboration incurred because the output has not been integrated into one report.

[8] There may be an inclination to take less responsibility, or take a more passive role, when the task has been divided up, or when there is a designated task head who will lead the integration of the report.

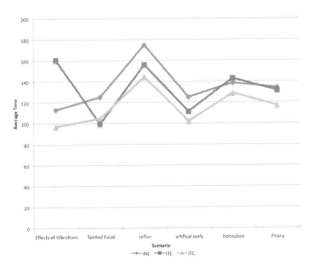

Fig. 12. Average time spent by individual analysts per topic for different modes of work (single, tacit, and open collaboration)

The primary source of citations in the report was the passage text encapsulated into the frames in the visual display (approx. 64%), with additional 11% coming from rapid results and the answer folders. In other words, 75% of cited material was selected from the passages directly offered by the system as relevant. The remaining 25% came from other parts of the retrieved documents, i.e., other passages than those explicitly displayed on the interface. This attests to the high-degree of precision of the COLLANE question answering component (Figure 14).

Finally, in Figure 15 we show selected performance statistics by the level of analytic experience. These results are averages over all sessions and work methods, and thus must be viewed only as illustrative (as discussed before, averaging scores in collaborative teams is not appropriate). Nonetheless, we note that while the most experienced analysts were in fact most effective and efficient in asking the right questions (their rate of productive questions[9] is very high at 71%), this effectiveness does not seem to translate into the higher report quality, as evidenced by the average cross-evaluation scores assigned to final reports (Scores columns in the chart).[10]

3.3 Summary of Evaluation Results

The preliminary results from this study suggest that COLLANE-supported tacit collaboration has the potential to produce significantly higher quality intelligence

---

[9] Productive questions (Good Questions in the chart) are defined as these that return at least one passage (citation) that is saved into the report.

[10] This may well be just a side effect of the current experiment, related to the selection of test topics. Further analysis, over a larger test sample is needed.

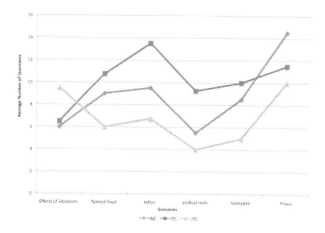

Fig. 13. Average number of questions asked by individual analysts working in different modes across all test topics

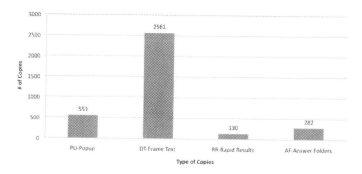

Fig. 14. Sources of citations in the reports

than would be possible by analysts working alone or in open collaboration groups. The results also show that to properly measure the impact of collaboration will require new evaluation techniques that go beyond the current metrics. This early prototype of COLLANE was well received by the analysts; nonetheless, more advanced research is required to exploit the full potential of collaborative technology on the analytic process. The workshop provided findings at three levels, as follows:

1. COLLANE Information Sharing and Tacit Collaboration are beneficial
   - *Information sharing in COLLANE improves the analytic process.* Analysts working in collaborative teams (open and tacit) are exposed to more topic-relevant information than analysts working alone. Therefore, we may infer that the collaborating analysts are using more evidence to arrive at more informed conclusions (example: Fig. 11).

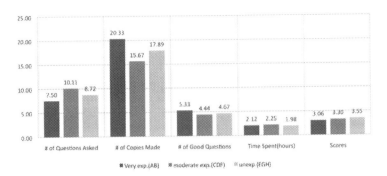

Fig. 15. Analytic effectiveness and efficiency by experience

- *The quality of the reported intelligence improves.* Qualitative assessment of the reports prepared by the collaborating teams suggests that better and more conclusions are drawn by collaborating analysts than by analysts working alone (ex. Fig. 7).
- *Current benefits of collaboration via COLLANE are only preliminary.* In order to obtain full benefits of tacit collaboration, COLLANE information sharing must be advanced to knowledge sharing capabilities.
2. Effects of Collaboration on the Analytic Process are positive
   - *Collaboration benefits the analytic process* but its effects vary between different forms. We found evidence that tacit collaboration is useful, but there was insufficient information to understand the effects of open collaboration.
   - *Tacit collaboration introduces a game-like competitive element.* Analysts working in tacit collaboration mode tend to pursue alternative or complementary interpretations of evidence. Unlike in the open collaboration, they are not being driven into consensus.
   - *Tacit collaboration improves analytic productivity* by inducing analysts to do more useful work per time unit. This is likely caused by increased information sharing and an element of competition. Tacit collaboration allows the analysts to benefit from the experience of others on their team without necessarily slowing them.
   - *Tacit collaboration helps to expose alternative interpretations* of the available evidence; the outcome is a portfolio of reports that facilitates the survival of sound alternatives.
3. New Evaluation Methodology is required to measure the full effects of collaboration
   - *Current evaluation methodology is inadequate* because it is geared towards a single output of the analytic process (report quality) and does not support the evaluation of multiple outcomes of collaboration.
   - *Evaluation of report quality is insufficient* because there is not enough time to do it in the current design and also there is no indepen-

dent assessment of the output by an external judge panel. The latter is particularly important for evaluating the output of tacit collaboration.

- **New metrics are required** to assess the relative value of a multi-faceted portfolio of reports covering a complex intelligence problem.
- **Revised evaluation design is required** in order to control for confounding factors such as level of experience, subject matter expertise, and analytic skills of the participants. In addition, we need ways of measuring effects of differently skilled participants on team performance.

3.4   Challenges

Not surprisingly, we found a number of challenges that would need to be addressed in future evaluations. Some of these challenges were already signaled in the preceding sections. Here we summarize them briefly:

- *Information sharing vs. knowledge sharing.* While information sharing currently facilitated by COLLANE is clearly beneficial, further improvements are expected by supporting knowledge sharing, i.e., the source information along with questions, annotations, and exploratory metadata left by the analysts.
- *Collaboration vs. competition.* Tacit collaboration seems an effective way of improving analytic effectiveness, but it also involves an element of competition, which needs to be taken into account. It requires further study to determine if this finding holds across further testing, and to determine if there are possible negative effects as well.
- *Collection vs. judgment.* Increased information sharing helps analysts to collect more supporting evidence but we need to provide tools to convert more information into better judgments.
- *"Correctness" of conclusion vs. soundness of argument.* We are interested in supporting sound arguments based on the available evidence, not simply the "correct" conclusion, which may be consistent with others' viewpoints.
- *Process quality vs. results quality.* We are interested in measuring both the process quality and the results quality, as well as in optimizing the connection between the two.
- *View rotations.* The current process does not support analysts' viewing each other's progress (e.g., evolving hypotheses); however such capability may be highly beneficial. Analysts expressed interest in being able to peek over "each other shoulders".

We have also identified challenges related to the mechanics of preparing an evaluation exercise. These include issues such as data collection, problem preparation, recruiting analysts, etc. Here are some of the key issues:

- *Data preparation*: The web mining method which depends on harvesting thousands of documents from the web through a series of rapid searches may not be adequate for creating data collections that can support analysis of highly complex topics; there is simply no guarantee that all relevant (and related but not relevant) aspects will be included. Potential remedies include human-in-the-loop interactive mining, mining more data, or using the open sources (e.g., internet). This last option has been frequently invoked by the participants; however, it raises a number of issues including stability of the experiment.

- *Time and scope*: The compressed nature of the experiment does not align well with typical analytical experience where analysts work on multiple topics but spend considerably more "clock" time on each. Analysts suggested that they should have an entire day to research a topic.
- *Access to the Internet*: Access to the open internet was not provided so as to maintain a controlled experimental environment. Nonetheless, analysts felt this limited their options too much, especially when dealing with unfamiliar and complex topics.
- *Access to specific data resources*: Some of the information needs raised by the tasks could be most naturally satisfied by searching specific data repositories (e.g., Government regulations on artificial reefs). Analysts thought that doing open-ended search for information that is readily available elsewhere made some aspects of the exercise unrealistic. We need to find a better way to balance the experimental needs with the realism of the evaluation exercise.
- *Access to other tools*: In addition to the above, access to other analytic tools was occasionally called for. Specifically, tools for organizing collected information by event date or release date was requested as an essential management tool.

4   Conclusions

In this chapter we described an advanced analytic system COLLANE and a process of conducting a task oriented evaluation with real users. COLLANE, which is in an early prototype stage, has been designed specifically to facilitate and support tacit and open collaboration among a group of analysts working on complex information problems. The preliminary evaluation described in this paper has led to a number of (indicative) observations, all of which need to be confirmed through further research:

1. Tacit collaboration is an effective means of improving analytic processes;
2. Tacit collaboration leads to better quality results;
3. Tacit collaboration does not drive analysts to a consensus; instead it exposes alternative approaches to complex problems;
4. Collaborative analysis is more efficient but also more demanding than working alone;
5. New metrics are required to adequately measure the full benefits of collaboration.
6. Information sharing may need to extend towards exchange of partially structured knowledge to further enhance the power of tacit collaboration.

Acknowledgements. We would like to thank Dr. Emile Morse for her assistance with the workshop planning and selection of evaluation instruments. LCDR Joseph Henriquez was instrumental in assembling the team of analysts who participated in the experiment and assisted in selection of the analytical tasks. This report is based on work supported by the IARPA CASE Program under the contract to SUNY Albany.

References

1. Allen, J., Core, M.: Draft of DAMSL: Dialog Act Markup in Several Layers (1997), http://www.cs.rochester.edu/research/cisd/resources/damsl/
2. Avouris, N., Margaritis, M., Komis, V.: The effect of group size in synchronous collaborative problem solving. In: Proceedings AACE Conf. ED-MEDIA (2004)
3. Dillenbourg, P., Traum, D.: Sharing solutions: persistence and grounding in multimodal collaborative problem solving. Journal of the Learning Sciences (2005)
4. Hardy, H., et al.: The AMITIES System: Data Driven Strategies for an Automated Dialogue. In: Speech Communication, vol. 48, pp. 354–373. Elsevier, Amsterdam (2006)
5. Hardy, H., Kanchakouskaya, V., Strzalkowski, T.: Automatic Event Classification Using Surface Text Features. In: Proceedings of the AAAI Workshop on Event Extraction and Synthesis, AAAI Press, Menlo Park (2006)
6. Kelly, D., Wacholder, N., Rittman, R., Sun, Y., Kantor, P., Small, S., Strzalkowski, T.: Using Interview Data to Identify Evaluation Criteria for Interactive, Analytical Question-Answering Systems. Journal of the American Society for Information Science and Technology, JASIST (2007)
7. Kirchhoff, K.: o and M. Ostendorf: Directions for multi-party human-computer interaction research. In: HLT-NAACL 2003 Workshop on Research Directions in Dialogue Processing (2003)
8. Lemon, O., Parikh, P., Peters, S.: Probabilistic Dialogue Modeling. In: Proceedings of the Third SIGdial Workshop on Discourse and Dialogue, Philadelphia (2002)
9. Morse, E.: An Investigation of Evaluation Metrics for Analytic Question Answering. In: Proceedings of AQUAINT Phase II, 6-month PI Meeting, Tampa (2004)
10. Ryall, K., Forlines, C., Shen, C., Morris, M.R.: Exploring the Effects of Group Size and Table Size on Interactions with Tabletop Shared-Display Groupware. In: CSCW '04, Chicago, Illinois, USA (2004)
11. Small, S., Strzalkowski, T., Hardy, H., Webb, N., Yamrom, B.: HITIQA: High-Quality Intelligence through Interactive Question Answering. Journal of Natural Language Engineering, Cambridge (to appear 2008)
12. Small, S.: An Effective Implementation of Analytical Question Answering. Doctoral Dissertation, Computer Science, SUNY Albany (2007)
13. Strzalkowski, T., et al.: Collaborative Analytical Workshop with COLLANE, Preliminary Report. Submitted to IARPA (2007)
14. Strzalkowski, T., Harabagiu, S.: Advances in Open-Domain Question Answering. Springer, Heidelberg (2006)
15. Strzalkowski, T., Small, S., Hardy, H., Yamrom, B., Liu, T., Kantor, P., Ng, K.B., Wacholder, N.: HITIQA: A Question Answering Analytical Tool. In: Intenational Conference On Intelligence Analysis (2005)
16. Strzalkowski, T., Small, S., Taylor, S., Lipetz, B.A., Hardy, H., Webb, N.: Analytical Workshop with HITIQA. A preliminary Report to IARPA (2006)

17. Strzalkowski, T., Small, S., Hardy, H., Liu, T., Ryan, S., Shimizu, N., Wu, M.: Question Answering as Dialogue with Data. In: Advances in Open-Domain Question Answering, pp. 149–188. Springer, Heidelberg (2006)
18. Traum, D.R., Andersen, C., Chong, W., et al.: Representations of Dialogue State for Domain and Task Independent Meta-Dialogue. Electron. Trans. Artif. Intell. 3(D), 125–152 (1999)
19. Traum, D.R., Rickel, J.: Embodied agents for multi-party dialogue in immersive virtual worlds. AAMAS, 766–773 (2002)
20. Wacholder, N., Kantor, P., Small, S., Strzalkowski, T., Kelly, D., Rittman, R., Ryan, S., Salkin, R.: Evaluation of the HITIQA Analysts' Workshops. Final Report (2003)
21. Wacholder, N., Kelly, D., Kantor, P., Rittman, R., Sun, Y., Bai, B., Small, S., Yamrom, B., Strzalkowski, T.: A Model for Quantitative Evaluation of an End-to-end Question Answering System. Journal of the American Society for Information Science and Technology (JASIST) (2006)
22. Webb, N., Hepple, M., Wilks, Y.: Dialogue Act Classification based on Intra-Utterance Features. In: Proceedings of AAAI workshop on spoken language understanding, AAAI Press, Menlo Park (2005)
23. Webb, N., Hardy, H., Ursu, C., Wu, M., Wilks, Y., Strzalkowski, T.: Data-Driven Language Understanding for Spoken Language Dialogue. In: Proceedings of the AAAI Workshop on Spoken Language Understanding, Pittsburgh, PA, AAAI Press, Menlo Park (2005)

Author Index

Brocki, Łukasz 273

Chen, Song 411
Cresswell, Tony 411
Cytowski, Jerzy 295

Ghandar, Adam 379
Gubrynowicz, Ryszard 273

Hajnicz, Elżbieta 211
Hardy, Hilda 411

Jankowski, Andrzej 3

Kimmel, Marek 359
Koržinek, Danijel 273
Kupść, Anna 241

Lipetz, Ben-Ami 411
Liu, Ting 411

Marasek, Krzysztof 273
Marciniak, Małgorzata 333
Matlatipov, Gayrat 83
Michalewicz, Zbigniew 379
Mykowiecka, Agnieszka 333

Piasecki, Maciej 163

Piskorski, Jakub 311
Pokrzywa, Rafał 359
Polański, Andrzej 359
Przepiórkowski, Adam 191

Rabiega-Wiśniewska, Joanna 61, 111
Radziszewski, Adam 163

Savary, Agata 111
Shaikh, Samira 411
Skowron, Andrzej 3
Strzalkowski, Tomek 411
Świdziński, Marek 143
Szalas, Alicja S. 43
Szałas, Andrzej 43
Szklanny, Krzysztof 273

Taylor, Sarah 411

Vetulani, Zygmunt 83

Webb, Nick 411
Woliński, Marcin 111, 143
Wu, Min 411

Zhan, Yu 411
Zurbruegg, Ralf 379